Dissecting the USMLE Steps 1, 2, and 3

Fourth Edition

Dissecting the USMLE Steps 1, 2, and 3

Fourth Edition

Francis Ihejirika, MD
Terra Caudill, MD, MS

Dissecting the USMLE Steps 1, 2, and 3

Fourth Edition

Copyright © 2009

Lulu Press

Authors:

Francis Ihejirika, M.D.

Terra C. Caudill, M.D., M.S.

Editors:

Jamison Strahan, M.D.

Amer Raza, M.D.

James Hwe, MS3

Illustrations:

Lindsey Jackson, M.D.

Francis Ihejirika, M.D.

Terra Caudill, M.D., M.S.

Book Cover Graphics:

Bob Fielder

Layout and Formatting:

Angel Editing

Dedication:

I dedicate this book to my parents, who instilled a strong desire in me to be excellent in life. I thank them for their love, their guidance, and their belief in me. I also owe much thanks to Dr. J.J. Shaw, a pediatric cardiologist at Saint Francis Hospital in Peoria, Illinois for being the one person who ignited this fire by showing me "how to understand" medicine rather than memorize it. I owe so much to my wife, Vickie, who has stood by me when I had nothing and I took my last paycheck to establish the PASS Program™. I also want to thank the countless students who have believed in me by attending the PASS Program™. I learn from them as they learn from me. And most importantly, I give all praise and thanks to God. For He is the one who guided me here and has blessed me with everything and everyone I need to be successful.

–Francis Ihejirika

To Phil, the love of my life, without whom I would not have made it through residency. To my parents, who encouraged me every step of the way. To my medical school study group: Jamie Strahan, Carlos Galindo, Steve Duffy, Steve Rath, Laura Hoelting, Kim Hennan, Richard Barber, Gina Bathurst, and Shawn Nishi – you guys are the best. Thanks for the all-nighters full of new mnemonics, clarification of concepts, and lots of laughs. Thanks also to my Step 2 study partner, Amer Raza, who made me study 25 hours a day and gave me a temporary addiction to espresso. Thanks to my Step 3 study partners, Al Taylor and Eros Sanchez, for all the innovative mnemonics and never-ending encouragement. Tons of thanks to my editors, Jamie and Amer, who are two of the smartest people I have ever known. Thanks to my illustrator, Lindsey, whose is an amazing artist, surgeon, and mother of triplets. Tons of thanks to James Hwe for his tremendous help in editing this edition and strengthening the Biochem chapters. And lastly, thank you to my God, who continues to guide my steps, hold my hand, and give me strength for each new day.

–Terra Caudill

Physician Assisted Student Success Program™

2302 Moreland Blvd, Champaign, IL 61822

http://www.passprogram.net

(217) 378-8018

Acknowledgements:

Gina Bathurst: ANP mnemonic

Will Chaviz: Newborns, An/Catabolic, Comma bugs mnemonics

Carlos Galindo: Meningitis mnemonic

Seunghee Oh: X-linked recessive diseases mnemonic

Matt Porac: Hypersensitivity mnemonic

Stephen Rath: Managing ventilators, Acid-Base methods

Eros Sanchez: Simplifying Micro Classifications, Palm/Sole Rash story

Malaika Scott: Low C_3 mnemonic

Jamison Strahan: Loop diuretics, Hb shift, Granulosa cell, Blots mnemonics

Al Taylor: All of the rest of the original mnemonics new to this edition

Kevin Woods: LITER mnemonic

Test Taking Strategies:

- Read the question at the end of the vignette *first*, then read the vignette to avoid wasting your time. There are times that the question can be answered without reading the vignette.
- You must answer the question *before* you enter the minefield of answer choices in order to avoid distractors.
- Read answers from bottom to top (e.g. E → A).
- Time yourself. Know what question you should finish by half-time.
- 3333Don't waste time on any one question. If you get stuck, pick an answer, mark it, and move on. It's better to finish the test than to get that one question right and end up missing many others, which negates all of your hard work.
- If you don't have a clue, listen to your gut feeling (even if you can't rationalize it) before you just guess. Then don't talk to yourself out of it!
- Another tip (if you don't have a clue) is to look at the answer choices, and you think there is more than one right answer, pick the lower one.
- If you are clueless about which drug is the correct answer, pick –mab. That tends to be the newer drug, and they like to put these on the test.
- What organ is the vignette talking about? Look for an answer that discusses the same organ.
- Look for a word in the question that has a similar word in the answer.
- For Biostats questions, always pick C, unless you know what the answer is.

Table of Contents:

Low Energy State:

"Without energy life would be extinguished instantaneously, and the cellular fabric would collapse."

–Albert Szent-Gyorgyi

Low Energy State:

This may be the most fundamental principle to understanding so much medicine. 98% of all illnesses fall into this category. When the body undergoes starvation, malabsorption, storage diseases, vitamin deficiencies, etc. due to a pathologic state; the body responds by conserving the energy it has to help the body heal, leading to a low energy state. This causes the following symptoms:

Brain: mental retardation, dementia
Heart: heart failure, pericardial effusion
Muscle: weakness, SOB, vasodilation, impotence, urinary retention, constipation, etc.
Primary Active Transport (ATPases): stop working → depolarization

<u>Rapidly Dividing Cells:</u>

- Skin: dry
- Cuticles: brittle (not nails b/c they are dead)
- Hair: alopecia
  - Female Alopecia Tx: Minoxidil
  - Male Alopecia Tx: Finasteride
- Bone marrow: suppressed → anemia, leucopenia, thrombocytopenia
- Vascular endothelium: breaks down
- Lungs: infection, SOB
- Kidney: PCT will feel the effect first
- GI: N/V/D
- Bladder: oliguria due to urinary retention
- Sperm: decreased
- Germ cells: predisposed to cancer – skin, GI, bone marrow (most rapidly dividing)
- Breasts: atrophic
- Endometrium: amenorrhea

Most common presenting signs: tachypnea and dyspnea
Most common presenting symptoms: weakness and SOB
Most common presenting infections: UTI and respiratory infections
Most common cause of death: heart failure
Most important complications: brain > heart> kidney

<u>Most Common Causes of Death:</u>
Most people: Heart failure
Cancer patients: Infection
Obstructive lung disease: Bronchiectasis
Restrictive lung disease: Cor pulmonale
Neuromuscular disease: Respiratory failure
SLE/Cervical CA/Endometrial CA pts: Renal failure

Most common cause of arrhythmias: Ischemia
Most common cause of bleeding into any cavity: Hypertension
Most common cause of atrial fibrillation: Long-standing hypertension

NOTES:

Vitamins/Minerals/ Trace Elements:

"Friendship is like vitamins, we supplement each other's minimum daily requirements."

–Anonymous

Vitamins:

Vit A:

- Night vision
- CSF production: Pseudotumor Cerebri → HA, papilledema
- Tx: serial LP/Acetazolamide
- PTH co-factor: moans, groans, bones, stones
- Liver stores fat soluble vitamins → Vit A/D overdose can kill you…

*Vit B: "**T**hese **R**eally **N**eat **L**ittle **P**ills **P**revent **F**unctional **C**omplications"*

Vit B_1 "Thiamine": dehydrogenases and transketolase (PPP) cofactors

- Beriberi => heart failure
- Wet: high cardiac output heart failure
- Dry: peripheral neuropathy
- Wernicke's encephalopathy: posterior temp lobe →ophthalmoplegia, ataxia, psychosis
- Korsakoff psychosis: mamillary bodies → anterograde amnesia, confabulation

Vit B_2 "**R**iboflavin" (source = milk, sunlight breaks it down): **FAD** cofactor

- *Got Milk? If not → angular cheilosis*

Vit B_3 "**N**iacin": **N**AD and dehydrogenases cofactor

- Pellagra: dermatitis, diarrhea, dementia, death
- Hartnup's: can't absorb Trp (niacin precursor)

Vit B_4 "**L**ipoic acid": Glycolysis

- No diseases known

Vit B_5 "**P**antothenic acid": AcetylCoA structure

- No diseases known

Vit B_6 "**P**yridoxime": transaminase cofactor (pyridoxyl phosphate) Ex: ALT, AST, GGT

- Neuropathy (brain uses transamination)
- Glossitis: painful, burning, swollen tongue
- Dermatitis
- Confusion
- Depression
- INH depletes body of Vit B_6

Vit B_9 "**F**olate"**:** Nuclear division of rapidly dividing cells

- Made by gut, and found in green leafy veggies *"foliage"*

- Megaloblastic anemia/Bleeding problems
- Hypersegmented neutrophils
- Fetal neural tube defects
- Beefy red tongue

Vit B_{12} "Cobalamin": cofactor for HMT (THF) and MMM (recycles odd carbon fats)

- The only water soluble vitamin absorbed in ileum, stored in liver
- Neuropathy: DCML/CS tracts (longest tracts, most myelin)
- Megaloblastic anemia
- Hypersegmented neutrophils
- Pernicious anemia
- Type A gastritis
- Diphyllobothrium latum tapeworm: easts B_{12}, found in raw fish
- Vegans

Vit C: collagen synthesis (Pro-OHase and Lys-OHase), anti-oxidant in GI tract

- Scurvy (bleeding gums and hair follicles, wooden leg = soft tissue hemorrhage)

Vit D: mineralization of bones, teeth

- Kids: rickets (lateral bowing of legs) *"genu varum"*
- Adults: osteomalacia

X-linked Dominant: Dad → daughter

- Vit D-resistant rickets (kidney leaks phosphorus): waddling gait
- Pyruvate dehydrogenase deficiency

Vit E: anti-oxidant in blood

- Decrease risk of Alzheimer's
- Hyporeflexia

Anterior leg bowing: Neonatal syphilis

Lateral leg bowing: Rickets

Anti-Oxidants:

- Vit E – in blood
- Vit C – in GI tract
- Vit A
- β-carotene

Diseases involving Oxidation:

- Cancers
- Alzheimer's
- Coronary artery disease
- Hemolytic anemia

Quick Reference:

If Ca and PO_4 decrease => Vit D deficiency

If Ca and PO_4 are different directions => PTH prob

- High Ca => hyperPTH
- Low Ca => hypoPTH

Vit K: γ-carboxylation => clotting (Factors 2,7,9,10; Prot C,S), made by gut flora

Gut Flora Production:

- Vit K
- Biotin
- Folate
- Pantothenic acid

Heparin: √ PTT
Warfarin: √ INR
LMW Warfarin: √ Factor

Anticoagulants:

IV Heparin^{+}: Cofactor for ATIII (start this before warfarin)

- $t_{1/2}$ = 6hrs, charged => can't cross placenta
- Heparin-induced thrombocytopenia (acts as a hapten on platelets), bleeding
- Activates hormone-sensitive lipase (↑triglycerides)
- Reverse w/ protamine sulfate^{-} (slower) or give FFP if actively bleeding
- Follow PTT (except SLE, then follow serum heparin levels)

LMW Heparin "Enoxaparin":

- Don't have to check PTT every day → follow Factor 10a activity
- Less thrombocytopenia
- Less bleeding complications

Oral Warfarin "Coumadin": inhibits Vit K-dep factors => can't make clots

- Fat soluble => crosses placenta
- If give w/ p450 inhibitors => bleed, skin necrosis (Fat: breasts, butt, belly, thighs)
- Initially: pro-coagulant until Factor 7 disappears => give w/ heparin until PT is therapeutic
- Keep pt in hospital 10 days (5 half-lives to reach steady state)
- Protein C decreases first b/c has shortest $t_{1/2}$; Factor 7 disappears next
- INR: 2-3x normal (Every lab has different controls)
  - If INR <5 → skip next dose
  - If INR 5-9 → stop warfarin
  - If INR >9 → give Vit K
- Reverse w/ Vit K (takes 6 hrs to work) + FFP (lasts 6hrs)

Clot Treatment:
Arterial: Aspirin
Venous: Heparin

Minerals:

β-carotene: anti-oxidant

- High: anorexia, DM, hypothyroidism

INR:
Measured PT/ Control PT

Biotin: Carboxylation

- Eggs (avidin) → ↓biotin → dermatitis, enteritis

Calcium (Ca^{2+}): Neuronal fxn, atrial depolarization, ventricle/SM contractility, bone/teeth

- Tetany (low Ca)

- Chovstek's sign: tap facial nerve => muscle spasm
- Trousseau's sign: take BP => carpal spasm
- **Copper (Cu^{2+}):** Collagen synthesis (Lys OHase cofactor), electron transport

Cu Deficiency – Kinky Hair disease (copper wire hair)
Cu Excess – Wilson's disease (AR): Ceruloplasmin deficiency

I) Autosomal Dominant: Structural => 50% chance of passing it on

II) Autosomal Recessive: Enzyme deficiency => 1/4 get it, 2/3 carry it (assume both parents are heterozygous)

III) If no family history => use **Hardy Weinberg**: $(p^2 + 2pq + q^2) = 1$

- $p + q = 1$
- Freq of carrier = p
- Freq of gene = q

Incidences:

Rare things: 1-3%
1 risk factor: 10%
2 risk factors: 50%
3 risk factors: 90%

Movement Disorders:

Huntington's (AD): triplet repeats, anticipation, no GABA, choreiform movements

- Cause of death: suicide (30 y/o), insurance will drop them upon diagnosis

Wilson's: ceruloplasmin deficiency

- Cu deposits in liver, eye (Kaiser-Fleischer rings), brain (lenticular nucleus)
- Choreiform movements
- Tx: Penicillamine (chelates Cu^{2+})

Trinucleotide Repeats: *"Fragile X **H**as a **F**reakishly **M**onstrous **P**air"*

Fragile X: low set ears, big jaw, short, retarded, big testes, mitral prolapse, aortic root dilation
Huntington's: dementia, choreiform movements
Friedreich's Ataxia: retinitis pigmentosa, scoliosis, spinal cord atrophy
Myotonic Dystrophy: cataracts, alopecia, bird's beak face, can't release handshake
Prader-Willi: hyperphagia, hypogonadism, almond-shaped eyes

Dinucleotide Repeats:

Hereditary Non-Polyposis Colon Cancer "HNPCC"

Uniparental Disomy:

Angelman's (only maternal genes):

- Ataxia, *"happy puppet syndrome"*

Prader-Willi (only **p**aternal genes) :

- Hyperphagia, hypogonadism, almond-shaped eyes

Iron (Fe^{2+}): Hb function, electron transport

Malnutrition:

Kwashiorkor: malabsorption
big belly (ascites)
protein deficiency

Marasmus: starvation
skinny
calorie deficiency

- If eat a bottle of multivitamins, worry about Fe
- Tx: Defuroxime

Magnesium (Mg^{2+}): PTH and kinase cofactor => need to check Mg levels before anesthesia

Zinc (Zn^{2+}): sperm, taste buds, hair

- Infertility, Dysgusia, Alopecia

Trace Elements:

Chromium (Cr): insulin function

Fluoride (F): hair and teeth, blocks enolase (glycolysis)

Molybdenum (Mb): purine breakdown (xanthine oxidase)

Manganese (Mn): glycolysis

Selenium (Se): heart function

- Se Deficiency: dilated cardiomyopathy
- Se Excess: garlic breath

Tin (Sn): hair

Breath Clues:

Garlic breath: high Selenium

Almond breath: Cyanide

Fruity breath: DKA

Minty breath: ASA overdose

Halitosis: Zenker's diverticulum

Ammonia breath: Budd-Chiari or Hepatic encephalopathy

NOTES:

NOTES:

Cellular Physiology:

"It is the cells which create and maintain in us, during the span of our lives,

our will to live and survive, to search and experiment, and to struggle."

–Albert Claude

Cell Death:

Necrosis: Non-programmed cell death = inflammation, nucleus destroyed first, explosion *(don't want to die...have to be drug off screamin' and hollerin'...)*
Apoptosis: Programmed cell death = no inflammation, nucleus destroyed last, implosion

Pyknosis: Nucleus turns into blebs *"pick blebs"*
Karyohexis: Nucleus fragments
Karyolysis: Nucleus dissolves

Types of Necrosis:
Coagulative necrosis: ischemic, cell architecture stable, most common
Ex: stroke, MI

Liquefactive necrosis: abscess formation (half liquid/half solid), anaerobic
Brain: most likely (already half solid/half liquid)
Lung: least likely (aerobic)

Hemorrhagic necrosis: organs w/ >1 blood supply **or** soft capsule, cell architecture lost
Ex: GI, lungs, brain, liver turn to mush

Fibrinous necrosis => fibrin deposition
Ex: collagen or vascular disease, renal failure

Fat necrosis => fat deposition, *"chalky"*
Ex: pancreas (sits on fat), breast

Caseous necrosis => TB, *"cheesy"*

Purulent necrosis => bacterial, *"pus"*

Granulomatous necrosis => T cells/MP (non-bacterial)

Irreversible Cell Injury:
Brain: 20 min (do CPR this long)
All Else: 6 hours (use t-PA by this time)

Signs of Irreversible Cellular Death:
I) Nuclear damage = *1st sign of irreversible cellular death*
- Mess w/ nucleus or chromosome → death
- If you don't die, things won't grow → microcephaly, micrognathy

II) Lysosome damage = *2nd sign of irreversible cellular death*
- I-cell Disease: failure of addition of mannose-6-phosphate to lysosomal proteins resulting in proteins not making it into the lysosome, then accumulate in inclusion bodies.

III) Mitochondrial damage = 3rd sign of irreversible cellular death, Mom → all kids

- Clue: "ragged red fibers"
- Leber's: atrophy of optic nerve
- Leigh's: chronic fatigue, poor sucking ability, loss of motor skills *"lay around"*

Monosomies: *less tissue*

Turner's (45, XO)

- Webbed neck (neck didn't develop)
- Shield-shaped chest (waist didn't develop)
- Coarctation of aorta (aorta didn't develop) => differential pulses/differential cyanosis
- Tx: Aortoplasty – cut out stricture
- Gonadal streak (gonads didn't develop): need in-vitro fertilization (risk=aortic rupture)
- Cystic hygroma = lymphatic fluid sac
- Rib notching (alternate blood route = intercostal aa. under ribs => ribs bulge out)
- Low-lying hairline
- NO mental retardation
- Newborn feet edema
- Cubitus valgus (elbows turn out)
- Growth hormone resistance (Tx: GH)
- ↑Risk of Hashimoto's

IQ Levels:

Superior: >130 (2SD)

Average IQ: 85-100 (SD=15)

Mild Retardation: <70 (can hold a job)

Moderate: <55 (needs a group home)

Severe: <40 (need supervision)

Profound: <25 (need caretaker)

Trisomies: *extra tissue*

Most common reason: non-disjunction => death

Most non-disjunctions *occur* in dad (he has many sperm)

Most *transmission* of non-disjunctions are from mom (eggs last a lifetime)

Trisomy 13 "**P**atau's": **p**olydactyly, high **p**alate, **p**ee problem, protruding abd (oomphalocele)

Trisomy 16: Most common cause of 1st trimester abortion

Trisomy 18 "Edward's": rocker bottom feet, *die within 2-3 months*

Trisomy 21 "Down syndrome":

- More babies born when Moms <35 y/o, ↑risk >35 y/o
- Highest incidence with Robertsonian translocation → 33% have it
- Simian crease, widely spaced 1st/2nd toes
- Hypothyroid (40%), macroglossia, male sterility
- Mongolian slant of eyes, Brushfield spots (speckled iris)
- Mental retardation (normal intelligence → mosaic)
- Early Alzheimer's (20-40 y/o), lowest depression rate
- No endocardial cushion (20%) → do Echo
- Duodenal atresia, Hirschsprung's
- High incidence of AML, C-spine instability

XXX syndrome (47, XXX): normal female w/two Barr bodies

XYY syndrome (47, XYY): tall aggressive male

Klinefelter's (47, XXY):

- Tall, gynecomastia, infertility, ↓testosterone, small penis/testes,
- Early BPH, ↑male breast cancer

Endocardial Cushion Defects:

ASD: septum primum didn't fuse w/ septum secundum
VSD: ventricular septum didn't fuse w/ endocardial cushion

Testicle Size:

Fragile X: short, big testes
Klinefelter's: tall, small testes
MD: bald, small testes
Prader Willi: fat, small testes
Kallman's: anosmia, small testes

Non-Cyanotic Cardiac Anomalies: *operate at 5y/o*

1) **VSD:** holosystolic murmur
2) **ASD:** fixed wide S_2 splitting
3) **PDA:** L pulm a → aorta): continuous murmur, wide PP, pink fingers/blue toes
   - Tx: Indomethacin to close PDA
4) **Coarctation of the Aorta:** rib notching, associated w/ Turner's, x-ray "3" sign
   - Pre-ductal: differential cyanosis
   - Post-ductal: differential pulses

CHEMOTHERAPY: affects rapidly dividing cells → skin CA (the most rapidly dividing)

1) **Induction: high dose to decrease malignant cells**
2) **Consolidation: get rid of residual malignant cells**
3) **Maintenance: hold fast…**

I) Alkylating agents: bind ds DNA ← *tx slow growing CA*
Busulfan – pulmonary fibrosis, tx CML
Cisplatin – renal failure, tx ovary cancers
Chlorambucil – tx CLL, tx membranous glomerulonephritis
Cyclophosphamide – tx Lupus nephritis; hemorrhagic cystitis ← prevent w/ Mesna
Dacarbazine – tx Hodgkin's, severe emesis ← prevent w/ Odansetron
Hydroxyurea – inhibits ribonucleotide reductase, ↑HbF, drug-induced SLE
Lomustine – crosses BBB
Melphalan – tx multiple myeloma
Metchlorethamine – tx Hodgkin's
Procarbazine – disulfiram reaction
Streptozocin – crosses BBB, causes DM (islet cell death)

Steroid-Resistant Tx:
"MAC"
1) **M**ethotrexate or
2) **A**zathioprine or
3) **C**yclosporine

II) Antibiotics:
Antimycin – inhibits ETC complex III
Bleomycin – pulmonary fibrosis
Dactinomycin – tx Wilm's tumor
Daunorubicin – cardiac fibrosis, tx AML
Doxorubicin "Adriamycin" – irrev cardiac fibrosis ← Dexrazoxane ↓cardiotoxicity
Mithramycin – decreases high Ca levels

III) Antimetabolites: replace nucleotides ←*tx fast growing CA* (acute, anaplastic)
5-FU – T analog, tx colon CA (Duke stage C) => mouth ulcers
6-mercaptopurine – purine analog, breaks down into uric acid
Azathioprine – purine analog, good for kidney transplants
Cladribine – purine analog, tx Hairy Cell Leukemia *"I'm clad with hairs"*
Cytarabine – inhibits DNA Pol, tx AML
Methotrexate – inhibits DHF reductase, anemia ← Leucovorin (folate analog)
Pemetrexed – tx malignant mesothelioma
Thioguanine – G analog

IV) Hormones:
Tamoxifen – vaginal bleeding, ↑risk of endometrial CA (E_2 antagonist)
Leuprolide – tx endometriosis (GnRH analog)
Flutamide – tx prostate CA (DHTr inhibitor)

V) Microtubule Inhibitors: affects cilia, villi, axon transport, epididymis
Etoposide – inhibits topo II, tx small cell lung CA
Paclitaxel – metaphase arrest, tx intraductal breast CA
Vinblastine – aplastic anemia *"blasts bone marrow"*
Vincristine – neuropathy

VI) Tyrosine Kinase Inhibitors:
Imatinib – tx CML

VII) Others:
Ondansetron (5-HT antagonist) – tx N/V/D
Megestrol (progesterone derivative) – tx appetite loss
Levamisole (immunomodulator) – stimulates NK cells
L-Asparaginase (nutrient depression) – anaphylaxis

Cell Anatomy:
Nucleolus: has ribosomal RNA
Free-floating Ribosomes: makes proteins that are used in the cytoplasm
Rough ER: makes proteins that need to be packaged (PrePro proteins)
Smooth ER: detoxifies

Labels:

- Pre → Rough ER (modifies collagen; N-acetylation of most)
- Pro → Golgi (modifies all other proteins), enter concave side, leave convex side
- Mannose-6-P → Lysosome
- N-terminal sequence → Mitochondria ← *HSP-70 folds proteins here*

NOTES:

NOTES:

Membrane Physiology:

"The art of medicine consists in amusing the patient while nature cures the disease."

–Voltaire

Cell Membrane Functions:

- Shape
- Protection
- Create concentration gradients
- Depolarization
- Selective permeability: Phospholipid bilayer is amphipathic (loves both)
- Hydrophobic – fat soluble
- Hydrophilic – water soluble

Fat soluble membranes: don't need transport proteins.

- Fat soluble compounds go through outer membrane and head for the nucleus because they have a nuclear membrane receptor.
- Concentration gradient is the only limiting factor. (Ex: steroid hormones – except cortisol)

Water soluble membranes: require a second messenger.

Fick's Principle:

(conc)(surface area)(flux) ← *easy to get through membrane*

(charge)(size)(thickness)(reflection coefficient) ← *hard to get through*

Reflection coefficient: # particles returned / # particles sent

- 0 => fat soluble
- 1 => water soluble

Unsaturated Fats: double bonds, easier to break down, more fluidity of movement

Saturated Fats: no double bonds, [NOT] allow temperature to escape, can't regulate body temp

Essential Fats: can only get from diet.

- Linolenic
- Linoleic – used to make arachadonic acid → prostagladins

The Drug Connection:

Ex: Statins are given once a day. So now you know….

- It has a long half-life
- It is a base
- Fat soluble
- Avoid in obese, elderly, babies (more adipose tissue)
- Easy to absorb
- Can get into your skin, pancreas, muscles, brain

- Increased CNS toxicity
- Metabolized by liver (not kidney)
- Hepatotoxic
- P450 (not GFR) dependant
- Increased volume of distribution

Now you have the Eyes Of Physio!....

Phagocytosis: entire membrane is used (need ATP, Ca^{2+}, actin, myocin)
Endocytosis: move nutrients in
Exocytosis: move waste out *(Lipofuscin: oxidized fat, protein → brown pigmented age spots)*
Pinocytosis: endocytosis of water and electrolytes *"cell drinking"* (Ex: skin)

Temperature Regulation:
Radiation – moving heat down its conc gradient (water traps heat, heat looks for cool surface)
- Ex: *"car radiator"*

Convection – environment moving past you (Ex: jogging)
- Ex: wind, movement *"open door of convection oven"*

Conduction – must be in contact (Ex: when you stop jogging → why you sweat more then)
- Ex: rolling over on cool sheets *"your conduct"*

Conductance – every membrane sits at -90mV except neurons and Purkinje fibers (-70mV)
- Ex: kid stuck finger in the socket => arrhythmia and seizures in next 24 hrs
- Most common cause of death = seizure
- Most common complication = can't clot due to fried BM (GP2b3a) => hemorrhage

Concentration gradient:
- More total Ca^{2+} (inside sarcoplasmic reticulum) inside the cells
- More total Mg^{2+} (on kinases) inside the cell
- More free Ca^{2+} and Mg^{2+} outside cell

Electrical gradient: "E" = Nernst number
- The E of any ion is the membrane potential at which the concentration gradient and electrical gradient are equal and opposite → no net movement
- Each ion wants to go to its own E

Ca^{2+}: +120 (greatest electrical potential, wins the race to the membrane)
Mg^{2+}: +120
Na^{+}: +65
Membrane potential: -90 (except neurons, Purkinje cells = -70mV)
Cl^{-}: -90 (least conductance at rest => no *net* movement)
K^{+}: -96

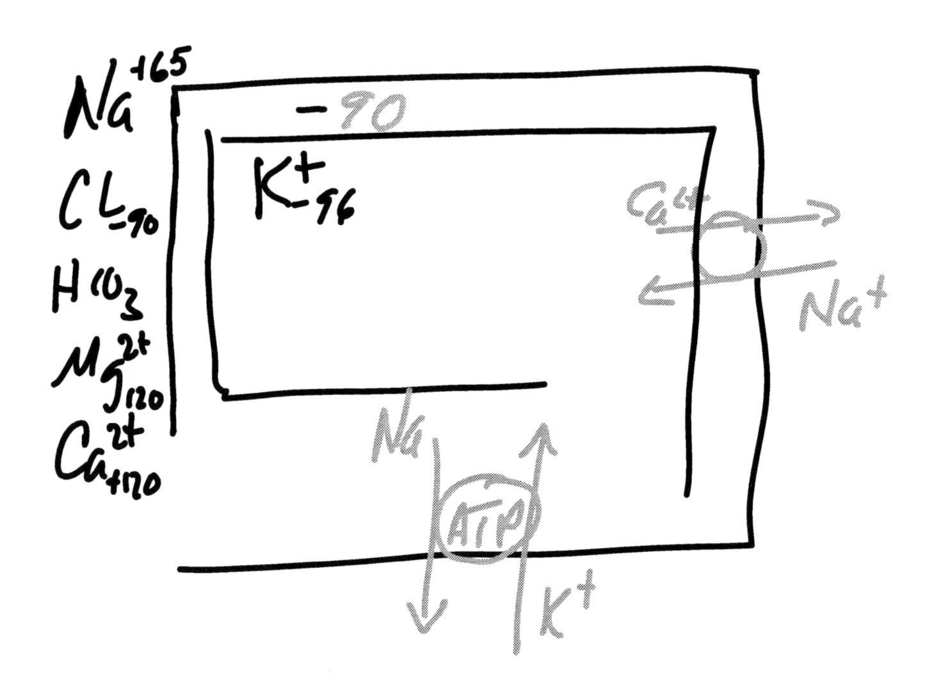

Driving force = $|E\ of\ ion - membrane\ potential|$ *(-90mV)*

Conductance (G) = movement across membrane (highest for K due to leak channels)

Permeability: *access across a membrane*
1) Channels – small ions (all except K are voltage regulated, only let ions into cell)
2) Pores – medium-sized molecules (Ex: sweat = NaCl, H_2O)
3) Transport proteins – all large molecules (Ex: glucose, HCO_3^-)

1° active transport: uses ATPase, *concentrates stuff*
2° active "facilitated" transport: uses Na gradient, *moves against gradient*
- Symport = co-transport
- Antiport = 1 in/1 out

II) Signal transduction – protein hormones (all receptors are glycoproteins, 7 TM spanning)
1) cAMP "sympathetic" pathway: *The Default Answer (the "aaaa" pathway)*
- Ex: SM relaxation

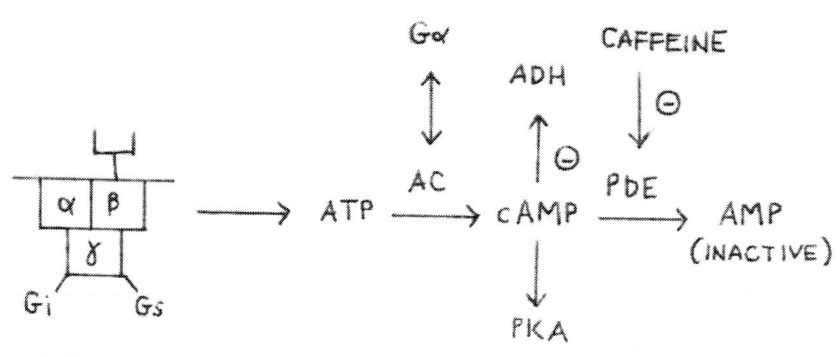

cAMP Diseases:

ETEC/Cholera: Gs on (via ADP ribosylation) *"turns the on on"*
Pertussis: Gi off *"turns the off off"*
Caffeine/Theophylline/Methylxanthines: inhibits PDE => ↑cAMP => ↑sympathetics

- Tx: Neonate central apnea, nocturnal asthma (bronchodilate)

Anabolic: active when dephosphoryated
Catabolic: active when phosphorylated

Low cAMP: sympathetic initially → after it crosses a certain unknown conc → becomes parasympathetic
High cAMP: parasympathetic initially → after crosses a certain unknown conc → becomes sympathetic

cAMP PDE Inhibitors:

- Caffeine
- Theophylline
- Methylxanthine

cGMP PDE Inhibitors:

- Sildenafil
- Vardenafil
- Tadalafil

2) IP_3/DAG pathway: *the "eeee" pathway…P,C,G,3*

- Ex: Hypothalamic hormones (except CRH – uses cAMP)
- Ex: SM contraction

3) Tyr kinase

- Ex: Insulin or GFs (Somatomedin, GPO, EPO, TPO)

4) Ca^{2+}-Calmodulin (4:1)

- Ex: SM contraction by distension (Ex: bladder, uterus)

5) Ca^{2+} alone

- Ex: Gastrin

6) NO pathway: NO → Guanylate Cyclase → cGMP → parasympathetic

- Nitrates: Nitroprusside, Nitroglycerin (dilate arteries and veins)
- Sildenafil, Vardenafil, Tadalafil
- ISDN: tx chronic CHF
- ANP
- Endotoxins

NOTES:

Inflammation:

"Pride is a powerful narcotic, but it doesn't do much for the auto-immune system."

–Stuart Stevens

Cardiac Glycosides = "Digitalis": *from foxglove plant*

1) Block Na/K pump => increase contractility
2) Stimulate vagus => slow down A/V nodes *(tx atrial arrhythmias)*

- Digoxin (renal excretion): given oral and IV (O/D Tx: Digibind, Digifab)
- Digitoxin (hepatic excretion): given oral
- Ouabain: only used in the lab

Ischemia: shut down Na/K pump => more positive => spontaneous depolarization

- K leaves => ST depression => 70% occlusion (Early ischemia; Ex: angina, TIA)
- Na enters => ST elevation => 90% occlusion (Late ischemia, Ex: MI, stroke)
- H_2O follows Na in => swelling (1st change of cellular death)
- Na leaves (b/c of concentration buildup)
- Ca enters => contraction

Ischemia (decreased blood flow) → Injury (hurt cells) → Infarct (dead cells)

Ischemia	Injury	Infarct
1) T wave peak	3) ST elevation	5) Q waves
2) ST depression	4) T wave inversion	

Angina Workup:

- Hospitalize for 24-hr observation
- Serial EKGs and cardiac enzymes
- Follow-up in 6 weeks:
    - Ca-Pyrophosphate scan: hot spot show dead calcified cells
    - Treadmill stress test: positive stress test if pain, ΔEKG, or ↓BP
    - Thallium stress test: cold spot shows ischemia
    - Dobutamine stress test: use dipyridamole

MI Treatment: *"MONA SHeM"*

- **M**orphine (pain)
- **O_2**
- **N**itroglycerin (dilates coronary aa. to ↑O_2 delivery to myocardium)
- **ASA** (irreversibly inhibits platelets) – GIVE 1st!
- **S**treptokinase or t-PA (breaks up clots), if <12hr and has not had prior CABG
- **He**parin (prevents new clots), unless pt has had surgery in past 2wk
- **M**etoprolol (decreases mortality by ↓myocardial O_2 demand) – GIVE 2nd!, limits infarction

Post-MI Heart Changes:

24 hr: Pallor

2-3 days: Mottling

4-7 days: Yellow-brown necrosis
7wk: Fibrotic scar

CABG Indications:

- Left main: 70% occlusion
- Three vessel disease

Inflammatory Response:

- <24 hr: Swelling
- Day 1: PMNs show up at 24hr
- Day 3: PMNs peak
- Day 4: MP/T cells show up
- Day 7: MP/T cells peak, Fibroblasts show up
- Day 30: Fibroblasts peak
- Month 3-6: Fibroblasts leave, fibrosis is complete

Likeliness To Depolarize:

This concept shows you how to predict what the side effects of any electrolyte state would be. For example, hypocalcemia is more likely to depolarize. Thus, they have an overall body state that can be described by the symptoms below.

More likely to depolarize: $\downarrow Ca^{2+}$, $\downarrow Mg^{2+}$, $\uparrow Na^{+}$ (early), $\uparrow K^{+}$ (early)
Brain: psychosis, seizures, jitteriness, insomnia
Skeletal muscle: muscle spasms, tetany, cramps
GI: diarrhea, then constipation (smooth muscle needs Ca for 2nd messenger system)
Cardiac: tachycardia, arrhythmias

Less likely to depolarize: $\uparrow Ca^{2+}$, $\uparrow Mg^{2+}$, $\downarrow Na^{+}$ (early), $\downarrow K^{+}$ (early)
Brain: lethargy, mental status changes, depression, delirium, sedation, coma
Skeletal muscle: weakness, SOB
GI: constipation, then diarrhea (Na to depolarize, then Ca-Calmodulin as 2nd messenger)
Cardiac: hypotension, bradycardia

Chronic ischemia => calcification

- Aorta calcification: "Monckeberg"
- Cancer calcification: "psamomma bodies"
- Dystrophic calcification = cellular injury
- Metastatic calcification = hyperCa

Acute diseases: swelling, $\uparrow$PMNs
Chronic diseases: fibroblasts, calcification

NOTES:

Endocrinology:

"War will never cease until babies begin to come into the world with

larger cerebrums and smaller adrenal glands."

–Henry Mencken

Steroid Hormones:

- Nuclear membrane receptors (*except cortisol – cytoplasm receptor*)
- Don't need a 2nd messenger
- Fat soluble

Steroids:

"PET CAD"

Progesterone

Estrogen

Testosterone

Cortisol

Aldosterone

Vitamin **D**

Protein Hormones:

- Cell membrane receptors
- Require 2nd messengers (*except T_4 – nuclear hormone receptor*)
- Water soluble

Definitions:

Endocrine: secretion into blood

Exocrine: secretion into cavities (Ex: pancreas)

Autocrine: works on *itself* (Ex: TPO)

Paracrine: works on its neighbor (Ex: SS, motilin)

Merocrine: cell is **m**aintained => exocytosis

Apocrine: **ap**ex of the cell is secreted (Ex: most sweat glands)

Holocrine: the **whole** cell is secreted (Ex: sweat glands in groin, axilla)

Hypothalamus Hormones: use IP_3/DAG

GnRH: stimulates LH, FSH

GRH: stimulates GH

CRH: stimulates ACTH (uses cAMP)

TRH: stimulates TSH

PRH: stimulates PRL

DA: inhibits PRL

SS: inhibits GH

Nomenclature:

Somatotrope = GH

Gonadotrope = LH, FSH

Thyrotrope = TSH

Corticotrope = ACTH

Lactotrope = PRL

Posterior Pituitary Hormones:

ADH: conserve water, vasoconstrict

Oxytocin: milk letdown, baby letdown

Anterior Pituitary Hormones:

GH: IGF-1 release from liver

TSH: T_3,T_4 release from thyroid

LH: Testosterone release from testis, E_2 and Progesterone release from ovary

FSH: Sperm or egg growth

PRL: Milk production

ACTH: Cortisol release from adrenal gland

MSH: Skin pigmentation

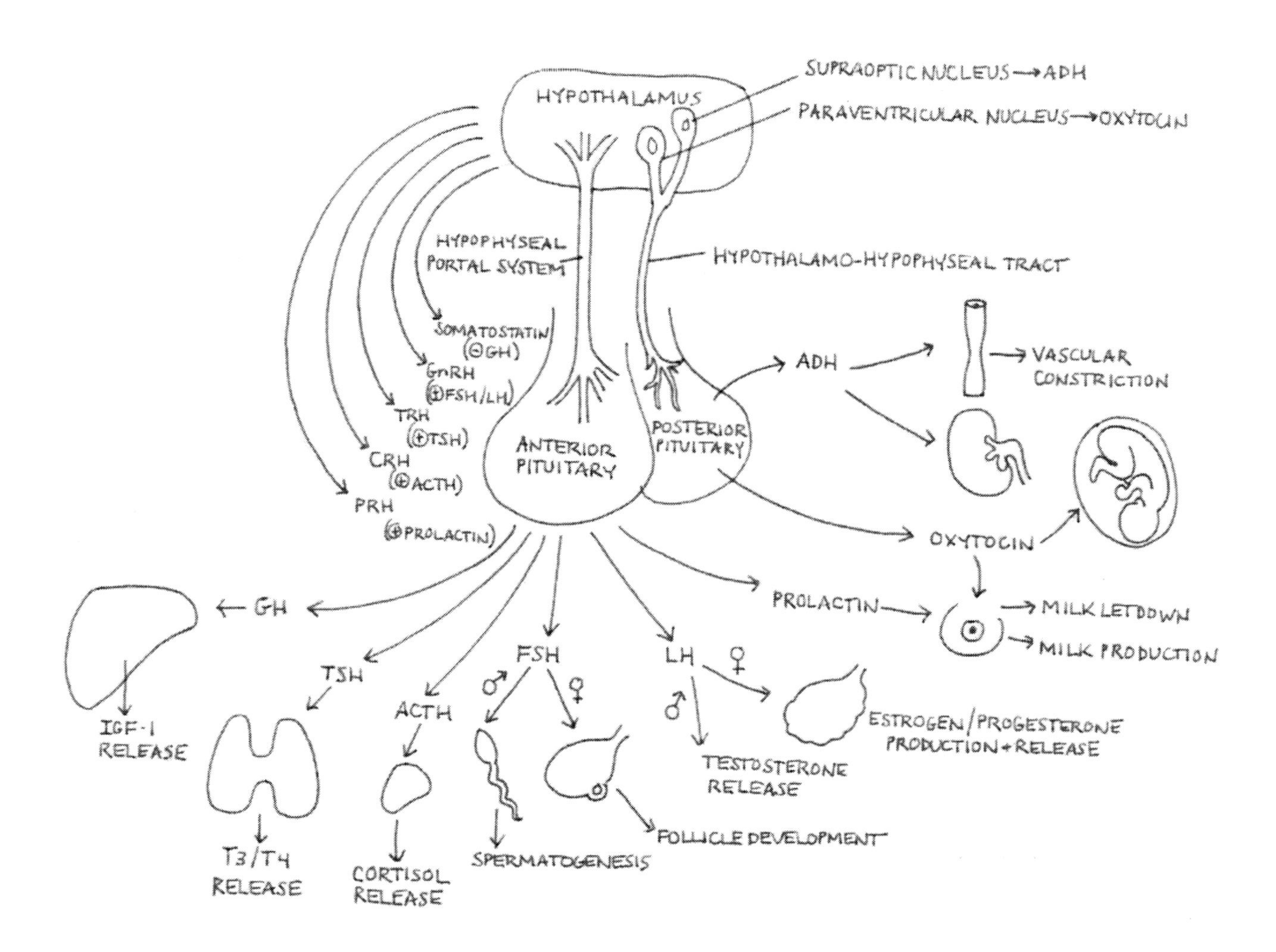

Hormone Overview:

ADH: conserves water, vasoconstriction
Aldo: reabsorbs Na (H_2O follows), secretes H/K in in kidney CD
ANP: inhibits Aldo, dilates renal artery (afferent arteriole)
Calcitonin: inhibits osteoclasts => low serum Ca^{2+} *"builds up bone-in"*
CCK: releases bile from GB/enzymes from pancreas *"like a WWE Smackdown!"*
Cortisol: gluconeogenesis by proteolysis => thin skin
Epi: gluconeogenesis and glycogenolysis in liver/adrenal cortex
Epo: erythropoiesis in bone marrow
Gastrin: stimulate parietal cells => IF, H^+ in stomach
GH: gluconeogenesis by proteolysis in liver => IGF-1 "somatomedin" to growth plates
GIP: enhances insulin action in pancreas => post-prandial hypoglycemia
Glucagon: gluconeogenesis, glycogenolysis, lipolysis, ketogenesis
Insulin: pushes glucose into cells everywhere except "BRICKLE"
Motilin: stimulates 1° peristalsis, MMC in duodenum
Oxytocin: milk ejection, baby ejection
Prolactin: milk production
PTH: chews up bone *"Phosphate Trashing Hormone"*
Secretin: secretion of bicarb, inhibits gastrin, tightens pyloric sphincter

No Insulin Required:
"BRICKLE"
Brain
RBC
Intestine
Cardiac, Cornea
Kidney
Liver
Exercising muscle

Somatostatin: inhibits secretin, motilin, CCK
Thymosin: helps T cells mature
VIP: inhibits secretin, motilin, CCK
Meet the Hormones: *Inhibition is the opposite of stimulation unless otherwise noted…*

RENAL HORMONES:

Hormone:	Function:	Second Mess:	Made:	Go:	Stimulated By:	Misc:
Epo	erythropoiesis	Tyr kinase	renal parenchy-mal cells	bone marrow	hypoxia (restrictive lung disease)	*polycythe-mia vera, renal cell CA*

Hormone:	Function:	Second Mess:	Made:	Go:	Stimulated By:	Misc:
ADH	opens water channels to dilute	IP_3/ DAG	hypothal-amus *(stored in posterior pituitary)*	kidney CD (V_2 receptors)	stress, high osmolarity	*responds in 30 min*

Diabetes Insipidus: too little ADH → urinate a lot → hypernatremia

- **Central:** brain not making ADH (Tx: DDAVP)
- **Nephrogenic** (XL-R)**:** blocks ADH receptor ← Li and Domecocycline
  - Tx: Thiazide diuretics *(paradoxical effect)*

DI: ↓ADH, dilute urine, ↑Na
SIADH: ↑ADH, conc. urine, ↓Na

Water Deprivation Test:
1) If urine concentrates → Psychogenic Polydipsia
2) If no response → Diabetes Insipidus → give DDAVP
   a) If urine concentrates >25% → Central DI (brain not making ADH) Tx: DDAVP
   b) If no response → Nephrogenic DI (receptors don't respond to ADH) Tx: HCTZ

SIADH: too much ADH → ↓serum Na (<115) → pee Na/retain H_2O

- Causes: Pain, Carbamazepine, Amph B, CA, Intracranial pressure, Restrictive lung dz
- Tx: Fluid restriction (turn IV down!) or Demecocycline/Li (ADH antagonists)
- Na<120 => seizures, arrhythmias (Tx: 3% NS)

Psychogenic Polydipsia: low plasma osmolarity, appropriate ADH levels

Hormone:	Function:	Second Mess:	Made:	Go:	Stimulated By:	Misc:
ANP	inhibits Aldo, dilates renal artery (afferent arteriole)	NO	wall of RA	kidney	stretch, high plasma volume	***ANP: Anti-Not Pee*** *(urinate) → heart failure causes polyuria*

ADRENAL CORTEX HORMONES:

Hormone:	Function:	Second Mess:	Made:	Go:	Stimulated By:	Misc:
Aldo	reabsorbs Na, secretes H^+/ K^+	None *(steroid hormone)*	adrenal ZG	kidney CD	low volume	↑*Aldo: Conn's tumor* ↓*Aldo: Addison's*

Adrenal Cortex: *"The deeper you go, the sweeter it gets"*
Zona Glomerulosa => Aldosterone "salt"
Zona Fasciculata => Cortisol "sugar"
Zona Reticularis => Testosterone "sex"

Steroid Synthesis Simplified:

17 →	Testosterone
21	
11 *"weak Aldo"*	
Aldosterone →	Cortisol

↓21-Hydroxylase → hypotension/ambiguous genetalia "CAH"
↓11-Hydroxylase → HTN/↑Na

Feedback Loop:
CRH → ACTH → Cortisol → ∅ACTH
POMC => proopiomelanocortin

- Opio part = β-endorphin => feel no pain
- Melano part = MSH => dark skin
- Cortin part = ACTH

Conn's syndrome: high Aldo due to tumor (↑Na, ↓K, ↑pH)

- HTN, H/A, muscle weakness, ↓renin
- Test: 24-hr urine Aldo on high-salt diet
- Tx: ACE-I (↑K), Spironolactone (blocks Aldo), Adrenalectomy

Hormone:	Function:	Second Mess:	Made:	Go:	Stimulated By:	Misc:
Cortisol	gluconeogenesis by proteolysis => thin skin	none *steroid hormone*	adrenal ZF	everywhere	hypoglycemia, stress	*responds in 2-4 hrs, lets all other processes continue under stress*

Low Cortisol: salt craving, hyperpigmentation, orthostatic hypotension

Addison's Disease "adrenal insufficiency": autoimmune destruction of adrenal cortex (↓Na, ↑K, ↑ACTH, ↓Cortisol)

- Test: Cosyntropin (ACTH) stimulation test: inject cortisol
  - No change → Addison's
  - ↑Cortisol → Pituitary dysfunction
- Tx: Hydrocortisone

Addisonian Crisis: fever, change in mental status, abdominal pain, orthostatic hypotension

- Tx: Hydrocortisone

Waterhouse-Friderichsen: adrenal hemorrhage ← N. meningitides

- Tx: emergent Methylprednisolone

Purulent Fallopian Tube:

To Adrenals: Waterhouse-Friderichsen (N. meningitidis)

To Liver: Fitz-Hugh-Curtis (N. gonorrhea)

High Cortisol: truncal obesity, moon facies, plethora, HTN, easy bruising, tan palmar creases, striae, buffalo hump

Cushing's syndrome: high cortisol (pituitary tumor, adrenal tumor, or small cell lung CA)

Cushing's disease: high ACTH (pituitary tumor)

Nelson's syndrome: pt with Cushing's had bilateral adrenalectomy, several years later, has hyperpigmentation + visual problems due to an unsuppressed pituitary adenoma → ↑ACTH

Tests To Order:

1) Morning cortisol
2) 24-hr urine cortisol
3) Low dose dexamethasone test
4) High dose dexamethasone test

Low-dose Dexamethasone Test (0.5mg q6 x 4) Dexamethasone = "Super Cortisol"

Suppress:

1) normal
2) obese (adipose tissue produces cortisol)
3) depression (stress produces cortisol)
Not suppress: have Cushing's => do high dose test

> **Survival Hormones:**
>
> Cortisol – permissive under stress
>
> TSH – permissive under normal

High-dose Dexamethasone Test (1mg q6 x 4)
"PAL"
Suppress: Pituitary tumor => ACTH (call brain surgeon)
Not suppress:
1) **A**drenal adenoma => Cortisol (call general surgeon)
2) **L**ung cancer (small cell) => ACTH, SIADH, Cushing's moon facies (call thoracic surgeon)

- Ectopic sites will produce much more ACTH than hyperstimulating the pituitary

Hormone:	Function:	Second Mess:	Made:	Go:	Stimulated By:	Misc:
Testo-sterone	external male genetalia	Tyr kinase	adrenal ZR	Testes	ACTH (utero) LH (after birth)	

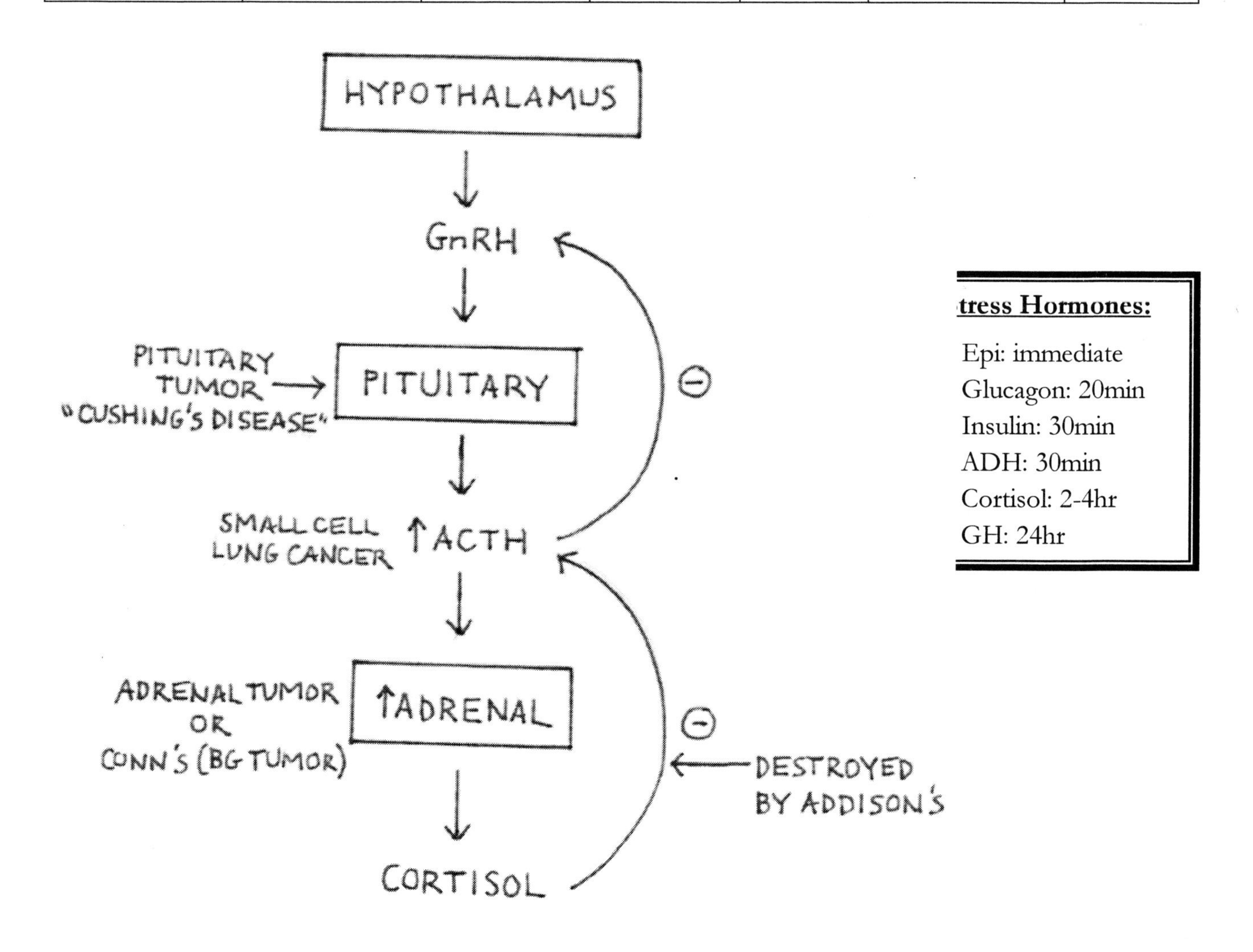

ADRENAL MEDULLA HORMONES:

Hormone:	Function:	Second Mess:	Made:	Go:	Stimulated By:	Misc:
Epi	gluconeogenesis, glycogenolysis	cAMP	adrenal medulla	liver, adrenal cortex	hypoglycemia, stress	*responds immediately, prefers β receptors*

Neuroblastoma: adrenal medulla tumor in kids; highest spontaneous regression rate

- abdominal mass
- dancing eyes "hypsarrhythmia"
- dancing feet "myoclonus"

Pheochromocytoma: adrenal medulla tumor in adults => sx come/go

- 5 P's: Palpitation, Perspiration, Pallor, Pressure (HTN), Pain (HA)
- Rule of 10's: 10% are malignant/calcify, familial, found in kids, bilateral, extra-adrenal
- Tests: Phentolamine (short acting α_{ns}-blocker => drop in BP), urinary VMA
- Tx: Phenoxybenzamine (irreversible α_{ns} blocker), follow Ca *"longer name acts longer"*
- Tx: No β-blockers! (unopposed α stimulation causes rapid ↑BP)
- Pre-op: 14d Phenoxybenzamine + fluid load (↑Na diet)
- Operative: Phentolamine (blocks α_{ns}) → Adrenalectomy → ↑NE/Epi → HTN crisis → β_1 blocker (everything's blocked except β_2) → depending on IVF to keep BP up…

PANCREATIC HORMONES:

Hormone:	Function:	Second Mess:	Made:	Go:	Stimulated By:	Misc:
Glucagon	gluconeogenesis, glycogenolysis, lipolysis, ketogenesis	cAMP	pancreas α cells	liver, adipose, adrenal cortex	hypoglycemia, stress (β_1 receptors)	*responds in 20min*

Hormone:	Function:	Second Mess:	Made:	Go:	Stimulated By:	Misc:
Insulin	pushes glucose into cells	Tyr kinase	pancreas β cells	everywhere except "BRICKLE"	hyperglycemia, stress (β_2↑insulin $\alpha 2$ ↓insulin)	*responds in 30 min*

Insulinoma: "Newborn Nesidioblastosis", high insulin => high C-peptide

↑C-peptide: Insulinoma
↓C-peptide: Insulin injection

Diabetes Mellitus: *ACE-I slow progression of nephropathy*

	Type I DM:	Type II DM:
Pathophys:	anti-islet cell Ab, GAD Ab, Coxsackie B	insulin receptor insensitivity
Insulin levels:	low insulin	high insulin (early)
Epidemiology:	thin Caucasian <20 y/o	obese Hisp,AA,NA >30 y/o
Genetics:	HLA-DR3 and HLA-DR4	genetic (95% twin concordance)
Toxicity:	DKA (must replenish K^+)	HONK
Symptoms:	polyuria, polydipsia, polyphagia	acanthosis nigricans
Notes:	90% islet cells destroyed => symptoms 10% hyperplasia => honeymoon period	organ damage, coma

Acanthosis Nigricans:
<50 y/o: DM type II
>50 y/o: Stomach CA

DKA: Kussmaul resp, fruity breath (acetone), altered mental status, follow anion gap
Tx: *"**I** **N**ever **K**ept **D**ogs"*

- Insulin drip: 7 units/hr (until ketones disappear, even if normal glucose)
- NS (to bring K down to normal)
- KCl: 20mEq (when K=5.5)
- D5 1/2NS (when Glucose=250 and Ketones=0)

DKA vs. HONK: calculate osmolarity…

HONK Tx:
1) 10L NS
2) Insulin drip
3) Replace electrolytes/bicarb

Foot Ulcer Risk Factors:

- DM/ Glycemic control
- Male smoker
- Bony abnormalities
- Previous ulcers

Dawn Phenomenon: 3am ↑glucose due to growth hormone

- Tx: ↑pm insulin

Somogyi Effect: 3am ↓glucose → 6am ↑glucose

- Tx: ↓pm insulin

Factitious Hypoglycemia: due to insulin injection (↑insulin, ↓C-peptide)

Erythrasma: rash in skin folds, coral-red Wood's lamp, micrococcus infection

- Tx: Erythromycin

Syndrome X = Metabolic Syndrome *"Pre-DM"*=> HTN, dyslipidemia, hyperinsulinemia

- Fasting glucose: 111-125 and
- Glucose challenge: 140-200 mg/dL

- Tx: Metformin

Tight glycemic control: ↓risk of retinopathy/nephropathy/neuropathy, ↓fetal anomalies
↓Glucose: caused by alcohol, exercise, pregnancy, delayed gastric emptying

DM Diagnosis:

1. Random glucose: >200 w/ symptoms or
2. Fasting glucose: >126 or
   a. repeat x 1
3. Glucose challenge:
   a. 75g: >200 @ 2hr
   b. 50g (if pregnant): >140 @ 1hr
   c. 100g (if pregnant): >125 @ 3hr

Hormones w/ Disulfides:

"PIGI"

- **P**RL
- **I**nsulin
- **G**H
- **I**nhibin

DM Retinopathy:

I) Background: weak capillary BM: leak plasma → retina

- Microaneurysms
- Hard exudates (sharp demarcation)
- Dot-blot hemorrhages
- Flame hemorrhages

II) Pre-proliferative:

- Edema of macula
- Infarction of retina (hypoxia) → soft exudates "cotton-wool spots"

Same α subunits:

1. LH, FSH
2. TSH
3. β-HCG

III) Proliferative:

- Neovascularization
- Ruptures
- Heal → collagen pulls → retinal detachment

DM Type I : (insulin does not cross placenta)
Hb_{A1c}: measures glycemic control over past 3mo = RBC lifespan

- <6% = good
- False + w/ hemolytic anemia

Short-acting:

- Lispro
- Regular
- Asparte

Intermediate-acting:

- NPH

Weight Gain DDx:

- Obesity
- Hypothyroidism
- Depression
- Cushing's
- Anasarca

- Lente

Long-acting:
- Glargine
- Ultralente

Initial Regimen:

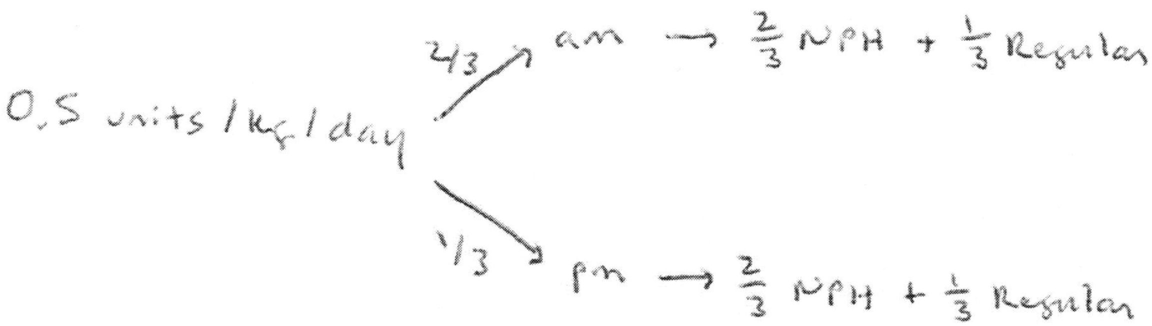

0.5 units/kg/day
- 2/3 → am → 2/3 NPH + 1/3 Regular
- 1/3 → pm → 2/3 NPH + 1/3 Regular

DM Type II:

1) Diet (low saturated fat), Exercise (weight loss upregulates insulin receptors)

2) Sulfonylureas: K^+ influx into islet cells → preformed insulin release *"let my insulin go!"*

1st Gen: weight gain, hypoglycemia
- Chlorpropramide – SIADH
- Tolbutamide
- Tolazamide
- O/D Tx: Dextrose, then Octreotide (inhibits insulin secretion)

2nd Gen: like 1st Gen, but inhibits gluconeogenesis and enhances peripheral glucose uptake
- Glipizide
- Glyburide – metabolized by kidneys *"y is lower like the kidneys"*

Overdose Tx:
1) Dextrose
2) Octreotide (inhibits insulin secretion)

Growth periods:
- 0-2 y/o
- 4-7 y/o
- Puberty

3) α-Glucosidase Inhibitors: ↓glucose **A**bsorption
- **A**carbose => flatulence (good for post-prandial hyperglycemia)
- Miglitol

4) BiGuanides: inhibits **G**luconeogenesis
- Metformin "Glucophage" – reacts with IV dyes, severe metabolic acidosis

5) Thiazolidinediones: ↑insulin sensitivity, check ALT/AST due to liver toxicity

- Pio**glit**azone "↑*sensitivity of the glit*"
- Rosi**glit**azone
- Tro**glit**azone – taken off the market b/c of liver failure

6) Incretin Mimetics:

- Exenatide "Byetta"

Hormone:	**Function:**	**Second Mess:**	**Made:**	**Go:**	**Stimulated By:**	**Misc:**
SS	inhibits secretin, motilin, CCK	cAMP	pancreas δ cells, duodenum	Nowhere *(paracrine function)*	high insulin, glucagon	**SSoma** => *steatorrhea, gallstones, DM*

GASTROINTESTINAL HORMONES:

Hormone:	**Function:**	**Second Mess:**	**Made:**	**Go:**	**Stimulated By:**	**Misc:**
Secretin	secretion of bicarb, inhibit gastrin, tighten pyloric sphincter	cAMP	Duodenum	Pancreas, Gall bladder	low pH	

Hormone:	**Function:**	**Second Mess:**	**Made:**	**Go:**	**Stimulated By:**	**Misc:**
CCK	digestion, bile release	IP_3/ DAG	duodenum	Pancreas => release digestive enzymes; GB => release bile	fat	*high pH inhibits*

Hormone:	**Function:**	**Second Mess:**	**Made:**	**Go:**	**Stimulated By:**	**Misc:**
GIP	enhances insulin action => post-prandial hypoglycemia	cAMP	duodenum	pancreas	low pH glucose	*insulin requires chromium*

Hormone:	Function:	Second Mess:	Made:	Go:	Stimulated By:	Misc:
VIP	inhibits secretin, motilin, CCK	cAMP	duodenum	nowhere *(paracrine function)*	high insulin, glucagon	**VIPoma** => *watery diarrhea*

Hormone:	Function:	Second Mess:	Made:	Go:	Stimulated By:	Misc:
Gastrin	stim. parietal cells => IF, H^+	Ca^{2+}	stomach antrum (G cells)	stomach body	high pH (>2)	ZE syndrome

Hormone:	Function:	Second Mess:	Made:	Go:	Stimulated By:	Misc:
Motilin	stimulates segmentation (1° peristalsis, MMC)	IP_3/ DAG	duodenum	duodenum *(paracrine function)*	low pH, distension	

PARATHYROID HORMONES:

Hormone:	Function:	Second Mess:	Made:	Go:	Stimulated By:	Misc:
PTH	chews up bone	cAMP	parathyroid chief cells	bone, kidney	low Ca, high Phos	

Parathyroid: Chief cells => PTH (Free Ca^{2+} = active)

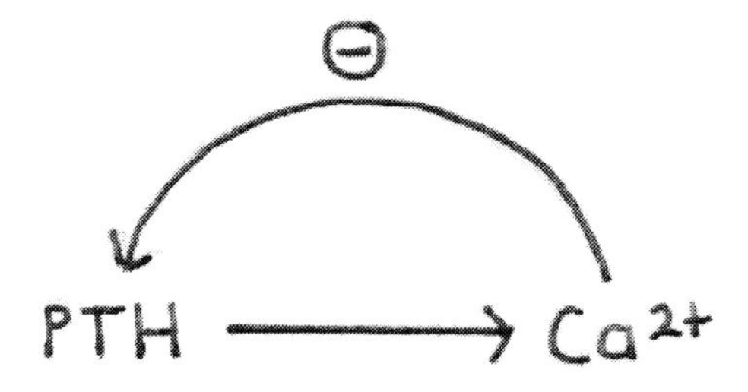

Chief Cells:
Parathyroid => PTH
Stomach => pepsin

Labs Simplified: *this little table is worth it's weight in gold…*

Ca:	Phos:	Disease:
↓	↓	↓Vit D
↓	↑	HypoPTH
↑	↓	HyperPTH

PTH: *"Pi/bone Trashing Hormone"*: ↓PO_4, ↑Ca^{2+}
To bone => osteoclasts (needs Mg, Vit A) => high Ca^{2+}
To kidney (PT) => waste phosphorus = low PO_4

To kidney (DT) => activate 1α-hydroxylase → activate Vit D
The Feedback Loop: PTH → ⊕Ca^{2+} → ∅PTH

Vit D: *"The Bone Builder"*: ↑PO_4, ↑Ca^{2+}
To GI: stimulates Ca-binding protein => high PO_4
To kidney (DT): activates Ca-ATPase => high Ca^{2+}
The Feedback Loop:

- Skin: "cholecalciferol" Vit D_3
- Liver: "calcifediol" 25-(OH) Vit D_2 ← 1α-hydroxylase ← PTH
- Kidney: "calcitriol" 1,25-$(OH)_2$ Vit D → ∅PTH

The Feedback Loops: whatever PTH does eventually come back to inhibit him (which is why he trashes PO_4 so it can't come back to haunt him)

HyperPTH:
1°: Parathyroid adenoma (MEN1,2): ↑PTH, ↑Ca^{2+}, ↓PO_4

- "bones" – osteoclasts
- "groans" – pancreatitis
- "stones" – kidney stones
- "moans" – psychosis
- Tx: ↓Ca: Furosemide, Calcitonin, Bisphosphonates
- Tx: Parathyroidectomy → ↓Ca → Bell's palsy

Basophil Prod:
"B FLAT"

- **F**SH
- **L**H
- **A**CTH
- **T**SH

2°: Renal failure (can't absorb/secrete => pseudohypo profile: ↑PTH, ↓Ca, ↑Pi)

- Tx: oral Ca (↓Pi to avoid metastatic calcification)

3°: Sarcoidosis

HypoPTH:
1°: Thyroidectomy (↓PTH, ↓Ca^{2+}, ↑PO_4) → intestine cells underexpress Ca transporters

Pseudo: Bad kidney PTHr (↑PTH, ↓Ca, ↑Pi)

- Sausage digits (short 3rd/5th digits), **Albright's osteodystrophy**, ↓urinary cAMP
- Cataracts, basal ganglion calcificaiton

PseudoPseudo: G-prot defect → no cAMP (↑PTH, ↑Pi, normal Ca^{2+})

Hungry Bone Syndrome: remove PTH → bone sucks in Ca^{2+}

Hormone:	Function:	Second Mess:	Made:	Go:	Stimulated By:	Misc:
Calcitonin	inhibits osteoclasts => low serum Ca^{2+}	cAMP	parafollicular "C" cells	bone	Ca^{2+}	*Medullary CA of thyroid*

Familial Hypocalciuria Hypercalcemia: ↓Ca excretion (urine Ca <200)

<u>MEN Tumors:</u>

MEN I "Wermer's": *"PPP"*

- **P**ancreas: gastrinoma/insulinoma/VIPoma (↑gastrin)
- **P**ituitary adenoma (↑PRL)
- **P**arathyroid hyperplasia (↑Ca)

MEN II "Sipple's": *30 y/o*

- Pheochromocytoma (↑VMA)
- Medullary thyroid cancer (↑calcitonin)
- Parathyroid hyperplasia (↑Ca)

MEN III "MEN IIb": *10 y/o*

- Pheochromocytoma (↑VMA)
- Medullary thyroid cancer (↑calcitonin)
- Oral/GI neuromas (Marfinoid habitus)

THYROID HORMONES:

Hormone:	Function:	Second Mess:	Made:	Go:	Stimulated By:	Misc:
T_3 = Liothyronine **T_4** = Levothyroxin	growth, differentiation *(brain grows for first 2 years)*	Tyr kinase	thyroid	everywhere	TRH → TSH	*Mom's T_3 lasts for 30 days after birth*

<u>Thyroid:</u> *Best test = TSH*

Thyroid: concentrates iodide → Thyroglobulin: I-Tyr-MIT(or DIT) in colloid

TSH → peroxidase → TG → 80% T_4, 20% T_3 (T_3 is the active form)

Liver: T_4 → T_3 via 5'-monodeiodinase

Sick: T_4 → rT_3 via 5-monodeiodinase (inactive storage when you're sick)

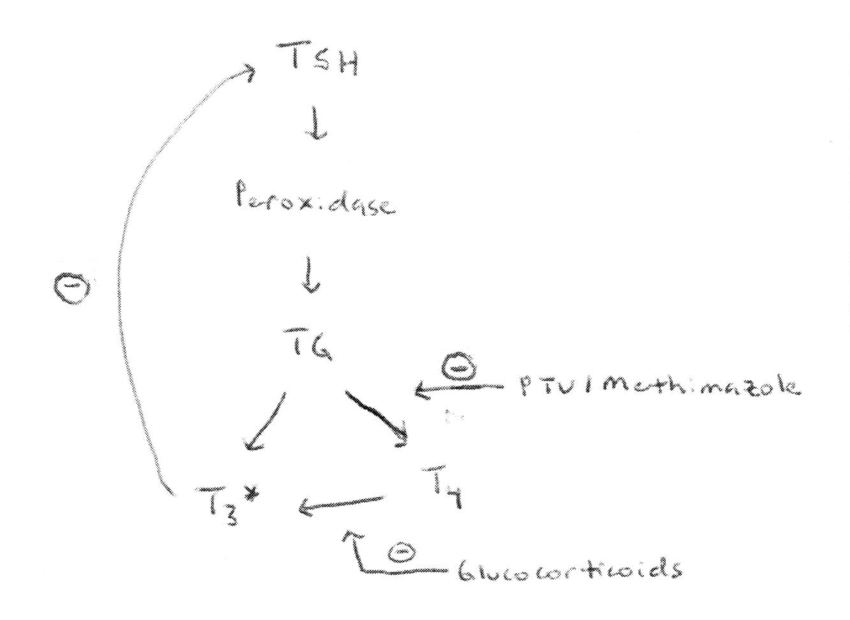

Exophthalmos:

- Grave's
- Histiocytosis X
- Cavernous sinus thrombosis

Enophthalmos:

- Horner's

Hyperthyroidism: speeds you up (or fatigue in elderly)

Grave's: TSHr Ab (the only stimulating Ab)

- Exophthalmos, pretibial (hard) myxedema, ↑risk of myasthenia gravis
- Tx: Radio-iodine ablation + Prednisone (for eyes)

DeQuervain's: subacute granulomatous thyroiditis (viral infxn invades thyroid => painful jaw)

- Tender/warm thyroid, lid lag, ↓uptake
- Tx: Propanolol/PTU or Methimazole

Silent thyroiditis: post-partum (immune system awakens after pregnancy → attacks thyroid)

- ↓radioactive iodine uptake

Thyroid Storm: fever, tachycardia, altered mental status

- Tx:

1) PTU or Methimazole
2) I or Li
3) Propanolol
4) Glucocorticoids: stops $T_4 \rightarrow T_3$

Hyperthyroidism:

Endogenous: High TG

Exogenous: Low TG

Plummer's disease: benign "toxic" adenoma in old people → ↑radioactive iodine uptake in nodule

1) Propanolol (slow heart so they won't die from heart failure)
2) Inhibit T_3, T_4 release:
   - PTU – also inhibits $T_3 \rightarrow T_4$ conversion, not cross placenta => agranulocytosis
   - Methimazole => scalp defects
3) Destroy thyroid w/ ^{131}I
4) Thyroid hormone replacement

Hypothyroidism:

1°: thyroid (measure TSH)

2°: pituitary (measure T_4)

3°: hypothalamus

5) No ASA (displaces T_4 from TBG)

Factitious Thyrotoxicosis: low TG

<u>Hypothyroidism</u>: slows you down (big tongue, ↓DTR)
Hashimoto's: CD4 kills thyroid

- anti-microsomal Ab "TPO"
- anti-peroxidase Ab
- anti-thyroglobulin
- Sx: coarse hair, slow speech, constipation, carpal tunnel, myxedema
- Tx: Levothyroxine "Synthroid" ← Ca/Fe inhibit absorption

Reidel's struma: "woody" CT in neck, die of suffocation, rule out thyroid CA

Cretin: congenital hypothyroidism

- Mental retardation, hoarse cry, big tongue, big fontanelle

Euthyroid sick syndrome: why you are tired after being sick

- "Low T_3 syndrome": ↓T_4 conversion to T_3 => hypothyroid sx

Myxedema coma: hypotension + coma

- Tx: may need ventilator
- 1) Levothyroxine "T_4" (OD Tx: PTU)
- 2) Hydrocortisone

Hyperthyroid: systolic HTN
Hypothyroid: diastolic HTN

Pregnancy: high total T_4 (T_4-TBG), nl TSH

Resistance to thyroid hormone: ↑T_3, ↑T_4, nl TSH

<u>RAIU:</u> *nuclear scan*
(+): Hyperthyroid/Adenoma
None: Thyroiditis
Low: Follicular

<u>TSH Tests:</u>
↑TSH: check free T_4
nl TSH: do FNA
↓TSH: do iodine scan → FNA (if cold)

Hormone:	Function:	Second Mess:	Made:	Go:	Stimulated By:	Misc:
TPO, Thymosin	help T cells mature	Tyr kinase	thymus	stays here *(autocrine function)*	T cells entering thymus	

MAMMARY HORMONES:

Hormone:	Function:	Second Mess:	Made:	Go:	Stimulated By:	Misc:
Oxytocin	milk ejection, baby ejection	IP_3/ DAG	Hypothal-amus (stored in post. pit.)	mammary glands	nipple stimulation	

Hormone:	Function:	Second Mess:	Made:	Go:	Stimulated By:	Misc:
PRL	milk production	Tyr kinase	pituitary acidophils	mammary glands	5-HT, TSH, nipple stimulation	*DA inhibits*

Pituitary:

Panhypopituitarism => hormones disappear in this order *(must replace cortisol + T_4):*
"Prolactin Goes First, Last Trails ACTH"

- **P**rolactin => no breast milk
- **G**H
- **F**SH/**L**H => 2° amenorrhea, male impotence
- **T**SH => hypothyroid *(need this to survive)*
- **A**CTH *(need this to survive)*

Acidophil Prod:
"GAP"
- **G**H
- **P**RL

Pituitary Tumors: *bitemporal hemianopsia, galactorrhea, amenorrhea*
Tx: cut off DA fibers, remove through sphenoid sinus (nose)
Expand laterally → CN 6
Expand upward → CN 2

- Non-functional tumors: 95% (Ex: chromophobic adenoma)
- Functional tumors: 5% (Ex: prolactinoma)
- Tx:

1)Bromocriptine
2)Cabergoline

Plummer-Vinson: esophageal webs
Plummer's: hyperthyroid adenoma

GROWTH HORMONES:

Hormone:	Function:	Second Mess:	Made:	Go:	Stimulated By:	Misc:
GH	growth, sends somatomedin to growth plates, gluconeogenesis by proteolysis	Tyr kinase	pituitary acidophils	liver => IGF-1 = "somato-medin"	GRH	*responds in 24 hrs, stress inhibits*

<u>Normal Growth Periods:</u>

- 0-2 yr
- 4-7 yr
- Puberty

<u>Too little GH:</u> kids w/ chronic disease are short
Pygmie: no somatomedin receptors
Midget: ↓somatomedin receptor sensitivity (proportional size)
Achondroplasia = Laron Dwarf: abnormal FGF receptors (extremities only)

<u>Too much GH:</u>
Gigantism: childhood presentation
Acromegaly:

- Adult bones stretch "my hat doesn't fit", coarse facial features, large furrowed tongue, deep husky voice, jaw protrusion, ↑IGF-1
- Tx: Octreotide (↓GH levels), Pituitary ablation

NOTES:

Rheumatology:

"Be temperate in wine, eating, girls, and sloth; or the gout will seize you and plague you both."

–Benjamin Franklin

Arthritis:

Rheumatoid Arthritis (HLA-DR4): pain worse in the morning (> 1 hr), autoimmune

- Attacks synovium → pannus (granulation tissue) in joint spaces
- Baker's cyst in semimembranous bursa (behind knee, rule out DVT)
- Swan neck deformities; Boutonniere's deformities (MCP, PIP joints)
- Subcutaneous nodules (fibrinoid necrosis w/ palisading epithelial cells)
- Synovial fluid: ↑protein (2-50k)
- Gets better with pregnancy

Synovial Fluid WBC:

>50k: Septic (N. gonorrhea)
2-50k: Rheumatoid arthritis/Gout/Lyme/Fungal/Virus
200-2k: Osteoarthritis/Hypothyroid (Tx: Acetaminophen)

Rheumatoid Factor:

IgM against IgG F_c

RA: 80% of pts

Sjogren's: 90% of pts

RA Tx Algorithim:

1) NSAIDs
2) + Hydroxychloroquine
3) + Methotrexate/Folate
4) + TNF-α Blocker:

- Etarnecept
- Infliximab
- Adlimumab

5) Glucocorticoid + Gold injection
6) Cyclosporine/Azathioprine

Juvenile RA: <16 y/o

- Pauciarticular: *knees/hips*
  - Type 1: ANA(+), iridocyclitis, good prognosis
  - Type 2: HLA-B27, bad prognosis
- Polyarticular: >5 joints involved → disability
- **Still's Disease:** fever at night, childhood death, salmon-colored rash (Tx: NSAIDs)

Polyarthritis: inflammation of eyes, ears, joints

Osteoarthritis: pain worse with activity, usually due to joint trauma

- Only non-inflammatory athritis
- "Gel phenomenon", crepitation of joints
- Heberden's nodes (DIP), Bouchard's nodes (PIP)
- Tx: Acetaminophen (not NSAIDs)

DIP Involvement:

- Osteoarthritis
- Psoriatic arthritis

Septic Arthritis: red, warm, swollen

- Dx: Synovial fluid WBC >50k (do biopsy/culture)
- Tx: Nafcillin + Cefepime

Tophi: gout crystals + giant cells
Podagra: big toe inflammation

Psoriatic Arthritis:

- Involves DIP joints asymmetrically
- HLA-B27
- Tx: Methotrexate + Etanercept (slows progression)

DeQuervain's tenosynovitis: occurs in Moms that lift babies

Pseudogout: Ca **P**yrophosphate crystals

- radiodense
- rhomboid crystals
- (+) birefringence
- usually in knee, wrist, shoulder (linear calcification of cartilage)
- ↑hemochromatosis, hyperPTH
- Tx: Thiazides

Gout: Uric acid crystals

- radiopaque
- needle crystals
- (-) birefringence
- usually in toes (toothpaste-like discharge)
- x-ray: rat-bite appearance

Acute Gout Tx: *no ASA or 5-FU (↓urate excretion)*

- Indomethacin: GI bleeding
- Colchicine: bad N/V/D, renal toxicity
- Intra-articular Corticosteroids
- Prednisone

Chronic Gout Tx: *Diet: No EtOH, Low purines*
Under-excretor Tx: ↑secretion of uric acid ← most common

- Probenicid *"Probes on the Excretor"*

Over-producer Tx: xanthine oxidase inhibitors

- Allopurinol – kidney failure
- Febuxostat – less side effects

Autoimmune Syndromes:
Systemic Lupus Erythematosis (SLE): young AA women

- Must meet 4 criteria: *"DOPAMIN RASH"*

Discoid rash: red, raised on face/scalp
Oral ulcers
Photosensitivity
Arthritis
Malar rash
Immunologic disorder: Anti-ds DNA, Sm, Cardiolipin Ab
Neurologic disorder: seizure or psychosis
Renal failure: *die of this*
ANA (+)
Serositis: pleuritis/pericarditis (Libman-Sacks endocarditis)
Hemolytic anemia

- Tx: NSAIDs, Steroids, Hydroxychloroquine, Methotrexate

SLE:

Type I: No renal dz
Type II: Mesangial dz
Type III: Focal proliferative dz
Type IV: Diffuse proliferative dz

Lupus Nephritis: anti-ds DNA Ab

- Tx: pulsatile Cyclophosphamide

Lupus Markers:

ANA: SLE sensitive
Anti-Smith Ab: SLE specific
Anti-ds DNA Ab: Lupus nephritis
Anti-neuronal Ab: Lupus cerebritis
Anti-histone Ab: Drug-induced SLE
Anti-SSA "Rho" Ab: Neonatal lupus (heart block)
Anti-cardiolipin Ab: recurrent abortions, false (+) VDRL

CREST:

*anti-**c**entromere Ab*
Calcinosis
Raynaud's
Esophageal dysmotility
Sclerodactly

Drug-induced Lupus: *"**H**HIPPPE" causes it, anti-**H**istone Ab*

- **H**ydralazine
- **H**ydroxyurea
- **I**NH
- **P**rocainamide
- **P**henytoin
- **P**enicillamine
- **E**thosuximide

Vasculitis:

High platelet count: Kawasaki
Normal platelet count: HSP

Scleroderma = Systemic Sclerosis: anti-**Scl**70 Ab

- Fibrosis, tight skin, esophageal stricture

Mixed Connective Tissue Disease: anti-RNP Ab

Raynaud's:

Tx: Nifedipine

- Scleroderma
- Takayasu's
- RA
- SLE

Takayasu's Arteritis "Aortic Arch Syndrome":

- Inflammation of aorta
- Asian women w/ weak pulses

Polyarteritis Nodosa (PAN): p-ANCA Ab

- Attacks gut and kidney, Hep B

Wegener's Granulomatosis: c-ANCA Ab

- Attacks **E**NT, **L**ungs, **K**idney *"ELK"*
- Sx: hemoptysis, epistaxis, hematuria
- Dx: biopsy

Goodpasture's: anti-GBM Ab

- Attacks lung and kidney => hemoptysis, hematuria
- Linear BM immunofluorescence, progresses to RPGN

Reiter's Syndrome: males, usually precipitated by Chlamydia

"can't see, pee, or climb a tree"

- Uveitis, Urethritis, Arthritis
- HLA-B27
- Keratoderma blennorrhagicum (look like a mollusk shell)
- Circinate balanitis (serpiginous penis plaques)
- Tx: Methotrexate + Tetracycline

Sjögren's Syndrome "keratoconjunctivitis sicca"**:** females, anti-SSB "La" Ab

- Dry eyes "sand in my eyes" or "Bitot spots"
    - Test: (+) **S**chirmer test →filter paper on eyelid
    - Tx: Pilocarpine (muscarinic agonist)
- Dry mouth, RA
- Dx: lip biopsy (specific)
- ↑Risk of Lymphoma

Vasculitis w/ Low C_3:

*"PMS in Salt Lake City"**

Post-strep GN

MPGN Type II

Subacute bacterial endocarditis

Serum sickness

Lupus

Cryoglobulinemia

BehÇet's Syndrome: recurrent oral and genital ulcers, uveitis

- Tx: Prednisone

Churg-Strauss: affects skin, kidney, nerves, GI, lungs, heart

- Asthma + Eosinophilia
- Tx: Zafirlukast

Felty's Syndrome: RA + leukopenia + splenomegaly

Kawasaki's disease "Mucocutaneous Lymph Node Disease":

"CRASH"

- **C**onjunctivitis
- **R**ash (palm/sole)

- **A**neurysm (coronary artery) → MI in kids (Echo every year)
- **S**trawberry tongue (like scarlet fever)
- **H**ot (fever > 102°F for at least 3 days + cervical lymphadenopathy)
- Tx: ASA, Flu vaccine (to avoid Reye's syndrome), IgG

NOTES:

NOTES:

Dermatology:

"Years wrinkle the skin, but to give up enthusiasm wrinkles the soul."

–Douglas Macarthur

Palm and Sole Rashes: *"TRiCKSSS"*

Let me tell you a little story about palm and sole rashes... Once upon a time a girl name Scarlet did some Tricks. She gets bit by a tick and gets Rocky Mountain Spotted Fever. Then, she sucks some Cox, jumps on her Kawasaki and crashes, leaving her with Scalded Skin. She's so crazy that when she went out on the town, she left her tampon in too long, and ended up with TSS and Syphilis.

- **T**oxic Shock Syndrome (Staph aureus) – tampon use (Tx: Clindamycin + 20L IVF)
- **R**ocky mountain (Rickettsia) – tender gastrocnemius (Tx: Doxycycline)
- **C**oxsackie A (Hand-Foot-Mouth disease)
- **K**awasaki
- **S**taph Scaled Skin
- **S**carlet fever (Strep)
- 2° **S**yphilis

Tx:
Face/Genitalia: 1% Hydrocortisone
Body: 0.1% Triamcinolone
Palm/Sole: Fluocinonide

Vesicular lesions:

Herpes: grouped vesicles on red base → painful solitary lesion

Varicella: red macule → clear vesicle on red dot → pus → scab

- Multiple stages present simultaneously

Dermatitis Herpetiformis:

- Ab to BMZ, IgA deposits in dermal papillae tips, assoc w/ celiac sprue

Red rash w/ mouth lesions:

Erythema Multiforme: target lesions (viral, drugs)

Stevens Johnson syndrome = Erythema Multiforme Major (mouth, eye, vagina)

Toxic Epidermal Necrolysis = Stevens Johnson w/ skin sloughing

Pemphi*gus* vulgaris: Ab against desmosomes

- Circular immunofluorescence, in epider*mis*, deposition b/w keratinocytes
- Oral lesions, (+) Nikolsky sign
- Tx: high dose steroids (fatal if untreated)

Bullous Pemphigoid: Ab against hemidesmosomes

- Linear immunofluoresence, in dermis, deposition on basement membrane
- "Floating" keratinocytes, eosinophils
- Tx: steroids

Allergic Rashes:

Atopic Dermatitis: associated w/ asthma, seasonal allergies (HS type 1)

Eczema: dry flaky atopic dermatitis in flexor creases (HS type 1) *"itch that rashes"*

- *Nummular dermatitis* = circular eczema
- *Spongiotic* = weeping eczema: scratching causes oozing *"like a sponge"*

- *Lichenification:* scratching => thick leathery skin

Goodpasture's disease: Ab against GBM => linear immunofluorescence (HS type 2)
Hives = Urticaria: "wheel&flare" (HS type 1) *if recurrent, rule out lymphoma*
Pityriasis Rosea: herald patch that follows skin lines (Tx: sunlight) *"C-mas tree pattern"*
Poison Ivy: linear blisters (HS type 4), type of contact dermatitis

Severe Pruritis:

1° Biliary Cirrhosis: bile salts, high alk phos, middle-aged female
Lichen Planus: autoimmune, **p**olygonal **p**ruritic **p**urple **p**apules
Scabies: linear excoriation "burrows" in webs of fingers, toes, belt line (Sarcoptes feces)

- Tx: Permethrin cream

Fitzpatrick Scale:
I-II: Burn
III: Tan after burn
IV: Tan
V: Brown

Other Skin Disorders:

Acne Tx:

- Retinoids: normalize follicular hyperkeratinization
- Minocycline: blue-grey pigmentation
- Isotretinoin: pseudotumor cerebri

Brown Recluse Spider Bite: painful black necrotic lesion (Tx: Dapsone)
Cellulitis: warm red leg, Staph aureus or Strep pyogenes (Tx: Cephalosporin or Vanc)
Cutaneous Anthrax: painless black necrotic lesion (eschar)
Decubitus Ulcer: "bedsore" (Tx: surgery)
DVT: blood clot in veins, associated w/ hypercoagulability (Tx: Warfarin)
Erysipelas: shiny red, raised, does not blanch, usually on face, assoc w/ Strep pyogenes

- Tx: Cephalosporins, Macrolides

Guttate Psoriasis: scaly stuff post-Strep
Ichthyosis: gradual lizard skin
Miliaria = "heat rash": burning, itching papules on trunk
Molluscum Contagiosum: fleshy papules w/ central dimple, pox virus (STD), Tx: Freezing
Psoriasis: silver scales on extensors, nail pitting, differentiated too fast, worse w/ stress

- Auspitz sign: remove scale => pinpoint bleeding
- Koebner's phenomenon: lesions at sites of skin trauma
- Tx: Topical coal tar, Calcipotriene, PUVA, UV-B, Steroids, Alephacet (anti-CD2 Ag)

Pyogenic Granuloma: vascular nodule at site of previous injury
Rosacea: blush, worse w/ stress/EtOH (Tx: topical Erythromycin, Metronidazole)
Seborrheic Dermatitis: dandruff in eyebrows, nose, behind ears
Seborrheic Keratosis (AD): rubbery warts with aging, greasy
Terry nails: white end of nails, liver cirrhosis
Thrombophlebitis: vein inflammation w/ thrombus (Tx: NSAIDs, warm compresses)
Vitiligo: white patches, anti-melanocyte Ab, assoc w/ pernicious anemia
Xeroderma Pigmentosa: bad DNA repair

The Red What?

Erythema **Chronicum Migrans:** Lyme disease (target lesion that spreads)
Erythema **Infectiosum:** Fifth disease "slapped cheeks" due to Parvovirus B19
Erythema **Marginatum:** Rheumatic fever (red margins)
Erythema **Multiforme:** Target lesions due to HSV, Phenytoin, Barbs, Sulfas
Erythema **Multiforme Major:** Stevens Johnson syndrome (> 1 mucosal surface)
Erythema **Nodosum:** Fat inflammation (painful red nodules on legs), sarcoidosis
Erythema **Toxicum:** Newborn benign rash (looks like flea bites w/ eosinophils)

NOTES:

NOTES:

Orthopedics:

"Sticks and stones may break our bones, but words will break our hearts."

–Robert Fulghum

Bone Remodeling:

1) Osteoclasts: dissolve bone (H^+, proteolytic enzymes)
2) Osteoblasts: build new bone (osteoid), occurs in periosteum
3) Mineralization (hydroxyapatite)

Kid Limps:

Legg-Calvés-Perthes: limp

- Avascular wedge-shaped necrosis of femur head
- Test: MRI

Osgood-Schlatter: "athlete's knee"

- Tibial tubercle inflammation
- Tx: NSAIDs, rest, ice

Slipped Capital Femoral Epiphysis: obese males w/ dull achy pain

- Test: Lateral x-ray "ice cream scoop falling off the cone"
- Tx: Emergent external fixation of hip w/ screws

Toxic synovitis: 2 y/o look normal w/ bad pain s/p URI

- Technetium scan uptake in hip joint

Bone Diseases:

Alkaptonuria "Ochronosis": kids w/ OA, black urine, homogentisic acid oxidase def.

Ankylosing Spondylitis: HLA-B27

- Ligament ossification: vertebral body fusion, kyphosis, stiffer in morning (better w/ exercise)
- *"Bamboo spine"* → x-ray sacro-iliac joint q3mo
- Uveitis, AR, pulmonary fibrosis, IgA nephropathy, syndesmophytes
- Tx: Indomethacin or COX-2 inhibitor or TNF antagonists

Cauda Equina Syndrome: *"saddle anesthesia":* can't feel butt, thighs, perineum

Costochondritis: painful swelling of chest joint-bone attachments, worse w/ deep breath

- Tx: NSAIDs, rest

Lumbar Disk Herniation: straight leg raise => pain, can't pee

- Tx: emergent disc decompression

Lumbar Stenosis: MRI *"hourglass"*, sciatica, pain relieved by sitting/leaning forward

Osteitis Fibrosis Cystica: inflammation of bone w/ holes, ↑PTH

Osteogenesis Imperfecta: blue sclera, multiple broken bones

Osteomyelitis: draining bones

- Most common: Staph aureus (Tx: Vancomycin)
- Sickle cell pts: Salmonella
- Drug users: Pseudomonas
- Test: MRI
- Tx: 6wk IV antibiotics

Osteopenia: lost bone mass

Osteopetrosis: ↓osteoclast activity => *marble bones* (obliterate own bone marrow)

Osteoporosis: *loss of bone matrix* (not calcification) => compression fractures

- Risk Factors: Old, Female, Thin, Sedentary, Smoker, Fam Hx, Alcoholic
- Type 1: Post-menopausal – increased osteoclasts
- Type 2: Senile (>75 y/o) – decreased osteoblasts
- Corticosteroid-induced – increased osteoclasts, decreased osteoblasts

Osteosclerosis: thick bones

Patella-Femoral Pain Syndrome: knee pain with long periods of sitting

Paget's Disease: "my hat doesn't fit", **p**aramyxovirus

- Fluffy bone, osteosarcoma, ↑osteoclasts/blasts, frontal bossing
- ↑CO heart failure, deafness
- ↑Uric acid (see stones on renal US), ↑alk phos, ↑urine Pro-OH
- Tx: Bisphosphonates

Rickets: soft bones (↓Vit D) => increased osteoid, waddling gait

- Craniotabes (soft skull)
- Rachitic rosary (costochondral thickening)
- Harrison's groove (soft ribs)
- Pigeon breast (sternum protrusion)

<u>Most Common Orthopedic Injuries:</u>

Ski/Football with lateral force: MCL tear
Dashboard injury: PCL tear
Clavicle injury: do angiogram to rule out subclavian a. injury (Tx: figure of 8 bandage)
Mom pulls kid onto curb: radial head displacement (Tx: reduce by supination)

Fall on hand: scaphoid fracture → tender anatomical snuffbox (Tx: 6wk tumb spica cast)
Fall on outstretched hand: radius/humerus shaft fracture (Tx: emergent pins)
Grab tree limb while falling: injure ulnar nerve → Klumpke's
Fall on elbow: humerus supracondylar fracture → "sail sign" → Volkmann's ischemia
Fall in shower: anterior dislocation of shoulder → lose round appearance → axillary n.
Tonic-clonic seizures: posterior dislocation of shoulder
Jump from building: fracture tarsal bone
Multiple bone fractures: fat emboli → confusion, upper body petechiae, eosinophils

__Ligament Tests:__ *do MRI*
ACL: anterior drawer "Lachmann" (feel knee "give way" or "pop" with forward pull)
PCL: posterior drawer (feel knee "give way" with backward pull)
MCL: tender at medial joint line
MM: McMurray test (feel "click" with knee twisting)

__Order To Fix:__
1) Dislocations
2) Open fractures
3) Nerve deficits

__Common Fracture Tx:__
Sling: clavicle fx
Cast: scaphoid/patella/ankle/calcaneus fx
Closed Reduction: humerus/talus fx, most kid fractures, dislocations
Open Reduction "ORIF": most adult fx, compound fx, severely displaced fx
External Fixation: contaminated wounds

__DVT Risks:__
1) Knee
2) Hip
3) Pelvis

__Scoliosis Tx:__ *pulmonary failure if untreated*
>20°: Brace
>40°: Spinal Fusion

T-score: -2.5 implies osteoporosis

__Osteoporosis Tx:__ *alk. phos. measures response to therapy*
1) Ca^{2+} (>1.5g/day)**, Vit D**

2) Calcitonin: nasal spray that decreases osteoclasts (tx bone pain)

3) Bisphosphonates: induces osteoclast apoptosis

- Alendronate "Fosamax" => esophagitis (stand 30min after), osteonecrosis of jaw (IV form)
- Risendronate
- Pamidronate: adjuvant therapy for metastasis
- Ibandronate "Boniva": take once-monthly

4) SERMs: Selective Estrogen Response Modulators

NOTES:

Ears/Nose/Throat:

"We have two ears and one mouth so that we can listen twice as much as we speak."

–Epictetus

Sinuses:

- Maxillary: only one that drains upward
- Ethmoid
- Sphenoidal
- Frontal

Infections:

Otitis Media: Strep pneumoniae

- Bulging tympanic membrane, immobile on pneumatic otoscopy
- Mastoid air cells → tympanic membrane perf → cholesteatoma (waxy granulation tissue)
  - Tx: Gentamycin ear drops
- Place tubes if >4 episodes/yr
- 1st Tx: Amoxicillin
- 2nd Tx: Amoxicillin-Clavulanate
- 3rd Tx: Tympanocentesis

Tongue Clues:

Big: Hypothyroid

Beefy Red: ↓Fe/↓Folate

Furrowed: Acromegaly

Strawberry: Kawasaki

Atrophy: ↓Vit B_{12}

Painful Swollen: ↓Vit B_6

Otitis Externa: Pseudomonas auregenosa

- "Swimmer's ear"
- Manipulation of auricle produces pain
- Tx: Ciprofloxacin + Ear wick with acetic acid solution
- Malignant: granulation tissue, DM, spreads to bone (Tx: Piperacillin-Ceftazidime

Mastoiditis:

- Otitis media + swelling behind ear
- Tx: Amp-Sulbactam (IV) + Myringotomy (drain the middle ear)

Parotiditis: Staph aureus

- Dehydrated/old patients
- Tx: Methicillin, I&D if not improved

Peritonsillar Abscess:

- Pain with swallowing, tilt head toward abscess, trismus, uvula deviation, 1 tonsil bigger
- Tx: I&D, Amp-Sulbactam

Retropharyngeal Abscess: Group A Strep

- "Epiglottitis" symptoms that occur over 2-3 days

Sinusitis: Staph aureus

- Face pain, bad breath, purulent discharge from nose

- Tx: Pseudoephedrine (or Amoxicillin if bacterial)

Viral Labyrinthitis: *dizzy, nystagmus, falls to one side*
- Tx: observe

Herpetic Gingivostomatitis:
- "Fever blisters"
- Red blisters that burst and leave ulcers behind in oral mucosa
- Lasts 1 week, will recur with stress

Ramsey-Hunt:
- Herpes zoster in ear, vesicles, vertigo

<u>Cavernous Sinus Thrombosis:</u>
- Sinus infection → clot in cavernous sinus
- CN6 palsy
- Exophthalmos/Ptosis
- Tx: Amoxicillin

<u>Vertigo:</u>

Acoustic Neuroma:
- Hearing loss, ringing, fullness
- HA/vertigo
- ↓sensation

Meniere's Disease:
- Triad: hearing loss, ringing in the ears, dizziness
- Tx: ↓Na diet, Triamterene (Na-losing diuretic)

Benign Positional Vertigo:
- Dizziness with head or eye movement, nystagmus
- Tx: Epley maneuver → Meclizine → Benzodiazepine

Motion Sickness:
- Due to stimulation of utricle
- Tx: Meclizine

<u>Nose Stuff:</u>

Most common location of nosebleed: anterior nasal septum
- Tx: nasal packing (3-4 days) + Dicloxacillin

Nasal Polyposis: Cystic Fibrosis

Nasopharyngeal Angiofibroma: heavy nosebleeds in teenage boys

Nasopharyngeal Cancer: Asians

NOTES:

Ophthalmology:

"Small is the number of them that see with their own eyes, and feel with their own hearts."

–Albert Einstein

Visual Acuity:

20/20 => can see bottom line of Snell chart 20ft away
20/100 => enlarge chart 5x to see bottom line of chart => legally blind

Terminology:
Amblyopia: difference in visual acuity (baby's head tilts toward bad eye)
Anisocoria: unequal pupils

- Newborns: genetic (AD)
- Adults: intracranial pressure

Blepharospasm: involuntary closure of eye with light (Tx: Botulim toxin)
Esotropia: cross-eyed

- Newborns: weak eye mm. for 1st 6mo (common) Tx: patch the good eye (3 y/o)
- >1y/o: eye turns in => intracranial pressure

Exotropia: eye turns out
Hyperopia: farsighted – need convex lens
Myopia: nearsighted – need concave lens
Presbyopia: loss of accommodation w/ aging (hold book at arm's length to read)
Stigmatism: defect in cornea
Strabismus: squint

Eye Diseases:
Blepharitis: eyelash inflammation
Conjuctivitis: red purulent eye discharge → keratitis (Tx: Erythromycin ointment)
Corneal abrasion: hazy light reflex due to foreign body
Corneal ulcers: white lesion due to contact lens
Cha*laz*ion: inner eyelid tumor *"it's laz-y and doesn't like to come out from under the eyelid"*
Hordeolum: stye, painful red lump near lid margin (Tx: warm compresses)
Pinguecula: yellow nodule on cornea, due to sun exposure
Pterygium: skin growth from conjunctiva to nasal side of cornea

Periorbital Cellulitis: pre-septal, ↓eye movements
Orbital Cellulitis: post-septal, proptosis (Tx: Amp-Sulbactam IV, I&D)

Cataract: monocular diplopia, will not recur after surgery
Macular degeneration: blind spots, decreased central vision (Tx: photodynamic therapy)
Retinal detachment: painless, flashes of light, floaters, amaurosis fugax (Tx: surgery)

Argyll-Robertson pupil: pupils accommodate but do not react, 3° syphilis
Kayser-Fleischer ring: ring of gold pigment around iris, Wilson's disease
Marcus-Gunn pupil: dilates with light (instead of constricting, CN 2 defect)

Parinaud's syndrome: bilateral paralysis of upward gaze, pineal tumor *"can't look up to pineal"*

Abnormal Eye Reflexes:

White reflex: Retinoblastoma → osteosarcoma

No red reflex: Cataracts – hard lens

- Newborns: sporadic *(may be caused by rubella virus)*
- Adults: sunlight, any disease with high sugar

Glaucoma:

Open-angle: overproduction of fluid causes high intraocular pressure

- Sx: Painless, ipsilateral dilated pupil, gradual tunnel vision, optic disc cupping
- Tx: Pilocarpine, β-blocker, Acetazolamide, Latanoprost (PG)

Closed-angle: obstruction of canal of Schlemm

- Sx: Sudden onset, pain
- Tx:
- 1) Induce miosis: Pilocarpine, α-agonist
- 2) Decrease aqueous production: β-blocker, Acetazolamide (CA inhibitor)
- 3) Laser iridectomy

Eye Clues:

Yellow color: Bilirubin

Yellow vision: Digoxin toxicity

Blue sclera: Osteogenesis imperfecta

Opaque: Cataract

Aniridia (no iris): Wilm's tumor

Iridocyclitis: Juvenile RA

Roth spot (central white hemorrhagic spot): Bacterial endocarditis

Cherry-red macula: Tay-Sachs, Niemann-Pick, Central retinal artery occlusion

Brown macula: Malignant melanoma

Retinitis Pigementosa (brown spot in retina): Friedrich's ataxia

Lisch nodules: Neurofibromatosis type I

Brushfield spots (speckled iris): Down's

Mongolian-slanted eyes: Down's

Almond-shaped eyes: Prader-Willi

Dislocated lens from top: Homocysteinuria

Dislocated lens from bottom: Marfan's

Yellow cholesterol emboli in retinal artery: Hollenhorst plaque

Hypertensive Retinopathy: neovascularization → *DM*

Grade I: arteriole narrowing, copper wiring

Grade II: A-V nicking

Grade III: retinal flame-shaped hemorrhages, cotton wool spots, waxy exudates
Grage IV: disc swelling

Central Retinal Vein Occlusion:

- "Blood and thunder" fundus, looks like swirls of bloody mass…
- Blurry vision worse in the morning

Central Retinal Artery Occlusion:

- Due to emboli
- Cherry red macula, pale retina
- Tx: Ocular massage

NOTES:

NOTES:

EKG/ Arrhythmias/ Electrolytes:

“”It’s the heart afraid of breaking that never learns to dance. It is the dream afraid of waking that never takes the chance. It is the one who won’t be taken who cannot seem to give. And the sould afraid of dying that never learns to live.”

–Bette Midler

Depolarize: *become positive from baseline*

Na channel:

m gates – outside

h gates – inside

- m gate opens => Na^+ leaks in => slow upstroke *"slow channels"*
- hit threshold potential => m/h gates open => Na^+ rushes in => ↑resting potential *"fast Na channels"*
- can't reach +65 b/c the higher you go => ↓Na^+ driving force, ↑K^+ driving force
- m gate closes => can't start another action potential = absolute refractory period

2) Repolarize: *become negative from a positive potential*

K^+ can actually reach its potential of -96 (unlike Na^+) b/c of K^+ leak channels

3) Hyperpolarize: *become more negative than baseline*

3 Na^+ in, 1 Ca^{2+} out channels reset the electrical membrane potential (1:1 in rvs direction)

3 Na^+ out, 2 K^+ in ATPase resets the concentration gradient (max activity at -96)

- Net positive → Na^+/Ca^{2+} pump
- Net negative → Na^+/K^+ pump

4) Automaticity: *resetting the membrane potential*

Charge movement: *All pumps that don't have an ATPase can be reversed…*

- Na/K pump: 3 Na^+ out, 2 K^+ in

→ net *negative* charge

- Ca/Na pump: 3 Na^+ in, 1 Ca^{2+} out

→ net *positive* charge

- Reverse Ca/Na pump: 1 Na^+ out, 1 Ca^{2+} in

→ net *positive* charge

Current (I) = change in membrane potential caused by *movement* of ions

Resting membrane potential: K channels are responsible for this

K^+: higher conductance/permeability

Na^+: greater driving force

Action potential: "all or none" => reach threshold, then fires (any extra = "overshoot")

Depolarize (Na) → Repolarize (K) → Automaticity (Na)

The Exception: SA/AV nodes (use Ca^{2+} to depolarize)

Extracellular Ca^{2+} => atrial depolarization

Intracellular Ca^{2+} => contractility

Cardiac Action Potentials:

Phase 0: Na^+ or Ca^{2+} in (depolarization)

Phase 1: K^+ out (initial repolarization)

Phase 2: Ca^{2+} in "plateau phase" (conduction to AV node) => contractility

Phase 3: K^+ out (repolarization)

Phase 4: Na^+ in (automaticity = hyperpolarization) – who can reset fastest => ↑slope

Absolute refractory period: all depolarization, some repolarization

Relative refractory period: rest of repolarization => need bigger stimulus to fire

Conduction Anatomy:

SA node: RA wall (near SVC)

AV node: interatrial septum

Purkinje: IV septum

Anatomy:

- RA/RV – more anterior *"Right behind chest"*
- LA – compresses esophagus

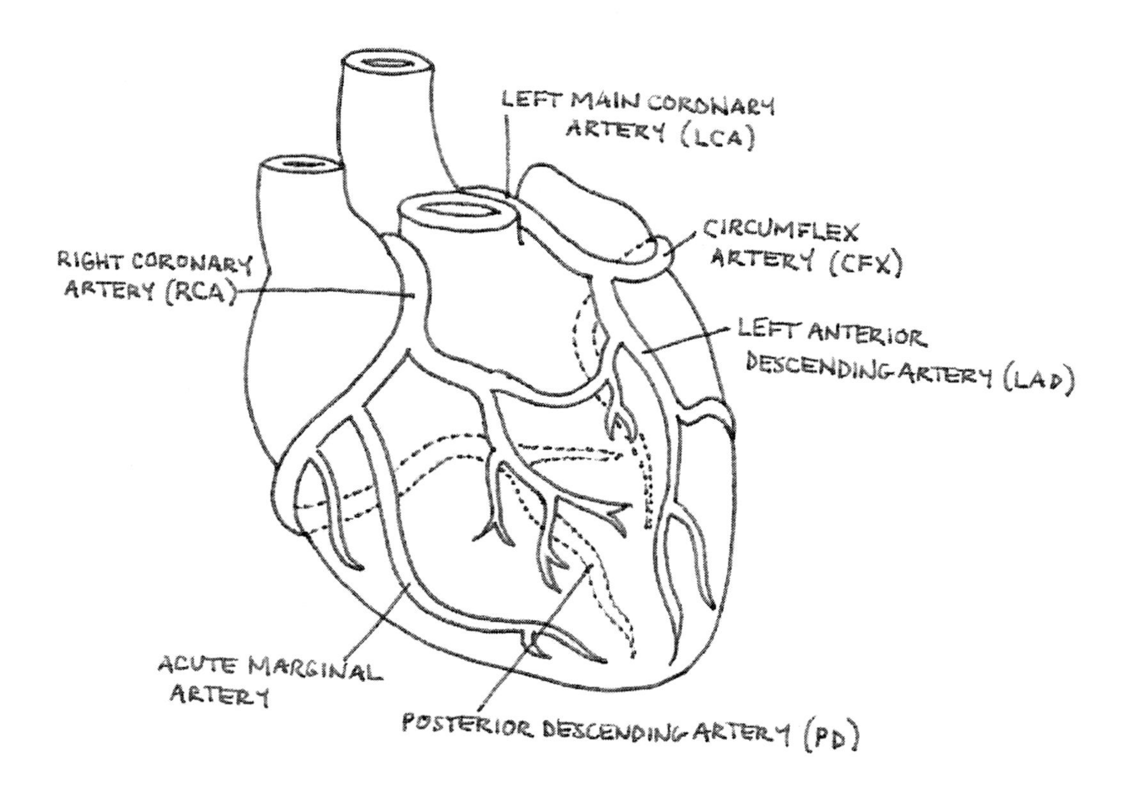

The dominant artery: the artery that *supplies the SA node* (usually the right coronary)

Blood supply: lose 40% blood supply => heart failure

85% of heart:

Aorta → Left main coronary a.→ L circumflex artery (LA) → marginal artery (LV)

→ LAD (ant. wall, septum, inf. wall, lower 1/3 post. wall)

15% of heart:

Aorta → Right coronary a. → (SA/AV, septum, top 2/3 post wall)→ R marginal (RV)
→ Posterior IV

Atria:

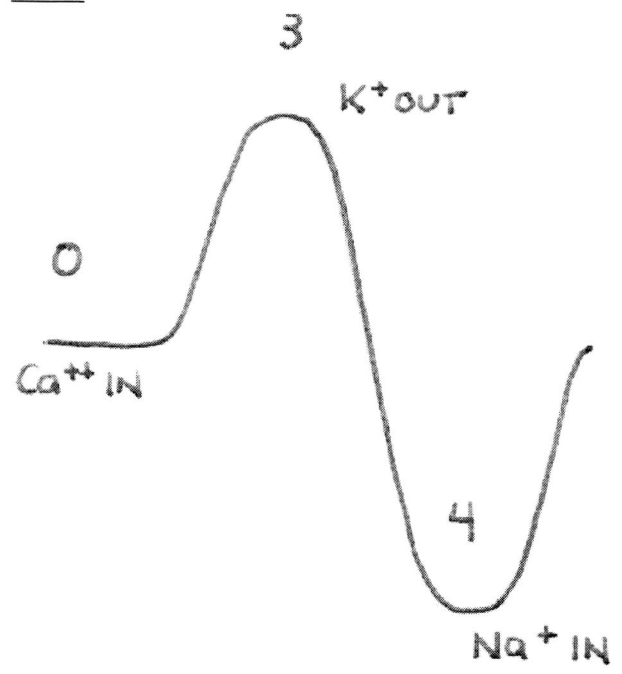

Atria Effects:

↑ Ca^{2+}: Arrhythmia

↓ Ca^{2+}: Heart Block

- Phases 0,3,4 only
- *Fires slow, but resets fast*
- Ex: Shock the heart to pause it so SA node (no phase 2) can take over
- Ex: Lidocaine attacks ischemic tissue only => silences ectopic site => SA node rules

Ventricles:

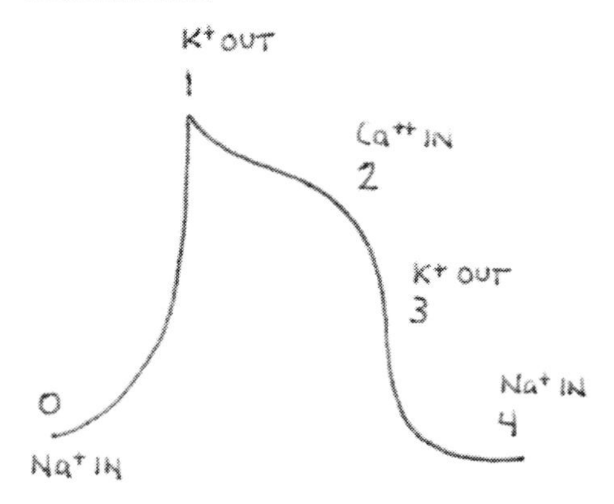

Ventricle Effects:

↑ Na^{+}: Arrhythmia

↓ Na^{+}: Arrhythmia

(due to Na/Ca channel)

- Now has Phase 1,2: lengthens absolute refractory period
  => why tetany (low Ca^{2+}) does not affect your ventricle
- *Fastest firing (Purkinje fibers), resets slow* => no control
- Anterior wall holds on to contraction longest => longest phase 2

EKG Waves:

P wave = Atrial depolarization (Phase 0)

- PR = Conduction from SA to AV node (Phase 2)

Q wave = Ventricle septum depolarization

R wave = Ventricle anterior wall depolarization

S wave = Ventricle posterior wall depolarization

- QRS = Total ventricular depolarization (Phase 0)
- QT = Ventricular depolarization/repolarization *(if too long, ectopic sites take over)*
- ST segment = Ventricle "plateau phase" (Phase 2)

T wave = Ventricle repolarization (Phase 3)

U wave = Ventricle automaticity (Phase 4)

MI EKG changes:

V_{1-2}: Septal

V_{3-4}: Anterior

V_{5-6}: Low lateral

I/AVL: High lateral

II/III/AVF: Inferior

Height = voltage:

Tall => enlarged ventricle

Short => small or compressed ventricle (or inflammation)

Width = duration:

Narrow => hypertrophy

Wide => dilated

Depolarization:

- SA node → AV node and LA → pause → IV septum → to RV → around apex to post side of heart
- Bachman's fibers – go from RA to LA so that atria will beat at the same time

Repolarization: starts on the posterior side of heart, opposite of depolarization

EKG Leads:

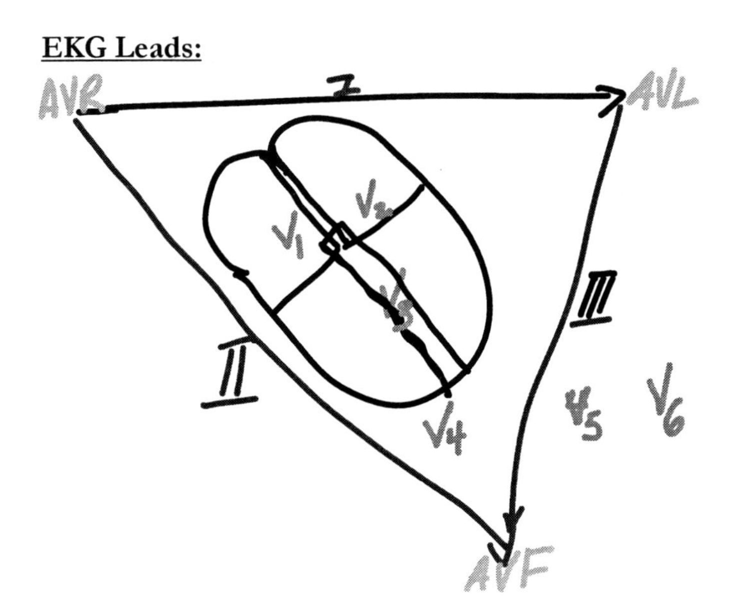

Bipolar Leads: have a negative and positive end
Lead 1: right arm to left arm, looking at heart from + electrode on L arm (sees **left side**)
Lead 2: right arm to left leg (sees **RV**)
Lead 3: left arm to left foot (sees **LV**)
AVR: on right arm (sees **RA**) *"The AV node sees the RA on its right"*
AVL: on left arm (sees **LA**) *"The AV node sees the LA on its left"*
AVF: left foot (sees **apex**)

Precordial leads: use V_1 and V_2 to see IV septum
V_1 – right upper sternal border (sees **RA**)
V_2 – left upper sternal border (sees **LA**)
V_3 – no anatomical site, half-way between V_2 and V_4 (sees **anterior wall**)
V_4 – left lower sternal border (sees **apex**)
V_5 – mid-clavicular line (sees **LV**)
V_6 – mid-axillary line (sees **LV**)

Electrodes only detect positive charges: see a wave coming toward it => + reflection
coming away from it => - reflection

EKG Interpretation:

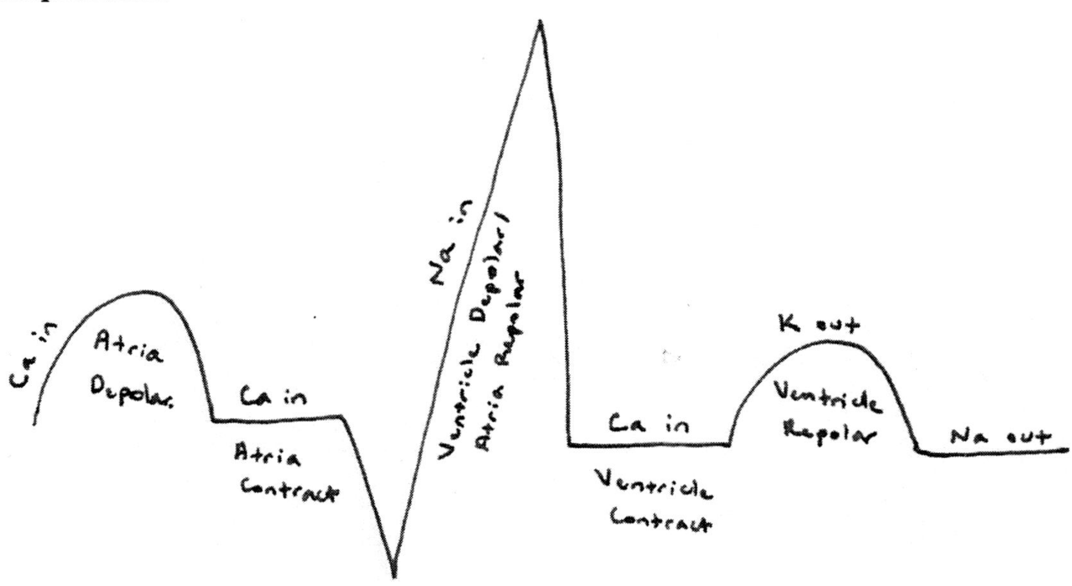

Summary:
↑**K:** peaked T waves
↓**K:** U wave
↑**Ca:** short QT
↓**Ca:** prolonged QT

Arrhythmia Tx: *Max heart rate = 220 – age*
Atrial arrhythmias: use Ca^{2+} to depolarize => use Ca channel blocker, then Warfarin
Ventricular arrhythmias: use Na^+ to depolarize => use Na channel blocker

Heart Block:

1st Degree: PR >5 small squares → bad SA node (Tx: exercise)

2nd Degree:

- o **Mobitz I:** PR lengthens "winks" until drops QRS → bad AV node (Tx: Pacemaker if sx)
- o **Mobitz II:** PR fixed, but some QRS are gone → bad His-P (Tx: Pacemaker)

3rd Degree: Regular P-P and R-R, but don't correspond → destroyed AV (Tx: Pacemaker)

Pacemakers:

Overdrive Pacemakers: use guidewire to get control away from ectopic site

On-demand Pacemakers:

1st letter = chamber location

2nd letter = chamber you are sensing

3rd letter = what you want pacemaker to do (I=inhibit)

PSVT Tx:

1) Monitored carotid massage

2) Adenosine

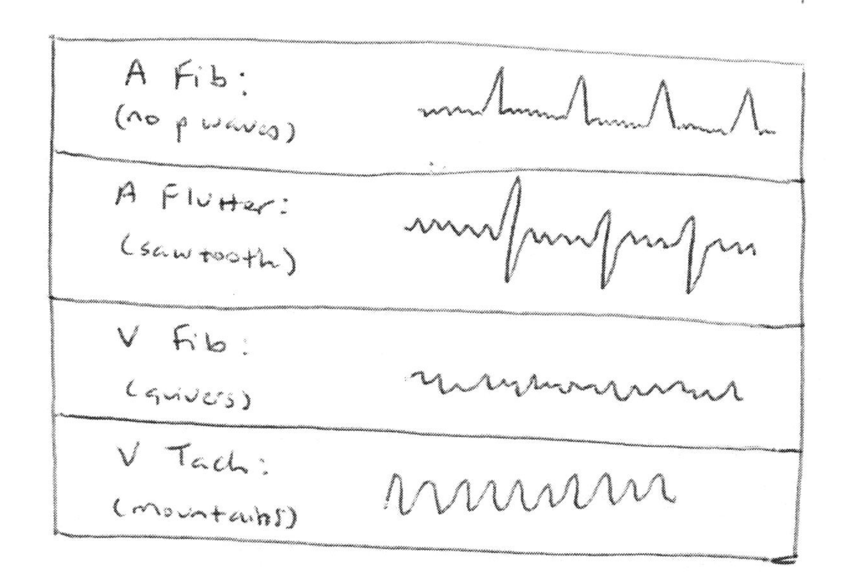

Premature Ventricular Complexes (PVCs):

Premature Beat: QRS has a pause after it

Bigeminy: PVC every other beat

Trigeminy: PVC every third beat

Ventricular Flutter: ribbon-like, "Torsade de Pointes"

- Tx: Mg, β-blocker

Ventricular Tachycardia: 3+ consecutive beats with HR>150, looks like mountains

- Tx: Amiodarone if BP normal, otherwise tx like V Fib

Ventricular Fibrillation: No recognizable QRS, looks like quivers

- Tx: Alternate Shock/Drugs (200→300→360J):

ACLS Reasoning:

1) Epinephrine – lowers threshold for cardioversion
2) Vasopressin "ADH" – holds H_2O to ↑BP
3) Amiodarone – blocks K, stops all cells
4) Lidocaine – blocks Na, stops ventricle, only acts on ischemic tissue
5) Procainamide – blocks Na, stops ventricle
6) Mg – makes cells less likely to depolarize

(+) Pulse: synchronize when you shock
No Pulse/V Fib: unsynchronized

Atrial Arrhythmias:

Premature Atrial Contraction: has a pause after it
Atrial Flutter: sawtooth pattern
Atrial Fibrillation: irregularly irregular, no p waves
Tx:
1) Slow rythym: Diltiazem/Metoprolol (for HR >120)
2) Increase contraction: Digoxin
3) Chemical Cardioversion: Amiodarone
4) Anticoagulate (for A Fib >48hr): Warfarin, Heparin

- Start 3 wk before cardioversion; stop 3wk after sinus rhythm

Electrolyte Imbalance Tx:

- Hyperkalemia means there is more K in the bloodstream (not the cell)
- Ca^{2+} and Mg^{2+} get to the door first in the race with Na^+
- Dilute with NS first… If all else fails, do hemodialysis

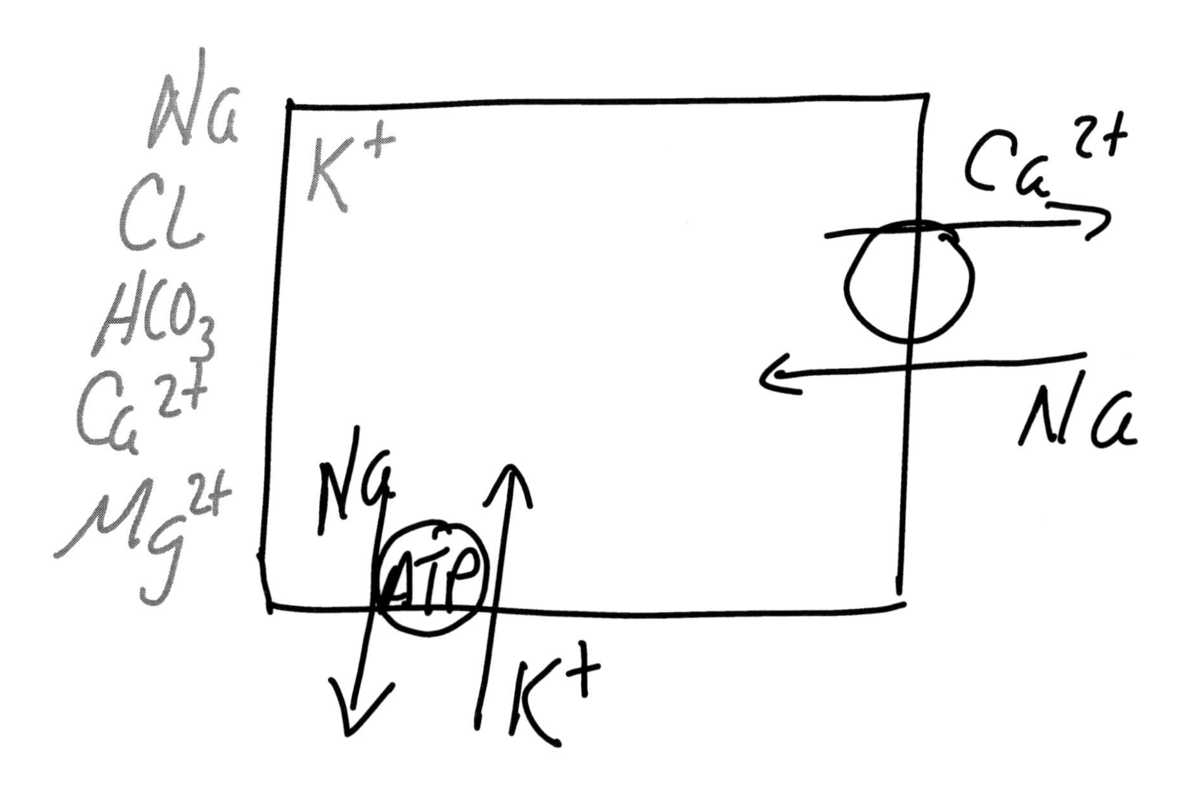

Low Mg: *more likely to depolarize → monitor DTR, outs, EKG, vitals q10min*

- Give Mg sulfate

High Mg: *less likely to depolarize*

1) Ca Gluconate (move Mg out of the way)
2) Furosemide (pee Mg out)

Corrected Ca:

0.8 (4-Albumin) + Ca^{2+}

Low Ca: *more likely to depolarize*

- Ca Gluconate
- ***Chovstek's sign:*** tap facial nerve => muscle spasm
- ***Trousseau's sign:*** BP check causes carpal spasm

Causes of ↑Ca:

- ↑PTH
- ↑Vit D
- Cancer

High Ca: *less likely to depolarize (except atrium)*

1) NS
2) Furosemide – pee Ca/Mg out
3) Calcitonin – intranasal
4) Pamidronate (bisphosphonate) – if Ca >16, takes 3 days to work
5) Mithramycin

Low K: *less likely to depolarize, narrow T waves (K leaves → negative cell)*

- Give K^+ *(<10mEq/hr)*

High K: *more likely to depolarize, peaked T waves*

"Can I Beat K?"

1) **C**a gluconate – save the SA node (unless pt is on digoxin)
2) **I**nsulin/Glucose (or Albuterol IV) – pushes K and glucose into cells (Na/glucose → Na/K pumps)
3) **B**icarb – make kidney pee K out
4) **K**ayexalate – cation exchange resin to poop K out

Low Na: *more likely to depolarize (Na goes out of cell → Ca flows in → + cell)*

- 0.9% NS *(<0.5 mEq/hr to avoid central pontine myelinolysis)*
- Seizures: 3% NaCl

High Na: *more (early) then less (late) likely to depolarize (Na in → + cell → Na leaves via Na/K pump)*

- 1/2 NS

Low PO_4:

- Give phosphate

High PO_4: *Refeeding Syndrome in anorexics*

1) Ca Carbonate (binds PO_4 in gut)
2) Insulin/Glucose (rapid PO_4 exchange)

Solutions:

Colloid: Albumin

Crystalloid: Na

Milk Alkali Syndrome:

- *Eat lots of Ca, short QT*
- Tx: IVF → Lasix (get rid of Ca)

ACE-I: ↑K

Albuterol: ↓K

NOTES:

NOTES:

Neuromuscular:

"Problems are to the mind what exercise is to the muscles, they toughen and make strong."

–Norman Vincent Peale

Central nervous system: brain/spinal cord (Cl^- in hyperpolarizes) – Oligodendrocytes

Peripheral nervous system: everything else (K^+ out depolarizes) – Schwann cells

Autonomic nervous system = automatic stuff
Somatic nervous system = moving your muscles

	Parasympathetic: *"sit and think"*	**Sympathetic:** *"get up and go"*
Function:	Rest-and-Digest => slows stuff down	Fight-or-Flight => speed stuff up
2nd Messenger:	cGMP	cAMP
Control:	Craniosacral: brain + below the belt	Thoracolumbar: above the belt
Preganglionic NT:	ACh (nicotinic receptor) Long fibers	ACh (nicotinic receptor): except sweat Short fibers
Postgangiolic NT:	ACh (muscarinic receptor): except skeletal mm, ganglia Short fibers	NE (α or β receptor): except sweat glands Long fibers
Side Effects:	*"DUMBBELS":* **D**iarrhea **U**rination **M**iosis *"constrict"* **B**radycardia **B**ronchoconstrict **E**rection *"point"* **L**acrimation **S**alivation	*Opposite of Parasympathetics:* Constipation Urinary retention Mydriasis *"eyes wide with fright"* Tachycardia Bronchodilate Ejaculation *"shoot"* Xerophthalmia (dry eyes) Xerostomia (dry mouth)

Stimulatory NT: depolarize
Tyr → (Tyr OHase) → DOPA → DA → NE → MAO (pre-synaptic) + COMT (post-synaptic)
Trp → (Trp OHase) → 5-HT (High: sleepy, Low: depression) → 5-HIAA
AcetylCoA + Choline → (Choline Acetyltransferase) → ACh → (AChase) → Acetate + Choline

Inhibitory NT:
Spinal cord: Gly
Brain: GABA

Catecholamines:
NE: NT
Epi: Hormone

COMT Inhibitors:

- Tolcapone

- Entacapone

Seratonin Agonists:

- Cisapride – tx GERD, not used due to Torsade
- Methysergide – tx headaches, die of MI, off market
- Sumatriptan – tx acute migraines
- Elatriptan

I) Adrenergic Receptors: *think sympathetic*

α_1 receptors: *vasoconstricts, uses Ca^{2+} (IP_3-DAG 2nd messenger system)*

- Sphincters => tighten
- Arteries => vasoconstrict
- Eye radial muscles => mydriasis w/o cyclopegia (freeze iris via radial muscles)

α_1 Agonists: *"Promote Enlarged Pupils"*

- **P**henylephrine – tx neurogenic shock
- **E**phedrine – OTC cold remedies
- **P**seudoephedrine – abused on street to make methamphetamine

α_1 Blockers: *"Decrease Prostate/Tame Tension"*

- **D**oxazosin – tx BPH/HTN
- **P**razosin – tx HTN only, priapism, 1st dose syncope *"Pass out"*
- **T**erazosin – tx BPH/HTN
- **T**amsulosin – only works on bladder/prostate, less side effects

α_2 receptors: *↓NE release*

- Decrease sympathetics
- Pancreas β cells (↓ insulin)

α_2 Agonists: *"Can Greatly Ameliorate (HTN)"*

- **C**lonidine – tx HTN, rebound HTN if stopped quickly
- **G**uanabenz – tx HTN
- α-Me-DOPA – tx HTN in pregnant women, hemolytic anemia

α_2 Blockers: *"Treat Your (Impotence)"*

- **T**olazoline – tx premie RDS
- **Y**ohimbine – tx impotence

α_{ns} receptors:

α_{ns} Blockers: *use for Epi reversal*

- Phentolamine – diagnose pheochromocytoma, short-acting, tx cocaine HTN
- Phenoxybenzamine (irreversible) – tx carcinoid, tx pheochromocytoma, long-acting

β_1 receptors: *Revs up the heart, ↑cAMP*
- CNS => ↑activity
- SA node => ↑HR and contractility
- JG => ↑renin
- Pancreas α cells => glucagon release
- Sympathetic => vasoconstriction

β_1 Agonists:
- Dobutamine (↑HR/↑contraction; no effect on BP) – tx CHF *"the* ***dope*** *in a class by itself"*

β_1 Blockers: *"A BEAM"*
- **A**tenolol (long acting)
- **B**utexolol – tx glaucoma
- **E**smolol (short acting) – tx thyroid storm
- **A**cebutolol
- **M**etoprolol

β_2 receptors: *Bronchodilate*

CNS: ↑activity

Ventricle: ↑contractility (not rate)

Lungs: dilation

Pancreas β cells: ↑insulin

Uterus: relax

Bladder: relax

Parasympathetic: vasodilation

β_2 Agonists: hypokalemia → check serum K

"FARTS"
- **F**ormoterol – use q12h
- **A**lbuterol – tx asthma (inhaler, use q4h)
- **R**itodrine – relax uterus to halt pre-term labor
- **T**erbutaline – relax uterus, bronchodilator (inhaler, use q4h)
- **S**almeterol – use q12h

β_{ns} receptors:

β_{ns} Agonists: ($\beta_2 > \beta_1$) *"Manage Inflated Lungs"*
- **M**etaproterenol – used as bronchodilator
- **I**soproterenol – tx heart block, tx bradycardia
- **L**evoproterenol

β_{ns} **Blockers:** *"TPN"*

- **T**imolol – tx glaucoma
- **P**ropanolol – tx tremor, tx panic attack, don't give with asthma
- **N**adolol – tx glaucoma

<u>α and β receptors:</u>
Direct Agonists:

- Epi – α_1/α_2 (high dose) β_1/β_2 (low dose) – tx bronchospasm, tx anaphylaxis => ↑pulse pressure
- NE – $\alpha_1/\alpha_2/\beta_1$ *"NE does NOT do β_2"* => blue digits (powerful vasoconstrictor)
  - Blue injection site Tx: Phentolamine
- DA – Low Dose: D_2 → perfuse kidney
  Int. Dose: β_1 → ↑contractility
  High Dose: α_1 → vasoconstrict, ↑afterload (high dose = 10μg/kg/min)

<u>Epinephrine Tx:</u>
Anaphylaxis: 1:1,000 (0.5mL q15min x 3doses → then 50mg Benadryl IV)
Cardiac Arrest: 1:10,000

Direct Blockers (α_1 + β_1): *"double treatment of HTN"*

- Labetalol – tx A Fib, most α properties
- Carvedilol – tx HTN crisis, tx chronic CHF

<u>II) Cholinergic Receptors:</u> Only anti-cholinergic, nonsympathetic effect = *hot, dry skin*

1) <u>Muscarinic:</u> *think parasympathetic actions*
Muscarinic Agonists: *"(Muscarinic Agonists) Can Promote Bladder Movement"*

- **C**arbachol – tx post-op urinary retention
- **P**ilocarpine – CF sweat test
- **B**ethanechol – tx post-op urinary retention
- **M**ethacholine – formerly used to diagnose reversible airway disease (asthma)

Indirect Muscarinic Agonists: inhibit AcetylCholinesterase
"PASS Program's Physio Educates Newbies"

- Parathion – organophosphate, irreversible "nerve gas" (Tx: Pralidoxime "2-PAM")
- Physostigmine
- Neostigmine – tx myasthenia gravis
- Pyridostigmine – tx myasthenia gravis
- Edrophonium – diagnose myasthenia gravis

Muscarinic Blockers: *think sympathetic actions*
"(Muscarinic Blockers) Do Block GMP; Almost Totally Imitating Sympathetics"

- **D**icyclomine – tx IBD sx (blocks Ach)

- **B**enztropine – tx dystonia/torticollis
- **G**lycopyrrolate – ↓pre-op pulmonary secretions
- **A**tropine – tx heart block, tx cholinergic crisis
- **T**rihexyphenidyl – blocks cGMP, tx Parkinson's tremor
- **I**pratropium – tx asthma, ↓cGMP
- **S**copolamine – tx motion sickness (patch)

2) **Nicotinic:** *think sympathetic*

Agonist:

- Varenicline – used for smoking cessation

Blockers:

"(Nicotinic Blockers) Have Near Sympathetic Tendencies"

- **H**examethonium
- **N**icotine – stimulates ganglia, then blocks (persistent depolarization)
- **S**uccinylcholine – flaccid paralysis *(the only depolarizing agent),* tx malignant hyperthermia
- **T**ubocurarine – histamine release (flushing, hypotension)

Malignant Hyperthermia Tx: Dantrolene → inhibits Ca release → MetHb/Cyanosis

- Cyanosis Tx: Methylene blue
- Next Surgery: use NO as anesthetic

Muscle Physiology:

All muscles use Na^+ to depolarize (except the atrium: Ca^{2+})

All muscles contract b/c of intracellular Ca^{2+}

- Ventricles/SM depend on extracellular Ca^{2+} to trigger that

Each neuron → multiple muscle fibers (1 motor unit)

Each muscle ← 1 nerve (only want 1 action)

3) **Skeletal Muscle:**

- Has motor units, uses recruitment (increase preload on muscle → increase recruitment)
- Electrochemically coupled => dependant on nerve for life and fxn
- T-tubule invaginations: depolarize → DHP → ranodine stimulation → SR Ca^{2+} release
- Use intracellular calcium for contraction
- **Rhabdomyolysis:** ↑serum K^+, Urine: 3+ blood/0 RBCs (b/c Mb is detected as Hb)
  - Tx: Bicarbonate (alkalinize urine to prevent precipitation)

II) Cardiac Muscle:

- Uses intracellular calcium for contraction
- Uses extracellular calcium to trigger intracellular release

- Acts as a *syncitium* => holds onto contraction until everyone contracts (need gap jxns)
- Has autonomics => don't need your permission to beat
- "Wall motion abnormalities": part of heart has died, and those cells won't contract

II) Smooth Muscle:

- Needs extracellular Ca^{2+} via Calmodulin for 2nd messenger system
- Uses intracellular Ca for contraction
- Has autonomics
- Acts as a *partial syncitium*, has autonomics (Ex: gut peristalsis)
- No sarcomeres => why it is smooth
- No troponin => actin and myosin are always bound = "latching" => bowel sounds
- No ATPase activity
- Has MLC kinase to phosphorylate; MLC phosphorylase to chop off

Neuromuscular Transmission:

Soma – makes and transports all proteins, NT
Kinesin – anterograde transport
Dynein – retrograde transport

Peripheral Nerve Injuries:

Neuropraxia: no axon injury (temporary loss of function)
Axonotmesis: loss of axon (grows 1mm/day)
Neurotmesis: loss of entire nerve

Spider Venom:

Black widow: red hourglass, Ach release, abd/back/thorax pain (Tx: Ca gluconate)
Brown recluse: violin-shaped band, Ca release, tissue necrosis (Tx: Dapsone)

Sequence of Events for Muscle Contraction:

- Depolarize
- Extracellular Ca^{2+} flows into T-tubule
- Ca^{2+} binds Troponin-C
- Troponin C releases *Troponin-T*
- Troponin T releases Tropomyosin
- Tropomyosin releases Actin binding sites
- Myosin head binds Actin
- *CONTRACTION:* no ATP used *(Ex: rigor mortis)*
- Myosin heads release ADP (from previous cycle)
- Myosin heads bind new ATP
- Myosin heads hydrolyze ATP → ADP +Pi (releasing 7300 cal)
- *RELEASE*

Don't Swim 30min After Meal:

- All blood in gut
- Skeletal mm. ran out of ATP

- Myosin head returns to start position
- Tropomyosin binds Actin
- Troponin T binds Tropomyosin
- Troponin C binds Troponin T
- Ca-ATPase pumps Ca into the SR
- Phospholambin inhibits Ca-ATPase when it's done

Massage:

- Induces skin inflammation
- Brings O_2/ATP to muscles

Heart Damage Rationale:

1) Subendocardial Infarct:

Ischemia blocks Na/K pump → K^+ leaks out → cell becomes negatively charged → ST depression (70% stenosis) → subendocardial damage *(Tx: Dilate to ↑blood flow)*

2) Transmural Infarct:

Na^+ rushes in → cell becomes positively charged → ST elevation (90% stenosis) → more likely to depolarize → transmural infarct *(Tx: Thrombolytics)*

4) Ventricular Fibrillation:

Na^+ drives all Ca^{2+} into cells → ↓extracellular Ca^{2+} → no p waves → SA node stops → vessels dilate, bladder and gut stop → ventricle can't contract but can depolarize → V Fib

Ischemia (decreased blood flow) → Injury (hurt cells) → Infarct (dead cells)

1) T wave peak	3) ST elevation	5) Q waves
2) ST depression	4) T wave inversion	

Cardiac enzymes: (rise→peak→last)

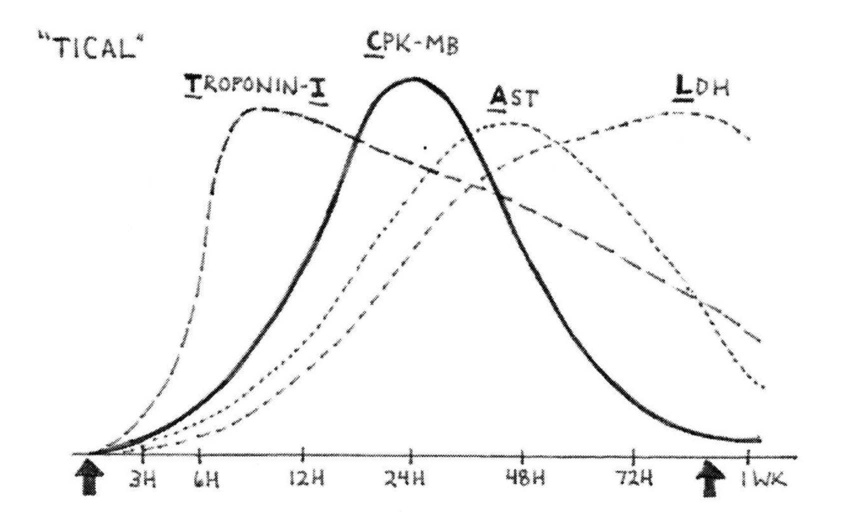

Myocardial Infarction: decreased O_2 → myocardial cell death

- Chest pressure or pain radiating to left arm, shoulder, jaw
- Epigastric pain radiating to back, scapula
- Sweating

- Sense of "impending doom"
- Nausea => inferior MI
- Silent MI => DM

CAD Risk Factors:

- Age (>45 male, >55 female)
- Fam Hx (Dad <55, Mom <65)
- Obese
- Smoker
- HTN
- DM
- Dyslipidemia

Post-MI Complications:

1. 2nd MI
2. Arrhythmias (most common cause of death)
3. IV/ pulmonary rupture
4. Aneurysm, heart failure
5. Dressler syndrome: Pericarditis 2-10wk post-MI => neck and pleuritic chest pain

(Tx: Steroids, NSAIDs)

Angina Workup:

- Hospitalize for 24-hr observation
- Serial EKGs and cardiac enzymes
- Follow-up in 6 weeks:
    - Ca-Pyrophosphate scan: hot spot shows dead calcified cells
    - Treadmill stress test: positive stress test if pain, ΔEKG, or ↓BP
    - Thallium stress test: cold spot shows ischemia
    - Dobutamine stress test: use dipyridamole to dilate vessel during test
    - 2-D Echo: evaluates anatomy of heart

Platelet ADP Receptor Blockers:

Dipyridamole – also dilates vessels
Ticlopidine – agranulocytosis, seizures
Clopidogrel – decreases clotting in high risk patients

Sarcomere: *place your thumbs up, interlace fingers => thumbs = Z lines*

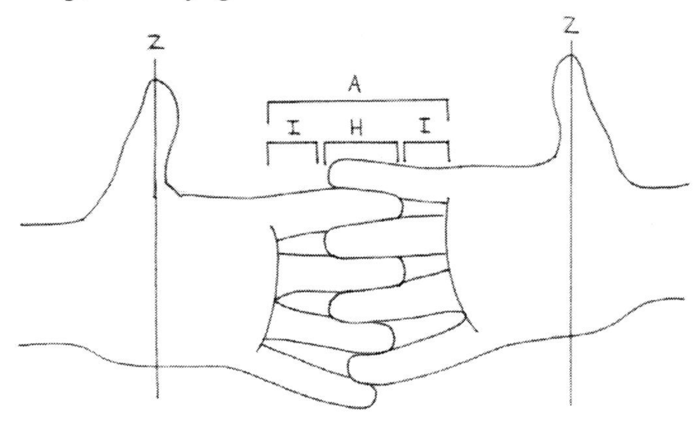

- Sarcomere = between Z lines, decreases during contraction
- *Light chain = actin*
- *Heavy chain = myosin*

- A band = length of myosin (will include some actin), *not change length*
- I band = actin only, decreases during contraction
- H band = myosin only, decreases during contraction
- T-tubules:
    - Cardiac muscle: Z line
    - Skeletal muscle: A-I junction
- Z-line: actin only
- M-line: myosin only, CPK located here

<u>Length-tension curve:</u> max overlap in sarcomere = 2.2 μm (I bands are touching)

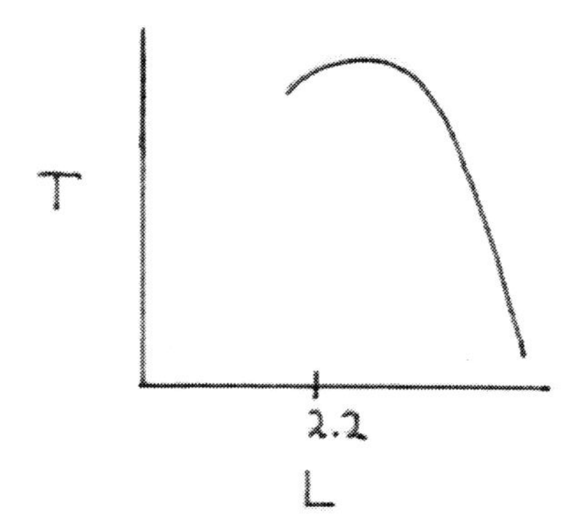

<u>Muscle Dz:</u>

Neurogenic: distal weakness + Fasciculations

Myopathic: proximal weakness + pain

Read from right to contract

Preload = tension on a muscle before work => increased time to cross bridge

Muscle cells hypertrophy by ↑number of cross bridges to handle increased preload

<u>Golgi Tendon Organs:</u>

Muscle will hold max weight for 1 sec → Golgi tendon lets go

Iso**tonic** – low weight, burns ATP when you release *"tones muscles"*

Isometric – build muscles → compresses arteries → ↑TPR → HTN

<u>Frank-Starling curve:</u> (heart dilates)

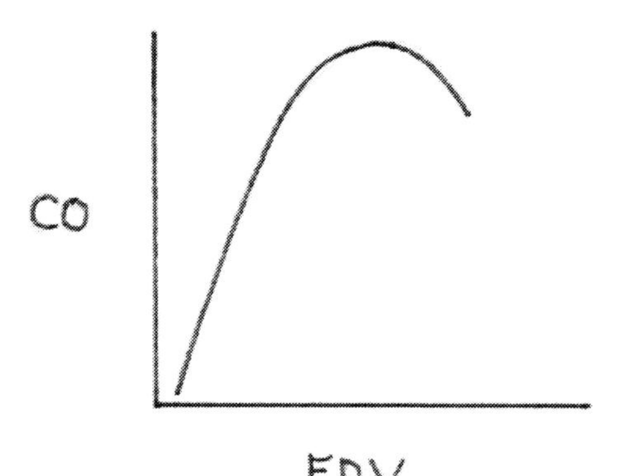

Sudden Death: EDV rises → increase CO → heart stops

Neuromuscular Diseases: *Cause of death = respiratory failure* (heart has autonomics)

I) Myositis: ↑ESR, ↑WBC, ↑AST, ↑ALT, ↑Aldolase, Myoglobinemia

Myositis: one muscle hurts

- Drug-induced: Rifampin, INH, Prednisone, Statins *"RIPS"*
- Bug-induced: Trichinella spiralis
- Endocrine disease: Hypothyroidism, Cushing's

Polymyositis: >1 muscle weak (hard to walk stairs, kneel)

- Elevated enzymes: CK, LDH
- Inflammatory cells: T-cell, MP
- Low EMG
- Dx: Muscle biopsy => inflammation

Dermatomyositis = myositis + rash

- Dysphagia of solids/liquids
- Rule out colon CA
- Heliotrope rash: purple periorbital edema
- Gottron's sign: scaly purple patches on MCP/PIP joints
- Tx: Steroids or Methotrexate

Anti-Phospholipid Ab Syndrome:

recurrent thrombosis, abortions

Type 1: False (+) syphilis

Type 2: Lupus anticoagulant (↑APTT)

Type 3: Anti-cardiolipin Ab

Fibrositis: pain w/ muscle movement

Fibromuscular Dysplasia: renal artery stenosis, child diastolic HTN

- CV bruit
- Angiogram "string of beads"

Fibromyalgia = 11 tender trigger points + axial skeletal pain, sleep disturbance, hurt all the time

- Rule out hypothyroidism
- Tx: Amitriptyline + Water aerobics

Polymyalgia Rheumatica: stiff, weak shoulders, pelvic girdle pain (can't comb hair/wave)

- Tx: Prednisone *(only disease where low-dose predisone improves sx <1 wk)*
- **Temporal Arteritis** "Giant cell arteritis": unilateral HA, blindness, thoracic aortic aneurysm
    - Age >60
    - ESR >60
    - Tx: Prednisone: 60mg now, then temporal artery biopsy (if sx)

II) Muscular Dystrophies:

Becker's (XL): *males get it, females carry it*

- Dystrophin missense mutation
- Symptoms >5 y/o, normal lifespan

Duchenne's (XL): dystrophin frameshift mutation => truncated protein, ↑CPK

- Pseudohypertrophy of calf (fat deposition)
- Gower sign – pt walks up legs to stand up
- Waddling gait – due to transferring torso on hips, toe stepping
- Die by age 30
- Tx: Prednisone

Myotonic Dystrophy (AD): *bird's beak face,* ↑muscle tone => can't let go of hand

- Triple repeat

Guillain-Barre: ascending paralysis *"Ground-to-Butt"*

- 2 wks after URI or *C. jejuni* infection
- Anti-ganglioside Ab
- No reflexes
- MP eat myelin off nerve axons → ↑CSF protein, segmental demyelination, ↓conduction velocity
- Polyradiculoneuropathy – many dermatomes involved
- Same presentation as tick bites, resolves spontaneously like MS
- Tx: Intubate if needed, IV Ig/Plasmapheresis

Transverse Myelitis:

- Guillain-Barre symptoms + back pain
- Post-viral
- URI → rapid myelopathy, urine retention, back pain
- Dx: MRI

Diabetes: sorbitol, glove&stocking neuropathy

3° Syphilis: Tabes dorsalis, Argyll-Roberston pupil, shooting/lancinating pain

Charcot-Marie-Tooth: fat muscle atrophy, stocking glove neuropathy, high arch foot

Myasthenia Gravis: post-synaptic Achr Ab => can't make an end-plate potential

- Middle aged female with ptosis, diplopia
- *Gets weaker as day goes by*
- Associated with thymomas → get CT chest
- Tests:

1) Repetitive stimulation EMG → weaker
2) Edrophonium "Tensilon": inhibits Achase → stronger
    - Repeat test after tx: get weaker → cholinergic crisis/get stronger → MG is worse

Myasthenia Gravis Tx:

- Anti-cholinesterases: Neostigmine, Pyridostigmine
- Immunosuppression: Prednisone
- Thymectomy

Myasthenic Syndrome = "Lambert-Eaton": pre-synaptic Ca^{2+} channel Ab →↓Ach release

- *Muscle contraction gets stronger as the day goes by*
- Associate with small cell CA
- Test: Repetitive stimulation EMG → stronger
- Tx: Immunosuppression

Multiple Sclerosis: anti-myelin Ab, symptoms come and go

- Middle aged woman with vision problems
- Optic neuritis => halo vision (can't see directly)
- Optic neuritis presentation → good prognosis
- Internuclear ophthalmoplegia: opposite eye won't go past midline
  - MLF lesion (connects CN 3 and CN 6)
  - Bilateral trigeminal neuralgia
- LP: myelin basic proteins, MRI (q3 mo): plaques
- Acute Tx: Methylprednisolone IV
- Chronic Tx: INF-β (can cause suicidal ideation), Glatiramer acetate

Metachromatic Leukodystrophy: Arylsulfatase deficiency, *childhood MS presentation*

III) Lower Motor Neuron Disease:

Amyotrophic Lateral Sclerosis (ALS) = Lou Gehrig disease:

- *Descending paralysis,* fasciculations in middle aged male
- Only motor nerves are affected
- CS tract and ventral horn
- Tx: Riluzole (↓pre-synaptic Glu)

Polio: *asymmetric fasciculations in a child,* 2 wks after gastroenteritis

Werdnig-Hoffman: *fasciculations in a newborn,* no anterior horns => no motor neurons

IV) Cerebellar Disease: affects depth perception, has intention tremor, dysdiodokinesis

Adrenal Leukodystropy (XLR):

- Adrenal failure
- Long chain fatty acids are not transferred via carnitine shuttle, stuck in mitochondria
- Rapid central demyelination
- Hyperpigmentation
- Seizures, death by age 12

Ataxia-Telangiectasia: spider veins, IgA deficiency
- DNA endonuclease defect
- Sx: ataxia, telangiectasias of skin/conjunctiva, recurrent sinus infxn, thymus hypoplasia

Friedreich's Ataxia: retinitis pigmentosa (brown pigment on retina), scoliosis
- Spinal cord atrophy – affects gracilis and cuneatus (ipsilateral)
- Sick sinus syndrome
- Triple repeat

<u>V) Cerebral Palsies:</u> permanent neuro damage <21y/o
Atonic cerebral palsy: no muscle tone => floppy
- Cause: frontal lobe tumor, stroke, AVM, anoxia

Choreoathetosis: *dance-like movements,* wringing of the hands, quivering voice
- Cause: kernicterus (bilirubin accumulation) → damage to basal ganglia
- Ex: Wilson's – Cu deposition
- Ex: Huntington's – caudate atrophy

Spastic Diplegia: midline cortical problem, leg problems, CMV infection

Spastic Hemiplegia: cortical problem on one side of the brain, herpes/toxoplasmosis infection

<u>Restless Leg Syndrome:</u> ↓*Fe* → ↓*blood flow to legs* → *irresistible urge to move legs*
- Clonazepam – sleepiness
- Pramipexole – DA agonist (contracts muscles to increase blood flow to legs)
- Ropinirole "Requip" – DA agonist

<u>Muscle Relaxants:</u>
- Succinylcholine: depolarizing blocker – use for intubation => hyperkalemia
- Tubocurarine: non-depolar (rvs w/ Edrophonium,Neostigmine) => histamine release
- Atracurium: NM blocker, degrades in plasma => OK for kidney, liver failure pts

<u>X-linked Recessive Deficiencies:</u> maternal uncle or grandpa had it
"Lesch-Nyhan went Hunting For Pirates and Gold Cookies"
- **L**esch**N**yhan (HGPRT def.) – self mutilation, gout, neuropathy
- **H**unter's (iduronidase def.)
- **F**abry's (α-galactosidase def.) – corneal clouding, attacks baby's kidneys
- **P**DH def.
- **G**-6PD def. – get infxns, hemolytic anemia
- **C**GD = chronic granulomatous disease (NADPH oxidase def.)

NOTES:

NOTES:

Vascular:

"Make no little plans; they have no magic to stir men's blood…

Make big plans, aim high in hope and work."

–Daniel H. Burnham

Autoregulation: *If systolic stays between 60-160 → perfusion is constant*

- Ischemic infarct: BP <60
- Hemorrhagic infarct: BP >160

Most common cause of bleeding into any cavity: HTN

Aorta:

- thickest layer of smooth muscle
- highest compliance = Δvolume/Δpressure (due to elastin)
- stratified squamous cell epithelium => made for abrasion due to speedy RBCs
- higher P_{linear} => blood rarely touches sides of aorta
- (Ex: don't hear marble in hula hoop as you speed it up)
- ↑ Pressure, little ↑ volume → thick aorta in elderly = wide pulse pressure
- Tx: ↓ Cholesterol intake, Ca^{2+} channel blockers (control calcification)

Arteries: sympathetic control, α_1 receptors
Monkenberg arteriosclerosis: old people
Arteriosclerotic occlusive disease: int. claudication, mm. pain w/ exercise, impotence
Hyaline arteriosclerosis: chronic intermittent HTN
Hyperplastic arteriosclerosis: malignant HTN, capillaries ready to burst

Arterioles: β_2 receptors

- do the most to protect BP
- most smooth muscle by cross section => "stopcock" to control BP
- "reactive hyperemia": cut nerve => arteriole dilation => sympathetic control
- malignant HTN => end organ damage
- job = to protect the capillaries
- low BP/ clot => pale infarct (1 blood supply), red infarct (2 blood supplies)

Capillaries:

- largest cross-sectional area
- thinnest walls => diffusion
- max filtration => push nutrients out
- any loss of capillaries will increase resistance
- largest cross-sectional area

Vessel Review:

Thickest layer of SM: Aorta
Most SM: Arteriole
Largest cross-sxn area: Capillaries
Highest compliance: Aorta
Highest capacitance: Vein/Veinules

Veins and Venules: parasympathetic control

- most capacitance => *60% of blood is pooled here*
- least smooth muscle, 1-way valves, blood flows superficial to deep

- venoconstrict (α_1 receptors) to mobilize this blood if skin gets cut (from GI, skin first)
- vein → venule → higher osmolarity → suck stuff into vessel

Cardiophysics: The Language of Pressure
Pulse pressure = volume rising in aorta, but blood is not flowing (Systolic – Diastolic)
Ex: Stiff aorta

$P_{Total} = P_{Transmural} + P_{linear}$
Transmural pressure = pressure on the sides of the vessel (↑ in collagen vascular dz)
Linear pressure = $\frac{1}{2}\ pv^2$ = pressure in the middle of the vessel (highest in the aorta)

Filtration forces:
Hydrostatic: think water (↑: CHF)
Oncotic: think protein; mostly albumin (↓: Cirrhosis)

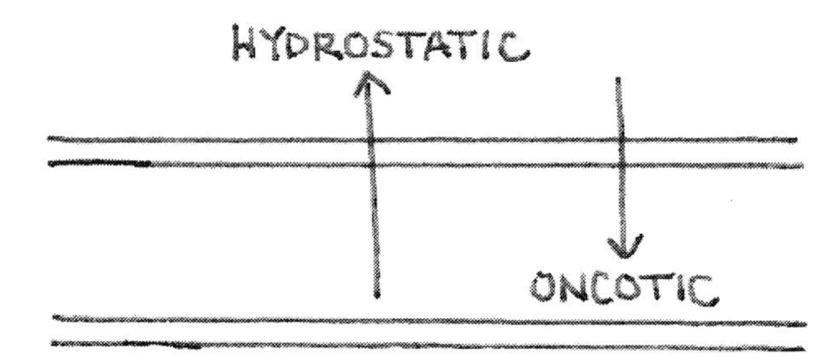

Flux: *Δhydrostatic – Δoncotic* (Δ= capillary – interstitial)
(think of this as "transudate – exudate")

Effusions:

Transudate:	**Exudate:**
Mostly **water**: specific gravity < 1.012	Mostly **protein** (>2g/dL): specific gravity >1.012
Low LDH (<200)	High LDH (>0.6 of serum or >200)
Too much water: • Heart failure • Renal failure *Not enough protein:* • Cirrhosis (can't make protein) • Nephrotic syndrome (pee protein) • Kwashiorkor (not eating protein) • Ménétrier's (GI losing protein)	*Too much protein:* • Purulent (bacteria) • Hemorrhagic (trauma, cancer, PE) • Fibrinous (collagen vascular dz, uremia, TB) • Granulomatous (non-bacterial) • Caseous (TB)

Systole = squish, ↓blood flow to coronary aa., more extraction of O_2 (Phase 1 Korotkoff)
Diastole = filling, ↑blood flow to coronary aa., less extraction of O_2 (Phase 5 Korotkoff)

Why does diastole have more blood flow?

1) less resistance (heart relaxed, not compressing coronary vv.)
2) aortic valve closed, coronary vv. ostea are open
3) more transmural pressure in aorta

A-VO_2 difference: *how metabolically active the tissue is*

- O_2 went out – O_2 came in *"give – take"*
- Heart has highest AVO_2 difference at rest (very high consumption)
- Muscle will have the highest AVO_2 after exercise
- GI will have the highest AVO_2 after a meal
- Kidney has the lowest AVO_2 all the time
- Brain will have highest AVO_2 after test

Flow: Q = ΔP/R

SVC → RA → (tricuspid) → RV → (pulmonic) → PA → Lungs → PVs → LA → (mitral) → LV → (aortic) → Aorta → (BC, LCC, LSC) → rest of body

- $R = \frac{1}{r^4}$ => radius is the *most important factor* in blood flow => cholesterol is bad!
- Poisselle Law: $Q = \frac{(\Delta Pin - Pout)r^4}{nL8\pi}$
- Change n (viscosity): ↓ via phlebotomy, ↑ via polycythemia or high glucose
- Change L (length of vv.): ↑ via obesity
- Change r (radius): ↓ via atherosclerosis

Resistance:

1) Resistance in series

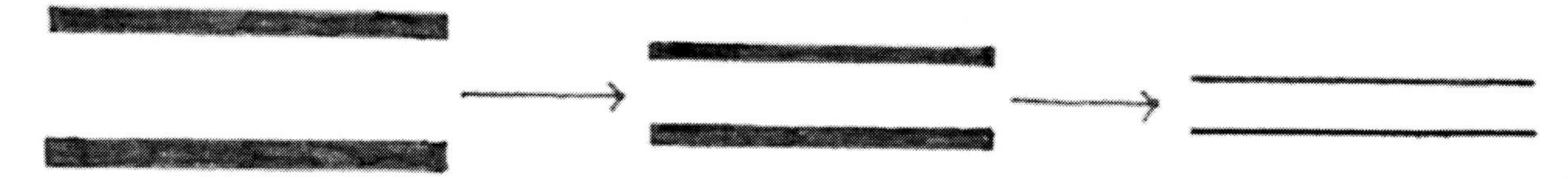

$R_{total} = R_1 + R_2 + R_3$

- As resistance increases, flow decreases and velocity increases
  - Ex: squeezing a garden hose
- Only two organs have this => higher pressure
  - Liver – detoxification (portal → hepatic vein)
- Slow stuff down to detoxify, then shoot it out…
- Macrophage "security men" need to find the junk
  - Kidney – filtration (afferent →efferent arteriole)
- Blood is sitting in "traffic jam"

2) Resistance in parallel – most of our organs have this

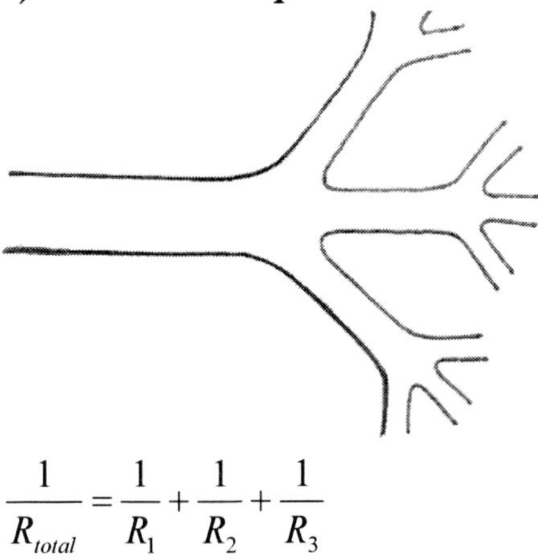

$$\frac{1}{R_{total}} = \frac{1}{R_1} + \frac{1}{R_2} + \frac{1}{R_3}$$

Ex: Remove an organ => increase resistance (transient ↑BP)

Ex: Add a baby => decrease resistance (why BP is low in pregnancy)

Organ:	**Vasodilators:**
Brain:	↑pCO_2, ↓pO_2 Ex: High altitude: low pO_2 => hyperventilation *(chronic => pulm vasoconstrict, secrete bicarb)*
Heart:	Adenosine Ex: ↑flow to SA node to prevent ischemia → stop arrhythmia → pause (Na channels reset) → restart heart
Lungs:	↑pO_2 Ex: hypoxia vasoconstricts blood vessels to lungs => polycythemia
Kidney:	PG-E, DA, ANP Ex: DA used in shock to increase blood flow to the kidneys Ex: NSAIDs will decrease blood flow to kidney => bad for elderly
GI:	Food Ex: blood rushes to stomach after meal
Skin:	Temp, ↑pCO_2 Ex: Face flushes in hot summer (builds up w/ walking, need blood flow to wash it out)
Muscle:	↓pH, ↑pCO_2 Ex: Exercise

When you stand up: BP ↓5-10mm Hg and Pulse ↑5-10 bpm

If pulse goes up >10 => hypovolemia

- CO = HR*(pulse)* x SV*(blood pressure)*

1. Early Shock: Pulse ↑ >10 when stand (orthostatic hypotension = compensated shock)
2. Late Shock: BP ↓ >10 when stand (uncompensated => low CO)

If pulse goes up <5 => Autonomic Dysfunction: *stand up and pass out w/o warning*

- Adults: **DM** (infarcts carotid sinus)
- Babies: **Riley-Day** *"Familial Dysautonomia"* (no reflexes)
- Elderly: **Sick Sinus syndrome** *(Tx: Pacemaker)*
- Parkinson's: **Shy-Dragger syndrome**

Carotid Body: Chemoreceptor
Carotid Sinus: Baroreceptor

The Low Volume State:

This is what happens whenever your body is depleted of volume. Your electrolytes become disorganized, and your body tries to fix it. This happens in a multitude of diseases states such as CHF, vomiting, sweating, aortic stenosis, anemia. Specifically, you end up with ↑total body Na, ↓serum Na (dilutional affect), ↓Cl, ↓K, ↑pH, and ↑TPR. This is due to the renin-angiotensin-aldosterone system.

Immediate: Carotid sinus baroreceptor – ↑sympathetics => high HR (reflex tachycardia)
Intermediate: Medulla – releases NE => vasoconstrict => high HR and TPR
Long-term: Kidneys – release renin => high TPR, ↑Na resorption and ↑K excretion. Excess water reabsorption leads to dilutional ↓Na, ↓Cl, ↓K. ↑K excretion leads to alkalosis.

1) Carotid sinus immediate response:

↓SV → ↓stretch → carotid sinus → ↓CN9 (aff) → ↓CN10 (eff) → ↑sympathetics → ↑HR, ↑BP
Thus, SV (stretch) goes the same direction as CN9,10 firing

Carotid sinus: baroreceptor that responds to stretch (volume) => regulate BP
Ex: Increase SV:

- Rub carotid sinus
- Vasovagal response (cough, sneeze, urinate) → replicate with tilt test

Ex: Decrease SV:

- Tonsillectomy (cut CN9)
- Stand up (blood drained down)
- Nitrates for angina (so… give β blocker first for MI pts to protect heart)

2) Medulla's intermediate response:

The nucleus tractus solitarius → NE release (higher affinity for α receptors) → constrict

3) Kidney's long-term (>20 min) response:

- NE also vasocontricts blood supply to kidney
- J-G apparatus in afferent arteriole of the kidney responds to flow and volume

Fix the pressure:

Mr. J-G releases renin → Liver (angiotensinogen) → ATI → Lungs (ACE) →
ATII (very potent vasoconstrictor) to efferent arteriole → ↑TPR → ↑BP
ATII constricts efferent more than afferent in order to re-establish GFR

Fix the volume:

ATII → Aldo → Na^+/K pumps in kidney DCT → ↑Na reabsorption → ↑total body water → drags in 3 waters with each Na → decreases serum Na → serum K decreases (secretion) → Aldo also secretes H^+ → ↑pH

Thus, Low Volume → Alkalosis

Acidosis Exceptions:

- Diarrhea (lose bicarb) => acidosis
- RTA type II (lose bicarb) => acidosis
- DKA (ketones) => acidosis

Fluid Review:

Vomiting => Alkalosis

Diarrhea => Acidosis

High Na Exceptions:

- Diabetes Insipidus
- Neglected elderly (eating salt, but not drinking pure water)

What is the Physiology of a Man in the Desert?...

Effect of Low Volume on Heart:

- ↓ Fluid volume
- ↓ Preload
- ↓ RA filling
- ↓ RV filling
- ↓ Ventricular contractility
- Softer S_1 (↓pressure)
- Narrow S_2 splitting (↓volume/↓pressure)
- ↓ CVP
- ↓ EKG amplitude
- ↓ Vascular pressure
- ↓ Pulmonary resistance
- ↓ PCWP
- ↓ LV filling
- ↓ Contractility of heart
- ↓ Stroke volume
- ↓ Cardiac output
- ↓ Blood flow
- ↓ Carotid stretch
- ↓ Firing of CN 9/10
- ↑ Heart rate

Effect of Low Volume on Kidneys:

- ↓ Firing of CN 9/10
- ↓ NE release

- ↑ TPR
- ↓ Blood flow sensed by J-G apparatus
- ↑ Renin release
- ↑ Angiotensinogen in liver
- ↑ Angiotensin I in lung
- ↑ Angiotensin II (coverted by ACE in lung)
- ↑ Vasoconstriction
- ↑ TPR more
- ↑ Aldosterone in the CD
- ↑ Na/H_2O reabsorption
- ↓ K/H

Effect of Low Volume on Urine:

- ↓ urine Na (↑Na/H_2O reabsorption)
- ↑ urine K
- ↑ urine pH (↓ urine H^+)
- ↑ FeNa

Effect of Low Volume on Serum:

- ↑ Total body Na (↑Na reabsorption by Aldo)
- ↓ serum Na (H_2O reabsorption by Aldo)
- ↓ serum K (K excreted by Aldo)
- ↑ serum pH (H^+ excreted by Aldo)

Effect on Low Volume on Lung:

- ↑ serum pH
- Harder to breathe in
- Restrictive lung disease
- ↓ pO_2 (hard to breathe in)
- ↑ Respiratory rate
- ↓ CO_2
- ↑ serum pH
- Respiratory alkalosis

Superficial Vein Thrombosis:

- Red cord
- Varicose veins
- Tx: warm compresses

Only Veins with No Valves:

- Brain
- Heart

DVT Prophylaxis:

Low Risk: SCDs

Moderate Risk: Heparin
High Risk (Ortho/Neurosurg/Previous DVT)**:** Warfarin with INR of 2-3
DVT Tx: LMW Heparin + Warfarin (can tx outpatient if no pulmonary sx or co-morbidities)

CVAs:
Stroke: infarcted cerebral tissue
TIA: neurologic deficits <1hr w/ no infarction
Dx: US → carotid stenosis
TEE → embolus

- Antiplatelets (asa, Clopidogrel)
- Anticoagulate (Heparin, Warfarin)
- Thrombolytics (t-PA)

Thrombus: artery occlusion
Embolis: moving thrombus

Strokes:
Carotid: anterior circulation

- TIA
- Amaurosis Fugax
- Aphasia
- Clumsy hands

Vertebrobasilar: posterior circulation

- N/V
- Syncope
- Ataxia
- Diplopia

IgA Deficiencies:

- Giardia
- Ataxia Telangiectasia

IgA Proteases:

- Strep pneumo
- H. influenza
- N. catarrhalis

Stroke Management:
1) CT: rule out hemorrhagic stroke
2) Treat based on time symptoms began:

- <3hr: t-PA (IV)
- 3-6 hr: t-PA (intra-arterial)
- 7-8 hr: Embolectomy
- 9+hr: Dipyridamole + asa

IgA Nephropathies:

- Alport's
- HSP
- Berger's
- Ankylosing Spondylitis

CVA Sequelae:
Agnosia: difficulty with comprehension
Agoraphobia: fear of not being able to escape *"can't go"*
Anosagnosia: can't understand that they are sick
Apraxia: can't follow commands
Aphasia: can't speak due to brain problem
Dysarthria: can't speak due to muscle problem
Agraphia: can't write

Rathke's Pouch =>

- Anterior pituitary
- Hard palate

Anomia: can't remember names
Prosopagnosia: can't remember faces

Vasculitis:

We all know that "–itis" means there is an inflammatory process, right? So the WBC count is going to be high. However, this is not a bacterial process – which means it has to be cell-mediated, which will have high levels of T cells and macrophages. These cells are going to be ripping RBCs and platelets, and you will see schistocytes in the peripheral smear. So now you know the whole CBC for every vasculitis: ↓RBC, ↓platelets, ↑WBC, ↑T cells, ↑MP, and schistocytes – without having to look it up! And, for collagen vascular diseases, the eosinophil count will also be elevated.

Just to tie this in… You have a vasculitis, and your RBC count is low. What do you need RBC for? Oxygen. And what do you need oxygen for? Making energy. So what state are we in? The low energy state. So now you know what you are going to see on their skin, lungs, GI, bladder, what their most common presenting signs and symptoms will be, and what they are mostly likely to die from. Now all you need to know is the CLUE that tells you which vasculitis it is…

Vasculitidies:

- Low volume state, low energy state, restrictive lung dz profile, cell-med inflammation
- Exposes BM → shows GP2b3a, which attracts platelets
- ↓RBC/platelets, ↑WBC/T cells/MP, schistocytes

Vaculitis:	**Clues:**
Ankylosing Spondylitis	Middle-age male w/ back pain, bamboo spine, + Schober test, HLA-B27, AR
Alport's	IgA deposition, deaf or cataract family
Behçet's	RA + oral and genital ulcers
Berger's	IgA deposition 2 wk after cold → renal failure (hematuria, edema)
Buerger's	Smoking Jews, necrosis of extremities
Churg-Strauss	Pulmonary eosinophils, asthma, p-ANCA
CREST syndrome	Mild scleroderma, anti-centromere Ab
Cryoglobulinemia	Acute non-bact inflamm(< 2mo), IgM
Diabetes mellitus	Glove-and-stocking distribution
Disseminated Intravascular Coagulation "DIC"	Sepsis, amniotic fluid emboli, abruptio placenta, D-dimers = fibrin split products
Felty's	RA + leukopenia + splenomegaly
Goopasture's	Lung and kidney, anti-GBM Ab, linear BM immunofluorescence
Hemolytic Uremic Syndrome "HUS"	Child renal failure 2 wk after eating raw hamburgers
Henoch-Schonlein Purpura "HSP" *Tx: Steroids/Immunosuppresants*	IgA deposition 2 wk after gastroenteritis, intussusception, normal Plt, purpuric rash from butt down "Hip-South Purpura"
Immune Thrombocytopenic Purpura	Purpura 2wk after URI, anti-platelet Ab

Tx: *1) Prednisone* *2) IV Ig* *3) Splenectomy*	
Kawasaki Disease *Tx: ASA + flu shot, IV Ig*	Strawberry tongue, red eyes/lips, palm/sole rash, cervical lymph nodes, kid MIs, high platelets, fever, "CRASH"
Leukocytoclastic Vasculitis	Drug allergy causes vasculitis
Mixed Connective Tissue Disease	Anti-RNP Ab
MPGN type I	C3 nephritic factor
MPGN type II	Kidney deposits, tram-tracks, low C3
Polyarteritis Nodosa "PAN" Test: Biopsy or Visceral Angiography	GI and kidney, P-ANCA Ab, HepB, mottled lace-like appearance
Post-strep Glomerulonephritis (strain 12)	2 wk after sore throat, *subepithelial deposit*
Progressive Systemic Sclerosis	Scleroderma that attacks organs
Rheumatoid Arthritis	Worse in morning, attacks synovium => pannus, amyloid => restrictive cardio, RF, attacks C1-C2 joint
Scleroderma	Anti-**Scl**70 Ab
Serum sickness	2 wk after vaccination (MMR)
Sjögren's	RA + sica (xerostomia/xeropthalmia), anti-SSB Ab, dental cavities
Still's disease = Juvenile RA	Iridocyclitis + RA
Subacute Bacterial Endocarditis	Septic emboli, Strep viridans
Syphilis	Aortitis, wrinkled intima, tree-bark appearance, obliterative endarteritis
Systemic Lupus Erythematosis "Lupus" *Tx: NSAIDs, Cyclophosphamide, Hydroxycholoroquine*	Ab: *dsDNA*/ Sm/ Cardiolipin, malar rash, die of renal failure, *subepithelial deposit,* IgG at derm-epid jxn, granular complement
Takayasu's = "aortic arch syndrome" *Tx: daily high-dose steroids*	Weak pulse, aortitis, Asian women, blind, Raynaud's, acute pain with cold
Temporal Arteritis	Unilateral temporal HA, jaw claudication, ESR >60
Thrombotic Thrombocytopenic Purpura "TTP" *Tx: Plasmapheresis*	Purpura + neuro symptoms, VWF esterase deficiency
Trousseau's	Migratory thrombophlebitis, assoc w/ cancer
Wegener's "**w***hole lot of* **c***rap*"	Lung, kidney, and sinus, **c**-ANCA Ab

Palm/Sole Rashes:

"TRiCKSSS"

Toxic Shock Syndrome

Rocky mountain spotted fever

Coxsackie A: Hand-Foot-Mouth disease
Kawasaki
Scarlet fever
Staph Scaled Skin
Syphilis

Vasculitis:
High platelet count: Kawasaki
Normal platelet count: HSP

HLA-B27 Diseases:

- **Psoriatic Arthritis:** attacks DIP joints, silver oval plaques on extensors, nail pitting, pencil-thin bones, gout, uric acid kidney stones
- **Ankylosing Spondylitis:** Ligament ossification, ↓lumbar curve, stiffer in morning, kyphosis, uveitis, AR, bamboo spine, (+) Schober test
- **Reiter's:** Uveitis, Urethritis, Arthritis

Pulmonary Eosinophilia:

- Aspergillus (Tx: Steroids)
- Loeffler's (Tx: Antibiotic or Anti-parasitic)
- Churg-Strauss (Tx: Zafirlukast)

ABI:
1.0: Normal
<0.9: PVD
<0.4: Severe ischemia

Low C_3:
*"PMS in Salt Lake City"**
Post-strep GN
MPGN Type II
Subacute Bacterial Endocarditis
Serum sickness
Lupus
Cryoglobulinemia

CREST:
Calcinosis
Raynaud's
Esophageal dysmotility
Sclerodactyly
Telangiectasia

Bacterial Endocarditis: *(Tx: Nafcillin + Gentamycin)*
Acute: Staph **A**ureus → attacks healthy valves
Subacute: Strep viridans → attacks damaged valves
Dx: TEE
Septic emboli: *ischemic stroke*

- Brain: Mycotic aneurysm
- **R**etina: **R**oth spots (central white hemorrhagic spots)
- Fingers/Toes: Osler nodes (painful red nodules) *"ouch"*
- Nailbed: Splinter hemorrhage

- Palm/**Soles**: Jane**way** lesions (painless dark macules)

<u>Cryoglobulinemia:</u> serum proteins (i.e. globulins) form a gel when exposed to cold
"I AM HE"
Influenza
Adenovirus
Mycoplasma
Hep B/C
EBV

NOTES:

Cardiac:

"The human heart feels things the eyes cannot see, and knows

what the mind cannot understand."

–Robert Valett

Systole:

S_1: M/T close → IC → Squishes blood out

Diastole:

S_2: A/P close → IR → Fills with blood

The Rules:

- Heart sounds are always made by valves closing
- L side has higher pressure/resistance => aortic and mitral valves close first

- Stand/Valsalva: ↓Regurg (↓venous return) Ex: ↑HCM, ↓AS
- Squat/Handgrip: ↑Regurg (↑venous return)

- Inspiration => hear R problems louder (↑blood volume on R side)
- Expiration => hear L problems louder

- Soft S_1 => M/T regurg (or mitral/tricuspid atresia - cyanotic)
- Loud S_1 => M/T stenosis (or ventricle contracting harder)

- Soft S_2 => A/P regurg (or aortic/pulmonic atresia - cyanotic)
- Loud S_2 => A/P stenosis (or high pressure in front of valves: systemic or pulm HTN)
- Wide S_2 splitting => ↑O_2, ↑RV volume, delay pulmonic valve opening
- Narrow S_2 splitting => ↓O_2,↓RV volume

- S_3 = volume (dilated)
- S_4 = pressure (hypertrophy)

Heart Sounds:

Mid-systolic click – hear valve buckling during systole

- Mitral valve prolapse (Tx: weight gain)

Ejection click – force the valve open during systole

- Aortic stenosis
- Pulmonary stenosis

Opening snap – force the valve open during diastole

- Mitral stenosis
- Tricuspid stenosis

Pulse = QRS (2+ = normal):

- Pulsus Tardus => AS
- Water-hamme**r** => A**R**
- Pulsus Alternans => DCM
- Pulsus Bisferiens => IHSS
- Pulsus Paradoxus => Cardiac Tamponade
- Irregularly Irregular => A Fib
- Regularly irregular => PVC

S_2 splitting: normal on inspiration (b/c pulmonic valve closes later)

- Right side has lower pressure => pulmonary valve stays open longer
- O_2 dilates pulm vv. => ↑flow => pulmonary valve stays open longer

Wide S_2 splitting:

- Increase O_2 (deep breath, ventilator)
- Increase RV volume (VSD, PR, lay down, dilated cardiomyopathy)
- Delay pulmonic valve opening (PS)

Fixed wide splitting of S_2

- ASD: L to R shunt

Paradoxical S_2 splitting: *aortic valve closes later*

- Aortic stenosis
- L BBB

Effects of Pulmonary HTN:

Cor pulmonale: RV failure

Eisenmenger's: Rvs L-R to R-L shunt

(become cyanotic, DVTs go systemic)

S_3: ***"SLOSH(S_1) –ing(S_2) in(S_3)"***

1) Dilated ventricle (estrogen stretches mm. apart, normal in teenage females)
2) Volume overload
3) Decompensation (heart gives out)

Radiating Sounds:

- to Neck: AS
- to Axilla: MR
- to Back: PS

S_4: ***"a(S_4) STIFF(S_1) wall(S_2)"***

1) Hypertrophied ventricle
2) Pressure overload
3) Compensation (Ex: *aortic stenosis* – most common, aging)

Estrogen Synthesis:

Ovary/Adipose/Placenta:

E_1: 20/**80**/0

E_2: **80**/20/0

The Estrogen Connection: Estrogen is a muscle relaxant => NM disease state

This concept is so simple, but explains so much. The key to understand here is that estrogen is a muscle relaxant. That means that any process where estrogen is increased will mimic a neuromuscular disease state. Thus, increased estrogen states (obesity, oral contraceptives, pregnancy, liver failure, p450 inhibition, etc.) will lead to the following:

- S_3, vasodilation, ↓BP, hemorrhoids
- constipation, urinary retention
- reflux, relax gall bladder → gallstones
- ↓osteoporosis, colorectal CA, LDL
- ↑breast CA, endometrial CA, DVT
- Proteins: ↑ESR, Lipoproteins, TBG, Angiotensinogen, Factor 1

Pre-eclampsia: ischemia to placenta => HTN (>140/90) + proteinuria (>300 mg/day)

- If <20 wks, think hydatidiform mole

- Mom gets cerebral hemorrhage/ARDS → dies
- HELLP syndrome: *hepatic injury*
  - **H**emolysis
  - **E**levated **L**iver enzymes
  - **L**ow **P**latelets
- Tx: delivery

Rule of 60's => immediate C/S

- HR below 60 bpm
- HR ↓ >60 bpm
- HR <100 for 60 sec

Eclampsia: HTN + seizures (shut down pump, Na is locked in cell but K can leak out)
Sx: HA, change in vision, epigastric pain
Tx: 4g Mg sulfate (seizure prophylaxis) → C/S

Pressure Rules: *Think: BP = resistance/ volume*

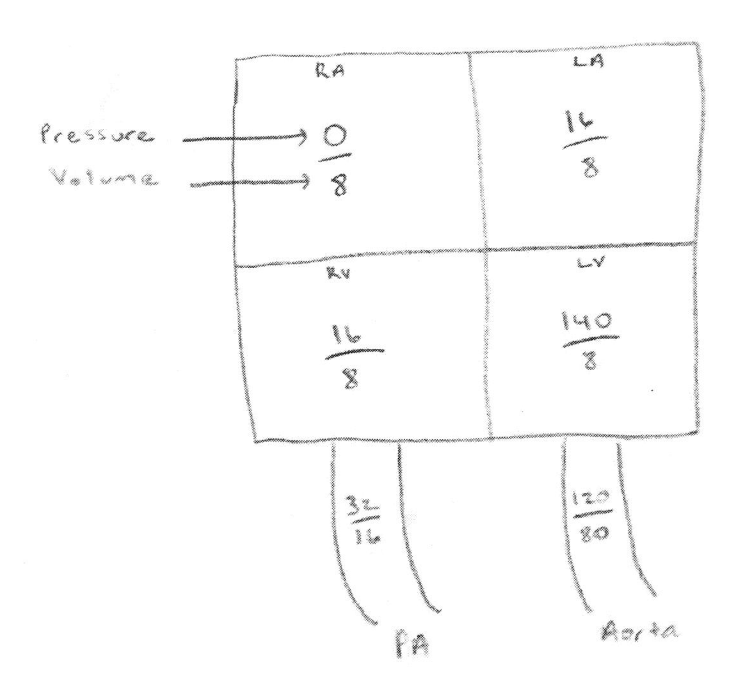

Lung has 2 blood supplies:

- Bronchiole artery (1/3) = 4cc
- Pulmonary artery (2/3) = 8cc

CVP = central venous pressure = average *RA pressure* (normal=3-5)

- high => heart failure, cardiac tamponade
- low => 1st sign of hemorrhage

PCWP = indirect measure of *LA pressure* due to volume in lungs (normal = 8+4 = 12)

- measured by Swan-Ganz catheter
- high => volume problem (Ex: pulmonary edema, CHF)
- low => resistance problem (Ex: hypoxia, fibrosis, Phen-Fen, ARDS)

Percent perfusion: (total = 5L/min)

- Brain: 20%
- Heart: 20%
- Kidney: 20%

O_2 saturation:

RA: 75%
RV: 75%
PA: 75%
LA: 100% (just came from lungs)
LV: 97% (thesbian veins drain deoxygenated myocardial blood into LV)

Cardiac Equations:
CO = SV x HR (Note: can measure HR via pulse, SV via BP)
CO = MAP/ TPR (MAP=BP)
CO = O_2 consumption/ AO_2 diff – VO_2 diff
SV = EDV - ESV

Blood Vessel:
Intima: endothelium
Media: elastin, SM
Adventitia: CT (w/ vasa vasorum)

CO = SV x HR: 5L/min
↑ SV: early exercise
↑ HR: late exercise (40-70% of max HR = fat-burning stage)

Max HR (pulse): 220-age *(higher => arrhythmia)*
Angina:
Stable => pain with exertion, relieved by rest (atherosclerosis) Tx: ASA, Metoprolol, NGN, Statin
Unstable => pain at rest (transient clots) Tx: Add Heparin, Eptifibatide, Plavix, Cardiac Cath
Variant *"Prinzmetal's"* => intermittent pain - wakes you up (coronary artery spasm) Tx: Diltiazem

Leriche syndrome: aorto-iliac claudication, butt hurts when they walk/during sex

Atherosclerosis:
Risk Factors: ↑LDL, HTN, DM, smoking
1) Fatty Streak: lipid foam cells
2) Fibrous Plaque: necrotic core with cholesterol crystals
3) Clot

HTN Terminology:
Mild Hypertension: >135/85
Moderate Hypertension: >155/100
Severe Hypertension: >175/115
Hypertensive Urgency: >200/110 (Tx: Slowly ↓BP over several days)
Hypertensive Emergency: plus end-organ damage (Tx: Nitroprusside to ↓BP by ¼)
Malignant Hypertension: plus papilledema

HTN Treatment: Can't use β-blockers and Ca-channel blockers together
First: Stop alcohol intake
AA/Hispanics: high salt diet → diuretics (↓SV)
Caucasians: stress → β blockers (↓HR)
Pregnant: Labetalol/Hydralazine/α-Me-DOPA

BPH: α-blocker
Angina: Nitroglycerin
MI: Esmolol
CHF: ACE-I + Spironolactone

Peripheral Vascular Dz: Ca channel blockers (↓SV, ↑TPR)
Atherosclerosis: Ca channel blockers (↓TPR) or thiazides

Osteoporosis: HCTZ (↑Ca^{2+})
Cocaine Users: Phentolamine
Opoid Withdrawal: Clonidine

Asthma: no β-blockers
Pulmonary Edema: Nitroglycerin or Furosemide
Renal Failure: ACE-I

DM: ACE-I
Gout: Losartan (pees out uric acid)
Pheochromocytoma: Phentolamine
Lupus: no Hydralazine
Scleroderma: ACE-I

Increase Digitalis Toxicity:

Sx: Yellow vision, SVT, AV block

High Ca: Na/Ca pump

Low K: binds Na/**K** pump

(more pumps for dig to bind)

Murmur Grades:
Grade 1: barely audible
Grade 2: easily audible
Grade 3: pretty loud
Grade 4: palpable thrill
Grade 5: hear with stethoscope off the chest
Grade 6: hear across the room without a stethscope

Cinchonism:

- Hearing loss
- Tinnitus
- Thrombocytopenia

MS: Rheumatic fever

MR: Endocarditis

Heart Murmurs:
Note: *Late murmurs → bad prognosis*
Bruit – turbulence in arteries
Murmur – turbulence across a valve (hole or stenosis)

Reynolds# >2500 => murmur

Anasarca DDx:

- CHF
- Cushing's
- Steroids
- Hypothyroid
- Low Albumin (kidney/liver failure)

SYSTOLIC MURMURS:
M/T are closed => ***Regurg*** = **holosystolic**
A/P are open => ***Stenosis*** = **ejection murmur**

Holosystolic:

1) Tricuspid regurg: endocarditis (IV drug abuser)
2) Mitral regurg: mitral valve prolapse/endocarditis, radiates to axilla, soft S_1
3) VSD: increase on expiration (LV contracts harder)

<u>Systolic Ejection:</u>
1) Pulmonary stenosis: congenital, carcinoid (local invasion), radiates to back

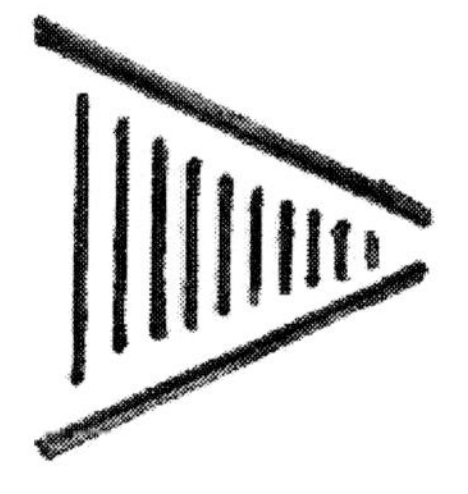

2) Aortic stenosis: aging (calcification) or bicuspid aortic valve

- Note: Don't give β-blockers or ↓Afterload drugs!
- 4-yr life expectancy; late murmurs are bad: *replace if area of valve < 1.5 cm³*
- Triad: syncope, angina, exertional dyspnea
- hear louder on exp, leaning forward, making fist, BP cuff
- hear less with Valsalva, squat
- pulsus tardus *"delayed carotid upstroke"* – takes the pulse a little while to get there
- radiates to neck, delayed carotid upstroke
- palpable thrill in suprasternal notch

3) Idiopathic Hypertrophic Subaortic Stenosis = IHSS = HCM
- sudden death in athletes, AD
- IV septum is thick on top, thin on bottom => flops down
- muscle fibers are disarrayed
- CXR: banana shape
- hear louder with Valsalva (less volume => hear flop louder)
- pulsus bisferiens – feel two peaks on pulse
- "closing (septum occlusion) following opening of aortic valve"
- Tx: β-blocker (HR/contractility), drink water (SV), no sports, Echo for family

Amyloid: ↑ESR

AA: **A**ny chronic disease

AB: **B**rain (Alzheimer's)

AB$_2$: β_2 microglobulinemia (renal failure)

AE: **E**ndocrine (medullary CA of thyroid)

AF: **F**amilial (MEN2)

Diastolic Murmurs:

A/P are closed => ***Regurg*** = **blowing**

M/T are open => ***Stenosis*** = **rumble**

Diastolic Blowing: *decrescendo*

1) Aortic regurg: aging or collagen diseases (Tx: ↓Afterload)

- Wide pulse pressure (↑systolic P, ↓diastolic P)
- De Musset's sign – head bobbing
- Quincke's pulse – see pulse in nail bed
- Water-hammer pulse – bounding "thumping" pulse
- **A**ustin Flint murmur (suction => 2° mitral regurg)

Disorganization:

- Muscle: HCM
- Bone: Paget's

2) Pulmonic regurg: congenital

- Graham-Steell murmur (suction => 2° tricuspid regurg)

Diastolic Rumble: thick atrium squeezes hard => whirlpool effect

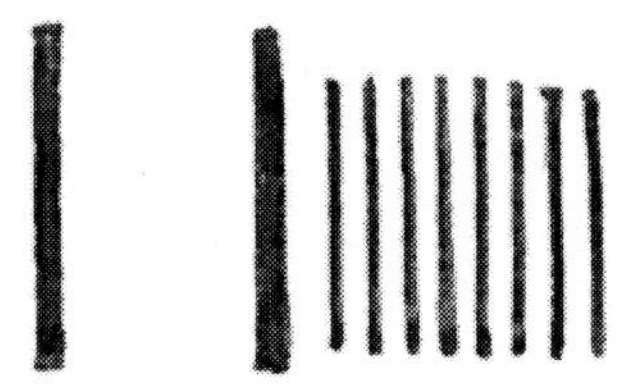

1) Tricuspid stenosis: rheumatic fever, carcinoid syndrome

2) Mitral stenosis: rheumatic fever => emboli, hemoptysis, loud S_1 *(Tx: No inotropics)*

- o Monitor progression of disease by murmur duration

Continuous murmur: "to/fro, machine-like"

A-V Fistulas: connection between an artery and vein

1) Congenital: PDA

- *Alprostadil (PG-E_1) will keep it open*
- *Indomethacin will close it*

2) Osler-Weber-Rendu: AVM in lung/gut/brain; sequester platelets/telangiectasias (Tx: embolize)

3) Von **H**ippel-Lindau: AVM in **h**ead/retina => renal cell CA risk *"can only see Hippo's eyes/head"*

4) Iatrogenic: dialysis fistula, stab femoral vessels

Friction "triphasic" Rubs:

Hear while breathing only => pleuritis

Hear when hold breath => pericarditis *(knife-like pain relieved by leaning forward)*

Murmur Review: *"HEB Rules…Mr. AS (holds me in his) ARMS"*

Systole: M/T close, A/P open	**H**olosystolic => **MR**/VSD
	Ejection murmur/click => **AS**/HCM
Diastole: A/P close, M/T open	**B**lowing => **AR**
	Rumble/Opening Snap => **MS**
Continuous:	AVM => PDA, OWR, VHL
	Friction rub => pleuritis, pericarditis

Cardiomyopathies:

Dilated: volume problem, systolic dysfunction

- Ex: Coxsackie B, Chagas, HIV, Doxorubicin, Alcohol

Restrictive: restricts actin/myosin, ↓filling, diastolic dysfunction

Collagen vascular diseases: fibrosis

Amyloidosis: *stains Congo red, Echo Apple-green birefringence, twisted β-sheet*

- 1° Amyloidosis (AD): big organs, ↑protein causes intracranial hemorrhage
- 2° Amyloidosis (chronic disease): Scleroderma, asthma, Wegener's

Hemochromatosis: Fe deposit in organs, Prussian blue stain, ↑transferrin (>50%)

- Triad: hyperpigmentation, arthritis, DM
- ↑ Infection w/ Fe-loving bugs: Listeria, Yersinia, Vibrio
- Tx: Deferoxamine + weekly Phlebotomy (16 units PRBC)
  - Bronze pigmentation: Fe deposit in skin folds
  - Bronze cirrhosis: Fe deposit in liver
  - Bronze diabetes: Fe deposit in pancreas
  - Hemosiderosis: Fe overload in bone marrow
  - Hemochromatosis: Fe deposit in organs, ligament calcification
- 1° (AR: HLA-A_3, A_6): duodenum absorbing too much Fe
- 2°: multiple blood transfusions (sickle cell anemia, thalassemias)

Constrictive:

Cardiac Tamponade: trauma, cancer

- Beck's Triad: distant heart sounds, JVD, hypotension
- Pressure equalizes in all 4 chambers, muffled heart sounds, no pulse or BP
- Kussmaul's sign, pulsus paradoxus (↓>10mm Hg BP w/ insp), pericardial knock
- EKG: electrical alternans
- Echo: small compressed heart
- Tx: Dobutamine (↑contractility), Pericardiocentesis, Pericardial window if recurrent

Hypertrophic: asymetric hypertrophy of IV septum, relatives have 25% risk

Ex: IHSS *(heart hypertrophies on the inside)*

Tx: Hydration/β-blocker

Li Effects:

Mom: Nephrogenic DI

Baby: Ebstein's anomaly

Cyanotic Congenital Cardiac Anomalies:

1) Transposition of the Great Arteries*: at birth*

- Aorticopulmonary septum did not spiral
- X-ray: egg-shaped heart
- Tx: Alprostadil (PGE_1 to keep PDA open until surgery)

2) Tetrology of Fallot: *>1 mo,* fatal without surgery

- Overriding Aorta: aorta sits on IV septum over the VSD and pushes on PA
- Pulmonary Stenosis "Tet spells" – *determines prognosis*
  - Turn blue when crying (reverse L to R shunt with exhalation)
  - Squat after running
- RV hypertrophy => boot-shaped heart
- VSD (L to R shunt)

3) Total Anomalous Pulmonary Venous Return: all pulmonary veins to RA, *snowman x-ray*
4) Truncus Arteriosus: spiral membrane not developed => one A/P trunk, mix blood
5) Ebstein's Anomaly: tricuspid sitting lower than normal, Mom's Li increases risk
6) Aortic atresia: blood can't get out of the heart
7) Pulmonary atresia: no blood to lungs
8) Tricuspid atresia: RA contracts harder, has FO/VSD
9) Hypoplastic left heart: small LV, low BP, weak pulse, ↑HR, AS, MS

Pressure-Volume loop:

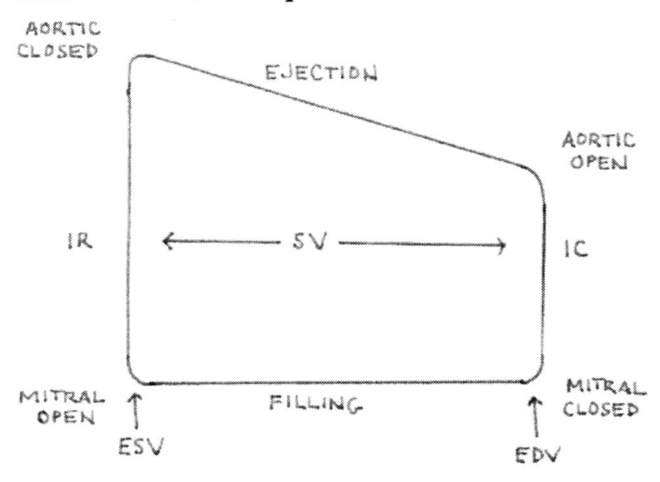

Mitral closes → (IC) → Aorta opens → Aorta closes → (IR) → Mitral opens

Isovolumetric Contraction (IC):

- needs to overcome aortic diastolic pressure of 80
- 81-120: blood enters the aorta
- 121: blood flows through the aorta (LV and recoil of aorta)

Stroke volume: EDV – ESV

Note: EDV and ESV always change in the same direction…

SV: how much you pumped out

EDV: total volume (↑ with volume or deep breath) "preload"

ESV: what's left after contraction (↓ by increased contractility: digoxin, dobutamine)

Ejection Fraction: SV/EDV

Normal: 50-80%

Low (< 45%): at least 40% of myocardium is dead

High: Athletes (low pulse => in good shape)

Pulse Pressure = Systolic – Diastolic
(40 = 120-80)

MAP = 1/3 systole + 2/3 diastole

Slow HR: spends more time in diastole

Fast HR: spends more time in systole (↑contraction → ↓flow in coronary aa. → clot&die

MAP can be approximated by: $\frac{systolic + diastolic}{2} = \frac{120+80}{2} = 100$

Wolff-Parkinson-White: δ wave

- Most common SVT in teenagers
- Bundle of Kent: accessory conduction that bypasses AV node

Pulm. Fibrosis: *"BBAT"*

Sx: insp crackles, ground-glass CT, low sats w/ walking

Tx: Steroids

- **B**usulfan
- **B**leomycin
- **A**miodarone
- **T**ocainide

- Tx: Procainamide, Phenytoin, Quinidine (block Na/Ca)
- No "ABCD": **A**denosine, β-block, **CCB**, **D**igoxin

Class I: Na^+ channel blockers – block ventricular arrhythmias

Class I_A: *"Queen Proclaims Disco" (prolongs AP)*

Quinidine – *blocks Ca^{2+}*, anti-cholinergic, cinchonism, platelet hapten, ↑p450, Torsade

Procainamide – *blocks Ca^{2+} and K^+ (via NAPA)*, NH_4^+ → GABA, SLE

Disopyramide

Class I_B: *"Lied To the Lung about the Mexican food that caused GI uPset" (shortens AP)*

Lidocaine – only acts on ischemic tissue, fat soluble, quickest $t_{1/2}$

Tocainide – **lung** fibrosis

Mexiletine – bad **GI upset**

Phenytoin – *blocks Ca^{2+}*, gingival hyperplasia, hirsutism, SLE, fetal hydantoin, ↑p450, hypotension if infused too quickly

Class I_C: blocks 90% Na^+ channels => die *"Flec (the alligator) Eats Props" (no effect on AP)*

Flecainide

Encainide

Propafenone

Torsade:

- Amiodarone
- Procainamide
- Quinidine
- Sotalol

Class II: β-blockers *(slows conduction)*

β_1: *"A BEAM"; begin with A-M (not L,C)*

- **A**tenolol – partially stimulates β receptor, long acting
- **B**utexolol – tx glaucoma *"Big Tex Tim treats glaucoma"*
- **E**smolol – tx thyroid storm (shortest acting) *"short Eskimos"*
- **A**cebutalol – partially stimulates β receptor *(Avoid w/ acute MI, angina, arrhythmias)*
- **M**etoprolol

β_{1+2}: *begin with N-Z (and L,C)*

- Timolol – tx glaucoma
- Propanolol – tx tremor, tx panic attacks (longest acting => not in kids or old)
- Nadolol – tx glaucoma
- Sotalol – also blocks K
- Pindalol – use for DM

Gray Skin:

- Chloramphenicol
- Amiodarone
- Deferoxamine

$\alpha_1+\beta_1$:

- Labetalol – tx A Fib
- Carvedilol – tx hypertensive crisis, tx chronic CHF (longer $t_{1/2}$)

Class III: K^+ channel blockers *(affects all of your cells; prolongs AP, slows conduction)*

- **A**miodorone – gray skin, cornea deposits, pulm fibrosis, ↓p450 *"kicks your lungs in the Ass"*

- Sotalol – also β_{ns} blocker
- NAPA
- Bretylium
- Dofetilide

Class IV: Ca^{2+} channel blockers => block atrial arrhythmias *(shortens AP)*
Verapamil – *"very"* cardioselective, digoxin toxicity
Diltiazem – cardioselective, tx A Fib, leg edema
Nimodipine – tx vasospasm after SAH *"Nemesis the spider hemorrhage"*
Nifedipine – vasoselective

Nicardipine
Amlodipine – ankle edema
Femlodipine

Open PDA: Alprostadil (PGE_1)
Close PDA: Indomethicin

Other Anti-Arrhythmics:
Adenosine – tx SVT, bronchospasm (slows AV node)
Digoxin – tx ↓HR, delirium (slows AV node - stimulates vagus, ↑contractility -Na/K pump block)
Epinephrine (↑HR, ↑contractility)
Magnesium – shuts down Na/K pump

Digoxin Toxicity:
- Amiodarone
- Spironolactone
- Quinidine
- Verapamil

p450-dependant drugs *levels rise if you inhibit p450*
"Women's DEPT"
Warfarin
Digoxin
E$_2$
Phenytoin
Theophylline

NOTES:

Hematology:

"I have nothing to offer but blood, toil, tears and sweat."

–Winston Churhill

C<u>**lotting Cascade:**</u> how you stop bleeding

Extrinisic: something cut you => Factor 7
Intrinsic: blood vv. pop (vasculitis, sepsis)

Bleeding Time: tells you it's a platelet problem or a vasculitis

Platelet problem => bleed from *skin and mucosal surfaces*
Clotting problem => bleed into *cavities* (↑ bleeding time)

- intracranial => herniation
- mediastinal => rips aorta
- pleural: hemoptysis => stops lungs from expanding
- pericardium => tamponade
- pelvis
- retroperitoneum
- thighs
- abdominal: hematemesis
  - Dark blood: melena
  - Bright red blood: hematochezia
  - **Cullen's sign:** bleeding around umbilicus => hemorrhagic pancreatitis
  - **Turner's sign:** bleeding into flank => hemorrhagic pancreatitis

Petechiae: dot hemorrhage
Papule: palpable
Ecchymosis: bruise

<u>**↑PTT and Bleeding Time:**</u>

- VWD
- SLE

<u>**Compartment Syndrome:**</u>
Tissue pressure >30: Rhabdomyolysis → Mb → renal failure

- Pain w/ muscle movement (first)
- Pallor
- Poikylothermia: cold
- Paresthesia
- Pulselessness: bad prognosis (last)

<u>**Hemolysis:**</u>

- ↑LDH
- ↓Haptoglobin

<u>**Virchow's Triad:**</u> *thrombosis risk factors*
1) Hypercoagulable (Ex: sepsis, trauma, amniotic fluid, cancer)
2) Stasis (Ex: A Fib, pregnancy, truck drivers, post-op)
3) Endothelial damage (Ex: vasculitis)

Intrinsic Pathway: measure PTT *"the PiTT is inside"*
Extrinsic Pathway: measure PT *"PeT the dog outside"*

The Coagulation Cascade Coalesced: *takes 2 hrs*

Inhibitors:	Intrinsic:	Factors:	Extrinsic:	*Inhibitors:*
		12		
	Kallikrein →	**11**		
		9		← *Warfarin*
Prot C →	**8 →**	**10**	**←7**	
Prot C →	**5 →**	**2**		← *ATIII* ← *Heparin*
		1		
		13		

Important Players:

Kallikrein: Released w/ inflammation → DIC
Factor 1: "Fibrogen", activated form is fibrin
Factor 2: "Prothrombin", actived form is thrombin
Factor 7: 2nd shortest half-life
Factor 8: Made by endothelium (only factor not made by the liver)
Factor 9: "Christmas factor"
Factor 11: Not increased by estrogen (E_2 loves to ↑Fibrinogen)
Factor 12: "Hageman factor"
Factor 13: Helps hold clot together
Protein C: shortest half-life
Protein S: Co-factor for Protein C
Type IV collagen (GP2b3a): anchors platelet to BM
von Willebrand Factor:

1) anchors Factor 8 to platelet
2) anchors platelets to endothelium

Vit K: Co-factor for γ-carboxylation => adds negative charges that are attracted to all that Ca^{2+}
Vit K dependant factors: Factors 2,7,9,10, Proteins C,S
Serine Proteases: Factors 2, 9-12

Platelets sit down and release…

- 5-HT => vasoconstricts only in the brain to protect it
- TXA_2 => vasoconstrict, platelet aggregation
- ADP => energy for platelet aggregation
- Ca^{2+} => attracts Vit. K-dep. factors (2,7,9,10,C,S) →*all dz with ↑Ca have ↑clotting, strokes*

Endothelium releases…

- PreKallikrein → (HMW Kininogen) → Kallikrein
- Prostacyclin: vasodilation
- Bradykininogen → BK (veno and vasodilator => increase blood supply => brings ATP)
- t-PA: activates plasmin to break up clots
- Makes Factors 5, 8, vWF

Form clot:

1) Platelets plug up the hole loosely
2) Fibrin tightens it all up, sticks platelets to endothelium

After clot is formed:

Platelets release PDGF: stabilize fibroblasts
Endothelium releases EDGF and bradykinin: dilates vessels to bring O_2 for energy

Pre vs. Post-mortum Clots:

Lines of Zahn: white clots that only occur in live people

Blood Clot:

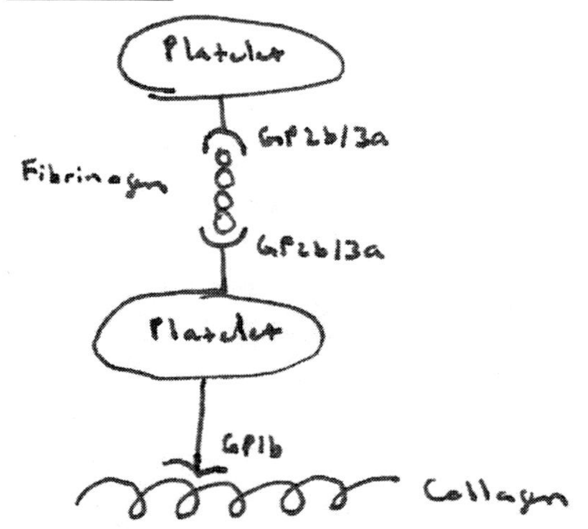

Platelets:
Normal:150-350K
<50K → steroids
<20K → bleeding

Abnormal Platelets:

Bernard-Soulier (AR): baby w/ bleeding from skin and mucosa, ***big** platelets* (low GP1b)
Glanzmann's (AR): baby w/ bleeding from skin and mucosa (low GP2b3a)
↑Platelets:

- Kawasaki
- Polycythemia Rubra Vera
- Essential Thrombocythemia

Idiopathic Thrombocytopenic Purpura "ITP":

- Anti-platelet Ab
- Child petechiae (better prognosis in childhood)
- Tx:

1) 60mg Prednisone
2) Ig
3) Rituximab
4) Splenectomy (should leave Howell-Jolly bodies, if not → have an accessory spleen)
5) Azathioprine/Cyclophosphamide
6) Platelets (if <20k or bleeding)

Thrombotic Thrombocytopenic Purpura "TTP":

- VWF Esterase Deficiency
- Young females w/seizures and big spleens
- Tx: Plasmapheresis

DO NOT GIVE PLATELETS TO: *just makes spleen eat more…*

- TTP
- HUS
- HIT

Bleeding Disorders:

Factor 13 deficiency: umbilical stump bleeding (1st time baby has to stabilize a clot) Tx: FFP
Factor V Leiden: Protein C can't break down Factor 5 => more clots
Protein C deficiency: skin necrosis with Warfarin use

Bleeding Treatment:

- **Mild Bleeding (nosebleed):** DDAVP = Vasopressin = Desmopressin → Factor 5, 8, vWF
- **Moderate Bleeding (heavy menstruation):** Cryoprecipitate – Fibrinogen, Factor 8, vWF
- **Severe Bleeding (shock):** FFP

Von Willebrand's Disease: *heavy menstruation*
Type 1 (AD): Decreased VWF production
Type 2 (AD): Decreased VWF activity (+ Ristocetin aggregation test)
Type 3 (AR): No VWF, ↓Factor 8

Hemophilia A (XR => uncle, grandpa had it): inactive Factor 8 (< 40% activity)

- Bleed into *cavities (head, abdomen, pleural, pericardial, etc.)*

Hemophilia B (XR): Factor 9 deficiency "Christmas disease"

- Bleed into *joints (knee, etc.)*

Hemophilia C: Factor 11 deficiency

Chronic DIC: migratory thrombophlebitis (due to lung, prostate, pancreatic, stomach CA)

Thrombolytics: plasminogen → plasmin → degrades fibrin → busts up clot

- t-PA – tx acute MI, strokes (<3 hrs)
- Streptokinase – also inhibits fibrinogen => more bleeding complications
- Urokinase – open clotted fistulas
- Alteplase – another form of t-PA
- Reverse w/ Aminocaproic Acid "Amicar"

Platelet Haptens:

- asa
- Heparin
- Quinidine

Thrombolytic Contraindications:

- Systolic BP >180
- CPR >10 min
- Recent sugery within 2wk
- A Fib/MS/Pericarditis
- Brain tumor/Head bleed history
- Pregnancy
-

Platelet Inhibitors: bleeding from skin, mucosa

COX Inhibitors:

- asa – irreversibly inhibits COX, cinchonism (don't give if gout, asthma, hyperthyroid)

Thrombin Blockers:

- Argatroban – blocks thrombin, use with HIT, liver metabolism
- Lepirudin – blocks thrombin, use with HIT, kidney metabolism

PDE Inhibitors:

- Cilostazol – dilates arteries
- Pentoxiphylline

Plasma: no RBC

Serum: no RBC or fibrinogen

GP2b3a Inhibitors:

- Abciximab – Ab against GP2b3a, rapid HIT (hours)
- Eptifibatide "Integrilin"
- Tirofiban

ADP Inhibitors:

- Clopidogrel "Plavix" – inhibits ADP, ↓stroke in high-risk pts (w/ asa)
- Ticlopidine – inhibits ADP needed for platelet fxn => agranulocytosis/seizures
- Dipyridamole – blocks ADP receptors, used in cardiac stress test (dilates vessels)

Heparin Induced Thrombocytopenia:

HIT-1: immediate → don't stop heparin

HIT-2: 4-7 days → use Argatroban/Lepirudin

Erythropoiesis: make new RBC

Yolk sac → Liver → Spleen/Lymph → Bone Marrow

(< 1mo) (1-2 mo) (2-4 mo) (> 4mo)

1y/o: if damage long bones, your spleen can make RBC => splenomegaly

Normal Labs:

Hb: **15**

Hct: **3xHb**

pO_2: **60-90**

Hemoglobin:

1. **HbA:** $\alpha_2\beta_2$ ← adults

2. **HbA_2:** $\alpha_2\delta_2$ ← minor component in adults
3. **HbF:** $\alpha_2\gamma_2$ ← fetal, gone by 6 months, high affinity for O_2
4.

Heme synthesis:

$SuccinylCoA_6$ + Gly → (δ -ALA synthase and ferrochelotase) → Heme

Hb Curve:

Shift to the Right: *loss of O_2*

"All CADETs face right"

- ↑**CO_2**
- ↑**A**cid/**A**ltitude (↑H^+ = ↓pH)
- ↑2,3-**D**PG
- ↑**E**xercise
- ↑**T**emp

Note: pH is usually the answer, 'cuz that is what students mess up on…

Shift to the Left:

- Mb
- MetHb
- HbF *(HbF holds onto O_2 until pO_2=40)*
- CO

Exercise:

- Blood flies through lungs (must sit still 0.75s to become fully saturated) → anaerobic
- ↑lactic acid → muscle pain
- The second wind => pO_2:60 → 40 (using Mb now)
- Breathing hard after race => repaying ATP to the body

Chronic Hypoxia:

More mitochondria in muscles

More EPO → ↑Hct

Microcytic Hypochromic => Low Hb synthesis *"FAST Lead"*

1) Fe deficiency (menses, GI bleed, hookworms, colon CA, bad diet, <6 mo – no switch to HbA)

- Angular cheilosis (lip cracks), Koilonychia (spoon nails), Pica (eat ice, clay)
- Earliest sign: ↑RDW
- Low Ferritin (Fe storer) from GI mucosa
- High TIBC (transferrin absorption) to get more Fe
- Low Gastroferrin absorption in stomach
- Low Lactoferrin in breast milk
- High retic count (peaks on day 7)

Anemia:

Acute: <1wk

Subacute: 1-3 wk

Chronic: >3 wk

normocytic until 2 mo

- Tx: Fe (Iron replacement: weight x Hb deficit)

2) Anemia of Chronic Disease: bone marrow shuts down; not replaced

- Plasma proteins kill off RBC (Tx: Epo)
- High Ferritin "Fe storer" (not being used => piles up)
- Low TIBC "Transferrin carrying capacity" (liver can't make transferrin)

3) Pb poisioning (>10µg/dL):

- Kids that ate paint chips, pottery artists, firing range workers, battery workers
- Hypospermia, miscarriages, foot/wrist drop
- Same labs as Anemia of Chronic Disease…
- Basophilic stippling
- Low ALA dehydrase and ferrochelatase => can't make heme
- "Free erythrocyte protoporphyrins" = porphyrin rings that you can't add Fe to
- Bruton's lines = blue gumline
- Child x-ray: lead lines in epiphyseal plates
- Test: Ca-EDTA challenge => lots of Pb excretion in urine
- Tx: <u>Hospitalize</u> (>30µg/dL):
- **P**enicillamine – pulls X^{2+} out of **p**lasma => anaphylaxis
- Dimercaprol "BAL" – pulls Pb out of bone marrow
- Succimer – binds Pb, oral agent
- EDTA – if Pb >45µg/dL

> **Ferritin:** Fe storer
>
> **Transferrin:** Fe limo

> **EDTA** = X^{2+} binder:
>
> Ca, Fe, Cu, Mg, Pb, Zn

4) Sideroblastic anemia: *low δ-ALA synthase ← Vit B_6 cofactor (Ex: INH)*

- Blood transfusions => Fe ppt => Sideroblasts, Pappenheimer bodies

5) Thalassemias: normal RDW (mostly small cells)

- α-thalassemia: Chr #16 deletion
- β-thalassemia: Chr #11 point mutation

> **<u>Enzymes that Need Pb:</u>**
>
> 1) ALA dehydrase
> 2) Ferrochelatase

<u>Hyperchromic:</u>

Spherocytosis (AD): defective spherin

- Microcytic **hyper**chromic
- Spherical RBC get stuck in spleen
- RBC have no central area of pallor
- Test: Osmotic fragility w/ hypotonic saline => cells burst b/c they don't have spherin to protect
- Tx: Folate (to produce erythropoiesis)

<u>Iron Transport:</u>

Ferritin: **stores iron**

Transferrin: **trasports iron**

TIBC: **transferrin absorption**

Disease:	Fe:	TIBC:	Notes:
Fe Deficiency	↓	↑	**High retics**
Anemia of Chronic Disease	↑	↓	***Just sittin' around…bone marrow shut down***
Pb Poisoning	↑	**nl**	***Don't need to bring in any more iron***
Sideroblastic Anemia	↑	↑	**Fe precipitates to form Pappenheimer bodies**
Hemochromatosis	↑	↑	**↑Ferritin >1000ng/mL, ↑Transferrin >50%**

RBCs: biconcave shape

Reticulocyte = Baby RBC

- Normal = <1%
- High => RBC being destroyed peripherally (bone marrow still functioning) → hemolytic anemia
- Low => bone marrow not working

RBC life = 120 days

Platelet life = 7 days

PMN life = 1 day

RBC Measurements: Big cells: normal MCH, low MCHC

RBC count: how many RBCs there are

Reticulocyte count: low => *dec production*, high => *RBC destruction*

Hb conc: how much Hb you have

RDW: RBC distribution width => *anisocytosis* (cell size variation)

Hematocrit: Add EDTA and centrifuge => RBC vol/Total vol

MCV: mean cell volume => hematocrit/RBC count => *micro/macrocytic*

MCH: mean cell Hb => *hypo/hyperchromic*

MCHC: mean cell Hb conc = #g Hb/ dL RBC => *hypo/hyperchromic*

Last names:

-penia => low

-cytosis => high

-cythemia => high

Heme Problems:

Acute Intermittent Porphyria:

- **↑Porphyrin production, urine δ-ALA, ↑urine porphobilinogen**
- **Abdominal pain, neuropathy, red urine (hemolytic anemia)**
- **Can be set off by stress (menses, Drugs: Barbs, Sulfas)**
- **Tx: 1. Fluids – wash away porphyrin ring**
  2. Sugar – break down bilirubin
  3. Opiates – stop pain (use Meperidine for abdominal pain)
  4. Hematin – inhibits δ-ALA synthase

Porphyria Cutanea Tarda: **Sunlight => bullae w/ porphyrin deposits**

- **Wood's lamp = orange-pink**
- **↑urine porphyrine**
- **Assoc w/ Hep C**
- **Tx: Plasmapheresis**

Erythrocytic Protoporphyria: **Porphyria cutanea tarda in a baby**

Globin Problems: **HbS polymerizes → sickling in kidney vasa recta**

Sickle Cell Disease (AR): **Homozygous HbS ($\beta_{Glu6 \rightarrow Val}$)**

- **Dactylitis (painful fingers/toes) at 6mo, short fingers**
- **Protects against malaria**
- "Crew haircut" on x-ray
- **↑Salmonella/Parvo B-19 infections → aplastic crisis**
- Vaso-occlusion: spleen infarction, avascular necrosis of femur (pain referred to knee)
- **Tx:**
  - **Hydroxyurea (↑HbF)**
  - **Folate (prevent aplastic crisis)**
  - **Pneumococcal vaccine**
  - **Sickle cell crisis Tx: O_2, PRBC**

Aplastic Crisis: **RBCs stop being made (no retics)**

Splenic Sequestration Crisis: **RBCs trapped in spleen (high retics)**

Sickle cell anemia: *sickles in kidney vasa recta*

- Salmonella, ParvoB-19 infxns => aplastic anemia
- Tx: O_2, hydroxyurea, pneumococcal vaccine

Stomach Rugal Folds:

Thick: Ménétrier's

Absent: Pernicious Anemia

Sickle Cell Trait: **painless hematuria, sickle with extreme hypoxia (can't be a pilot, fireman)**

- **Heterozygous HbS**

HbC Disease: **($\beta_{Glu6 \rightarrow Lys}$), still charged => no sickling**

PRBC:

1 unit (500mL)

only transfuse if sx...

- **↑ Hb by 1-2g/dL**
- **↑ Fe by 3-4g/dL**

Thalassemias:

- **Hb α subunit – 4 genes**
- **Hb β subunit – 2 genes**

- **α-thalassemia** (Chr.16 deletion) – AA, Asians
- 1 deletion => nl
- 2 deletions "trait" => microcytic anemia
- 3 deletions => hemolytic anemia, Hb H = β4 (ppt w/ brilliant cresyl blue)
- 4 deletions => hydrops fetalis (stillborn baby), Hb Bart = γ4

- **β-thalassemia** (Chr.11 point mutation) – Mediterraneans
- 1 deletion "β minor" => more HbA_2 and HbF
- 2 deletions "trait/intermedia/major" => only HbA_2 and HbF => hypoxia at 6 mo

Cooley's Anemia: **see w/ β thalassemia major (no HbA => excess RBC production)**

- **Baby making blood from everywhere → frontal bossing, hepatosplenomegaly, long extremities**
- **Tx: Total body transfusion q60-90 days → hemochromatosis (Tx: Defuroxamine)**

__Anemia__: **(Hb <11 for everybody)**, less Hb => less O_2 carrying capacity
<9 => moderate
<7 => severe

↓O_2 transport → dilate arterioles → ↓blood return to heart → ↑CO, ↑PP
- Skin atrophy
- Fatty cardiac myocytes
- Neuron degeneration

__Thrombosis Risk Factors:__
"Virchow's triad"
1) Turbulent blood flow *"slow"*
2) Hypercoaguable *"sticky"*
3) Vessel wall damage *"escapes"*

__Hypoxia:__
Acute => SOB
Chronic => Clubbing of fingers/toes (furthest from blood supply)

__Cyanosis__ => 5g Hb fully desaturated at this moment
- An anemic person will rarely be cyanotic b/c 5g will be a bigger percentage of their Hb
- Easier for polycythemic pts to become cyanotic b/c they have more grams of Hb available

__Hemolysis:__
Intravascular: RBC destroyed in blood vv. → low haptoglobin (binds free floating Hb)
- Ex: Vasculitis

Extravascular: RBC destroyed in spleen (problem w/ RBC membrane) => splenomegaly
- Ex: Hemolytic anemias

__Megaloblastic__: gets bigger, can't divide → hypersegmented neutrophils
1) Vit B_{12} deficiency: tapeworms, vegans (↓pain and temp, + Romberg test due to neuro sx)
R-Vit B12 → binds IF (in ileum), need pancreatic enzymes to cleave R off...

Type A Gastritis: atrophic gastritis
- Anti-parietal cell Ab

Pernicious anemia: s/p terminal ileumectomy, ↑gastric CA, absent stomach rugae
- Anti-IF Ab, (+)Schilling test, ↑MMA, ↑Homocysteine
- MMCoA mutase – recycles odd carbons
- Homocysteine Me-Transferase – makes THF (need for nucleotide synthesis)

__Schilling Test:__ *label Vit B_{12}^**
- \+ IF reabsorption → pernicious anemia

- + Antibiotic reabsorption → bacterial overgrowth
- + Pancreatic enzyme reabsorption → pancreatitis

2) Folate deficiency: due to eating old, overcooked food => glossitis (beefy red tongue)

- Drugs: Phenytoin, Phenobarbital, Bactrim

3) Alcohol: Fetal EtOH syndrome: smooth philtrum, stuff doesn't grow, mental retardation

4) Anticonvulsants:

- Phenytoin
- Carbamazepine
- Valproic acid
- Ethosuximide

> **Philthrums:**
>
> Smooth: Fetal EtOH
>
> Long: William's

<u>**Hemolytic Anemias:**</u> *macrocytosis*

<u>**1) Intravascular:**</u> ***IgM***

G-6-PD Deficiency (XR): jaundice, dark urine, sudden drop Hb (>3g/dL), unconjugated hyperbilirubin

- Mechanism: HMP shunt → G-6-PD → GSH (protects RBC against oxidation)
- <u>Drugs:</u>
    - Sulfa drugs: Dapsone
    - Naphthalene moth balls
    - Fava beans
- Tx: IVF to protect kidneys (no asa)

> **<u>RBC Haptens:</u>**
>
> *"PAD PACS"*
>
> - **Penicillamine**
> - **α-MeDopa**
> - **Dapsone**
> - **PTU**
> - **Antimalarials**
> - **Cephalosporins**
> - **Sulfa drugs**

Cold Autoimmune: RBC agglutination, fixes complement, active in distal body parts

- Post-mycoplasma infection
- Mononucleosis infection

Paroxysmal Nocturnal Hemoglobinuria:

- The only hemolytic anemia caused by an acquired cell-membrane defect
- (+) Ham's acid hemolysis, Sugar-water test
- Flow cytometry: CD55/CD59
- Hemolytic anemia, thrombosis, red urine
-

<u>**Extravascular:**</u> ***IgG***

Warm Autoimmune: anti-Rh Ab, active at body temp

- Drugs: The RBC Haptens
- Tx: Steroids, Splenectomy

> **<u>Finger Abnormalities:</u>**
>
> **Sausage digits:** Pseudo hypoPTH
>
> **6 fingers:** Trisomy 13
>
> **2-jointed thumbs:** Diamond-Blackfan anemia
>
> **Painful fingers:** Sickle cell anemia

Paroxysmal Cold Autoimmune: fixes complement

- Donath-Landsteiner Ab
- Massive bleeding after cold exposure

Production Anemias:

1) Diamond-Blackfan: kid is born without RBCs (high adenosine deaminase), 2-jointed thumbs

2) Aplastic Anemia "pancytopenia": ↓RBC/↓WBC/↓Platelets

- T-cell induced apoptosis of progenitor cells (autoimmune → thymoma)
- Virus: ParvoB-19, HepC (transfusions), HepE (pregnant women), EBV
- Drugs: *"ABCV"*
  - **A**ZT
  - **B**enzene
  - **C**hloramphenicol
  - **V**inblastine
- Tx: Immunosuppresion or Bone marrow transplant

Funny Shaped RBCs:

Basophilic stippling: Lots of immature cells => ↑mRNA, stain blue (Pb poisoning)

Bite cell: Hemolysis

Burr cell = Echinocyte: Pyruvate kinase deficiency, Liver dz, Post-splenectomy

Cabot's ring body: Vit B_{12} deficiency, Pb poisoning

Doehle body: PMN leukocytosis
#1 cause: Infection
#2 cause: Corticosteroids
#3 cause: Tumors

Drepanocyte = Sickle cell: Sickle cell anemia

Helmet cell: fragmented RBC (Hemolysis: DIC, HUS, TTP)

Heinz body: Hb denatures and sticks to cell membranes (G-6PD deficiency)

Howell-Jolly body: Spleen or bone marrow should have removed nuclei fragments
#1: Spleen trauma
#2: Hemolytic anemia
#3: Cancer

Pappenheimer bodies: Fe ppt inside cell (Sideroblastic anemia)

Pencil cell = Cigar cell: Fe deficiency anemia

Rouleaux formation: Multiple myeloma

Schistocyte: Broken RBCs (DIC, artificial heart valves)

Sideroblast: MP pregnant with iron (Genetic or Multiple transfusions)

Spherocytes (AD): Old RBCs

Spur cell = Acanthocyte: Lipid bilayer dz

Stomatocyte: EtOH, liver dz

Target cells: less Hb (Thallasemias)

Tear drop cell = Dacrocyte: **RBCs squeezed out of marrow (Hemolytic anemia, Bone marrow cancer)**

NOTES:

NOTES:

Gastrointestinal:

"If you greatly desire something, have the guts to stake everything on obtaining it."

–Brendan Francis

Average Weight:

>20% => obese

Females:

5ft: 100lb. + 5lb/in

Males:

5ft: 106lb + 6lb/in

Medium Frame: +15lb

Large Frame: +30lb

Embryology:

Foregut = lips to duodenum (ligament of Treitz) ← Celiac artery
Midgut = duodenum to transverse colon (splenic flexure) ← SMA
Hindgut = transverse colon to rectum ← IMA

I) Cephalic Phase: CN10 = sensory input
Limbic System: controls urges

Cerebrum: can override the limbic system

Pineal Gland: tells you what time of day it is; responds to 5-HT (made of Trp)

- ↑melatonin at night, ↓during day due to bright light
- Warm milk, turkey: Trp → Melatonin *(takes 4 days for pineal to reset => jet lag)*

Rhythms of the Day:

- 1st 8 hours: Catabolism => exercise in the morning to burn most fat
- 2nd 8 hours: Mixture of catabolism and anabolism
- 3rd 8 hours: Anabolism, melatonin is high

BMI :

weight (kg)/ surface area(m^2)

<18: underweight

>25: overweight

>30: obese

>40: gastric bypass surgery

Hypothalamus:

- Progesterone: similar to the amino acid sequence of testosterone =>↑hunger
- Sympathetics => can't eat *"GI Signs of Dating"*

Hunger center: (lateral)

- NE and 5-HT stimulate this center 20% of the time
- Low glucose/ sight of food → increased firing of hunger center
- Lesion: Anorexia Nervosa (20% below normal weight) Tx: Amitriptyline

Satiety center: (medial)

- NE and 5-HT stimulate this center 80% of the time
- High glucose/ gastric stretch → increased firing of satiety center
- Ex: Gastric bypass → stomach stretches sooner to produce satiety
- Ex: CA on stomach wall stretches it resulting in weight loss
- Lesion: Bulimia => knuckle abrasion, loss of teeth enamel, metabolic alkalosis, hypokalemia
- Lesion: Prader-Willi => hyperphagia

II) Oral Phase:

Salivary Glands:

How Saliva is made:

1° saliva: take up *isotonic* plasma (salivary duct is impermeable to water)

2° saliva: *hypotonic*, more K (Na/K pump), more HCO_3^- (Cl/HCO_3 pump)

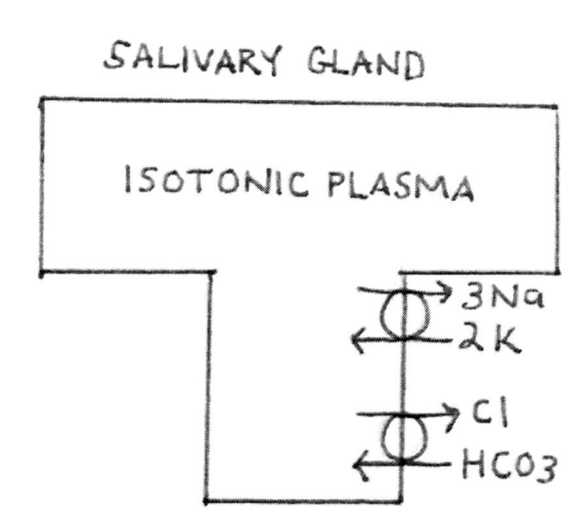

Watery saliva → Lingual > Parotid > Submandibular > Sublingual ← *Mucous saliva*

- *Lingual*: CN 7 (under tongue)
- *Parotid*: CN 9 (Stenson's duct: behind first top molar) ← *Mumps*
- *Submandibular*: CN 7 (in jaw)
- *Sublingual*: CN 7 (on side of mouth)

Saliva Components:

- Lysozyme: detergent to impair adhesion to teeth
- IgA: protection against encapsulated bacteria
- Salivary amylase: carbohydrate digestion
- Bicarb: neutralizes acidic food

Solid Dysphagia:

- Schatzki's rings
- Cancer

Teeth:

10 mo: Incisors → cut (2 bottom, 2 top)

15 mo: Bicuspids → chop (4 top, 4 bottom)

18 mo: Molars → grind (4 top, 4 bottom)

28 mo: Decidiua done

8 y/o: Permanent teeth – need fluoride and calcium

20 y/o: 3rd Molars *"wisdom teeth"*

Solid+Liquid Dysphagia:

- Esophageal spasm
- Scleroderma
- Achalasia

Muscles of mastication: from 1st brachial arch, CN 5

- Masseter: closes mouth, moves cheek
- Temporalis: closes mouth, moves jaw back and forth
- Medial pterygoids: closes mouth
- Lateral pterygoids: Lowers jaw => opens mouth

- Buccinator: slides food sideways (not mastication)

Tongue:

Sensory (somatic):
Front: CN V_3
Back: CN9

Sensory (taste):
Front: CN 7
Middle: CN 9
Back: CN 10

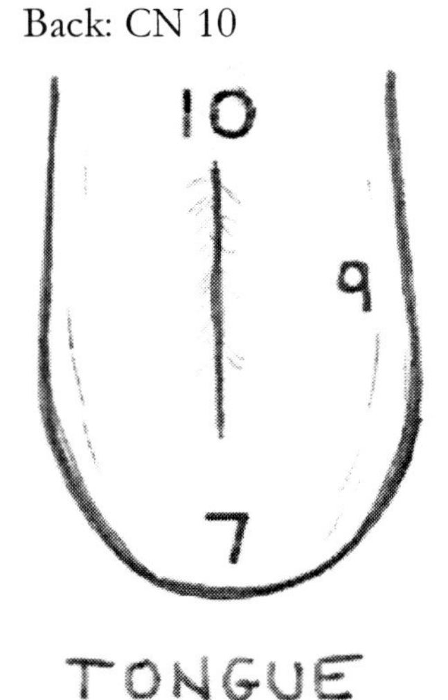

X-rays:
Corkscrew: Esophageal spasm
Apple core: Cancer
Stacked coin: Intussusception
Thumbprint: Toxic megacolon
Abrupt cutoff: Volvulus
Barium clumping: Celiac sprue
String sign: Pyloric stenosis
Bird's beak: Achalasia

III) Pharyngeal Phase:
Swallow:

- Tip of tongue goes up
- Side of tongue forms gutter
- Tongue tip goes up to hard palate to let gravity start rolling bolus down
- Epiglottis closes off glottis to protect trachea => don't talk while eating
    - CN9 – upper pharynx pain
    - CN10 – lower pharynx pain
- Bolus goes over epiglottis to posterior pharynx to esophagus
- Post pharynx comes down and medial
- Soft palate lifts via LVP => opens UES
- UES relaxes and bolus rolls down

IV) Esophageal Phase: esophagus is located slightly on left side
Upper 1/3: Skeletal muscle (stratified squamous epithelium)
Middle 1/3: Mixed
Lower 1/3: SM (tall columnar)

UES: Nucleus ambiguous (gag), lesion this => *NPO* (will aspirate food → feeding tube)

- Stylopharyngeus muscle – CN9
- Pharyngeal constrictors – CN10

LES: CN 10 dorsal motor nucleus (peristalsis), lesion this => *can feed by mouth* (via Ca-Calmodulin)

1° peristalsis: UES

- Contract: CN10 (IP_3/ DAG)
- Release: Auerbach's plexus (VIP inhibits CN10)

2° peristalsis: entire esophagus (Ca/Calmodulin – SM contraction by distension)

<u>Stratified Squamous Epithelium:</u> *protects against abrasion*

- Skin
- Upper esophagus
- Rectum/Anus
- Aorta
- Urethra
- Upper Vagina

<u>Double Bubble:</u>

- Duodenal Atresia
- Annular Pancreas

<u>Esophagus Diseases:</u>

Test Approach: 1)Esophography, 2)Endoscopy

Achalasia: lose Auerbach's plexus in LES => ↑esophageal tone (*choke on solid food*)

- Barium swallow: "bird's beak" = "string sign" = "up-side-down-ace-of-spades"
- Kids: Congenital (choke w/ feeding at 4 mo)
- Adults: Chaga's (it eats ganglia)
- Test: Manometry
- Tx: NGN, local Botox, sphincterotomy

Barrett's esophagus: chronic acid causes metaplasia (squamous → tall columnar), ↑risk of AdenoCA

Boerhaave's: perforate all layers of esophagus → L pleural cavity → acid eats lungs

- Vomit → sudden left chest pain
- Left pleural effusion, left pneumothorax *"bores into lungs"*
- *Hamman's sign:* subcutaneous emphysema => air escaping under skin (crunching sound)
- Dx: Gastrographin swallow (water soluble)
- Pleurocentesis: low pH, high amylase

Choanal atresia: membrane b/w nostrils and pharynx => *blue when fed, pink when cry*

Cloaca (persistant): one opening for rectum/bladder/vagina

Diverticula: pouch → cough undigested food, malodorous halitosis

- Mechanism: uncoordinated swallow
- Dx: Barium swallow
- **Zencker's Diverticulum:** above UES (Tx: Excise + Cricopharyngeal myotomy)
- **Traction Diverticulum:** mid-esophagus (only true diverticulum)
- **Epiphrenic Diverticulum:** above LES

Duodenal atresia: *bilious vomiting after 1st feeding*, double bubble sign, Down's syndrome

Esophageal atresia w/ TE fistula: blind pouch esophagus, *vomit first feeding*, gastric bubble

Esophageal spasm: hypoactive neurons, "*corkscrew*" *barium swallow*

- Tx: NGN, anticholinergics

Esophageal varices: vomit blood everywhere, portal HTN

- Tx: Band ligation or Sclerotherapy (more complications)

Gastroesophageal Reflux Disease "GERD": heartburn

- Tx: Elevate head of bed, H_2 blockers, PPIs

Choanal Atresia:

- blue when feed
- pink when cry

Tetrology of Fallot:

- blue when cry
- pink when stop

Hirschsprung's: lose Auerbach's in rectum => constipation (narrowed segment is involved)

Mallory-Weiss: tear mucosa of LES at GE junction (vomiters: EtOH, bulimia), ↑hiatal hernia

Plummer-Vinson syndrome: upper esophageal webs, spoon nails, Fe-def. anemia

Schatzki rings: lower esophageal webs (hot food causes it to bleed)

Tracheoesophageal fistula: milk drips into trachea, cough and choke with each feeding

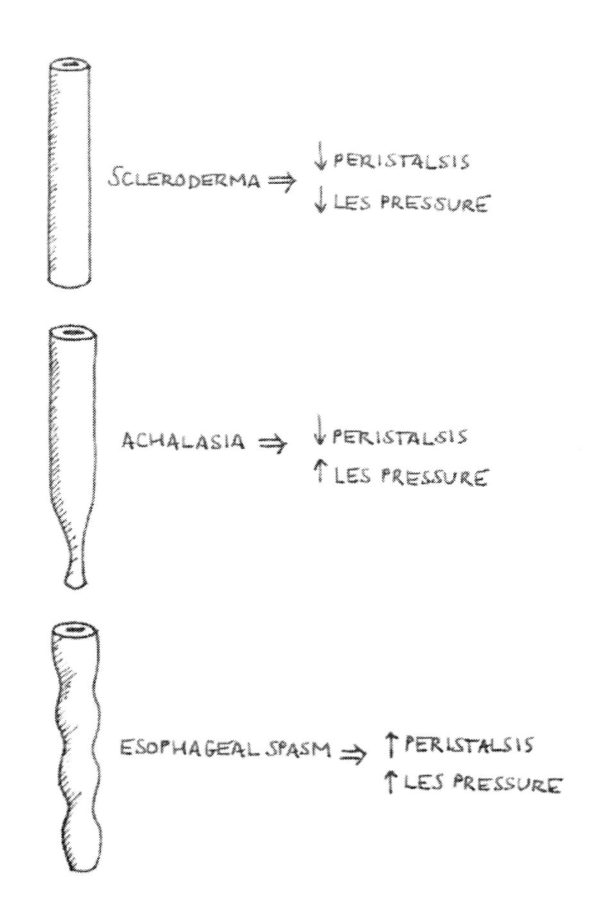

V) Upper GI Phase:

1) Stomach: pH=1-2 => kills all bacteria (except H. pylori – due to urease)
Peristalsis: begins in middle of stomach body
Antrum: *G cells* (stim by high pH) => gastrin
Body: *Parietal cells* (stim by gastrin) => IF, H^+ (via carbonic anhydrase)
Body: *Chief cells*: Pepsinogen → Pepsin to digest protein via H^+

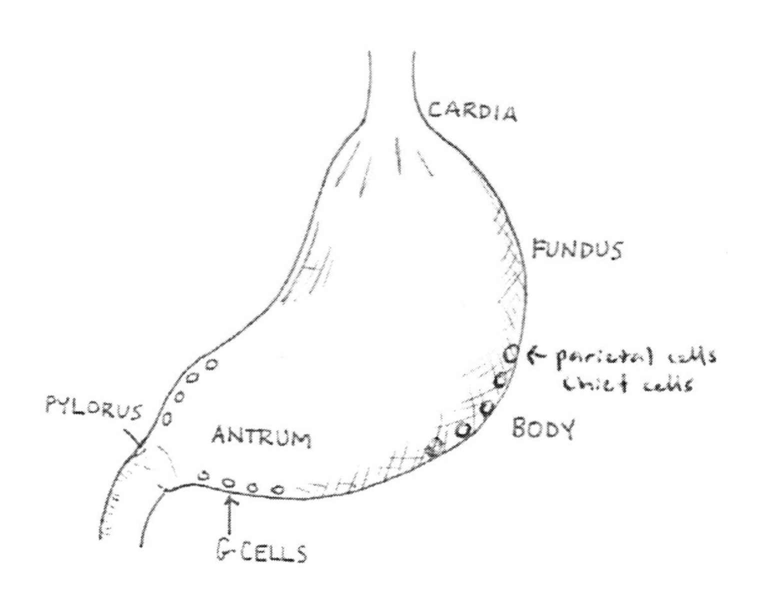

GI pH:
Stomach: pH = 1-2
Duodenum: pH = 3-4
Early Jejunum: pH = 5-6
Late Jejunum: pH = 7-8
Ileum: pH > 9

GI Protection Against Acid:

1) Goblet cells => mucus
2) Alkaline tide => bicarb by-product from parietal cells
3) PGs => mucus release (*NSAIDs, Prednisone causes peptic ulcers by ↓mucus thickness)*

Stomach Diseases:

Bezoar: mass of hair and vegetables => gastric antrum obstruction

Chronic Gastritis: upper GI bleeding, pain less w/ antacids, vomiting

Type **A**: fundus

- **A**nti-parietal cell Ab
- **A**trophic gastritis
- **A**denocarcinoma risk

Stress Ulcers:

1) Parasympathetics: ↑H^+
2) Sympathetics: vasocontrict → ↓blood flow

Type **B**: in stomach antrum (more benign)

- Related to meds, spicy foods, H. pylori

Hiatal hernia: due to obesity (increased abdominal pressure) or restrictive lung disease

Sliding type (90%): fundus slides from esophageal hiatus to thorax => sucks acid into thorax

- Tx: weight loss, H_2 blockers, Nissen Fundoplication => lose ability to belch

Rolling type (10%): fundus through diaphragm hole, strangulates bowel *"rolls through a hole"*

- Tx: surgery

Right Masses:

RUQ olive mass: Pyloric stenosis

RLQ sausage mass: Intussuseption

Ménétrier's disease: lose protein, thick rugal folds (stomach)

Tx: H_2 blockers/ Anticholinergics

Peptic Ulcer Disease:

Gastric (antrum): protection barrier broken down => PPIs will relieve sx, but not fix problem

- Pain worse: *during meal* (increased acid production)
- Type A blood, NSAID or steroid induced
- Pseudolymphoma: lymphocyte infiltration
- 80%: due to H. Pylori => MALT lymphoma (Tx: Antibiotics, not surgery!)
- 20%: predisposed to cancer
- Tx: Misoprostol (PG-E analog) => causes abortions

Duodenal (2nd part): too much acid; perforation => pancreatitis, free air under diaphragm

- Pain worse: changing pattern of pain => perforation, pain relieved by food
  - *30 min after meal* (takes that long to get to duodenum)
  - *at night* (no food to buffer acid at night)
- Type O blood; no risk of malignancy
- 100%: due to H. pylori => urease (breaks down mucus)

- Perforation: guaiac (+), erodes into gastroduodenal artery
- Dx: UGI series w/ Gastrograffin (water-soluble contrast)
- Tx: H_2 blockers or PPI

Helicobacter Pylori:

- Dx: Urease breath test (eat *C-urea → *CO_2 in breath) or fecal Ag test
- General Tx: 2Abx + PPI (can add bismuth to suffocate bacteria)
- Tx: Omeprazole + Amoxicillin + Clarithromycin **or**
- Tx: Ranitidine + Metronidazole + Tetracycline + Bismuth *"4 drugs for 4wks"*

Indications for Ulcer Surgery: *"IHOP"*

- **I**ntractable pain (meds don't help)
- **H**emorrhage
- **O**bstruction from scarring
- **P**erforation

Malabsorption DDx:

- Pancreatitis
- Celiac sprue
- Whipple's dz
- Lactase deficiency

Peptic Ulcer Surgery:

Bilroth I: antrectomy/hook stomach to duodenum → G hyperplasia → more ulcers

Bilroth II: antrectomy + hook stomach to jejunum
1) Dumping syndrome: palpitations after eating (Tx: small fatty meals)
2) Blind Pouch syndrome: 1ft of duodenum has nothing, causing bacterial overgrowth
3) Afferent Loop syndrome: eating causes cramps, vomit brown stuff
4) Reactive hypoglycemia (↑GIP → ↓glucose)
5) Pernicious anemia, steatorrhea
Tx: eat small high fat meals to slow it down (fat takes longest to digest)

Selective Parietal Cell Vagotomy: least symptoms

Pyloric Stenosis: thickening of pyloric muscle

- Projectile vomiting (3-4 wk old)
- Olive sign: feel an "olive mass" in RUQ
- Dx: String sign (barium trickling down)
- Tx: pyloric myotomy (split the muscle fibers)

Microsteatosis:

- Pregnancy
- Reye's syndrome
- Acetaminophen O/D

Macrosteatosis:

- EtOH

2) Small Intestine:

Most common cause of Small Bowel Obstruction: Adhesions (x-ray: multiple air-fluid levels)

Hormones:

Secretin (stim by low pH) → Pancreas: secrete bicarb, tighten pyloric sphincter, ↓gastric emptying
CCK (stim by fatty food) → squeeze Pancreas/Gallbladder → enzymes/bile
Motilin: Peristalsis

- 1°: Segmentation
- 2°: MMC (Migrating Motor Complex), squeezes every 90min

GIP: enhances insulin (stimulated by glucose => reactive hypoglycemia)
Enterokinase: activates the first trypsin → activates everybody else
SS: inhibits secretin, motilin, CCK
VIP: inhibits secretin, motilin, CCK, relax SM
Amylase: breaks down carbs

- Lactose → glucose + galactose
- Sucrose → glucose + fructose
- Maltose: 2 glucoses w/ α -1,4 branching
- α-dextrin: 2 glucoses w/ α -1,6 branching

Absorption: most occurs in "Big Daddy Jejunum"
Duodenum/Jejunum/Ileum (%):
Fe^{2+}: **90**/10/0
Fat: 10/**80**/10
Vit A,D,E,K,B_{12}, bile salts: 0/0/**100**
All else: 10/**90**/0

Beginning of Digestion:
Carb digestion: Mouth (amylase)
Protein digestion: Stomach (pepsin)
Fat digestion: Small intestine (lipase)

Small Intestine Diseases:
Osmotic Gap: *stool osmolarity – 2(stool Na +K) = <50*
Constipation: < 3 bowel movements/ wk
Diarrhea: > 200g/day poop *(who measured that?)*

Bloody Diarrhea:
"CASES"
Campylobacter
Amoeba (E. histolytica)
Shigella
E. coli
Salmonella

Osmotic diarrhea: watery

- Cl^- excretion causes H_2O to leave
- Increased osmotic gap
- Ex: Celiac sprue, lactose intolerance, sorbitol

Secretory diarrhea: laxatives, non-invasive microbes

- Increased osmolarity → increased water → dilutes Na → can't pump
- Normal osmotic gap
- Ex: ZE syndrome, carcinoid syndrome, cholera

Inflammatory diarrhea: blood, pus

- Ex: UC, Crohn's, dysentery

Non-infectious diarrhea:

- Pellagra
- Carcinoid
- Glucagonoma

- VIPoma (watery diarrhea)

Steatorrhea:
- Cystic Fibrosis
- Celiac Sprue
- Zollinger-Ellison
- SSoma
- Gallbladder CA

Celiac Sprue:
- Jejunum, wheat allergy, villous atrophy, unmasked by gastric bypass
- Oxalate stones, osmotic diarrhea, ↑T-cell lymphoma
- Fe-deficiency anemia, vitiligo, ↓Vit D
- Dermatitis herpetiformis *(Tx: Dapsone)*
- Anti-gliadal Ab
- Anti-reticulin Ab
- Anti-endomysial Ab
- Tx: Gluten-free diet (no wheat, rye, barley)

Dapsone Tx:
- Dermatitis Herpetiformis
- Brown recluse spider bite
- Leprosy
- Toxoplasmosis

Tropical Sprue: ileum celiac sprue (Tx: Erythromycin)

Mesenteric Ischemia: plugged blood supply, pain out of proportion to exam
- Dx: Angiography, spiral CT

3) Liver (normal=6-12 cm): fenestrated endothelial cells => free flow of serum across

Functions:
- Synthesis of plasma proteins, bile acid, coagulation factors, lipids (Ito cells)
- Stores vitamins
- Detoxification

Cirrhosis Diet:

Early: high protein

Late: low protein

(avoid encephalopathy)

Zone 1: Periportal
Zone 2: Intermediate
Zone 3: Central – contains P450 system

Liver Enzymes:
- PT: acute liver injury
- LDH: hemolysis, MI, PE, tumor
- ALT: viral hepatitis or MI
- AST: alcoholic hepatitis *"A Scotch & Tonic"*
- GGT: alcoholic hepatitis
- Alkaline Phosphatase: made by bone, liver, placenta

Hep B Association:
- Polyarteritis Nodosa
- Membranous Glomerulonephritis

- Bilirubin: increased production, decreased excretion, or the liver ain't doing it's job…

Total Bilirubin: Direct (liver) + Indirect (hemolysis)

Hep C Association:

- Lichen planus
- Cryoglobulinemia
- Porphyria cutanea tarda

Child-Pugh:
Labs: Albumin, Bilirubin, INR
Exam: Ascites, Encephalopathy

1° Biliary Cirrhosis: autoimmune destruction of bile ductules in liver,

- Xanthelasma, no jaundice, pruritis at night
- Anti-mitochondrial Ab
- Tx: Cholestyramine, Ursodeoxycholic acid, Phenobarbital, Terfenadine

1° Sclerosing Cholangitis: bile duct inflammation

- US "beading", "onion skinning"
- Assoc w/ Ulcerative Colitis
- Staging: liver biopsy (bad prognosis if Alk Phos >125)
- Diagnosis: ERCP or PTC
- Tx: Cholestyramine, Ursodeoxycholic acid (Sjö grens, dark skin, pruritis), Abx

Ascending Cholangitis: common duct stone gets infected => dilated ducts w/ pus

- Tx: Emergency decompression

Budd-Chiari: hepatic vein thrombosis (poor prognosis) → fetor hepaticus (NH_4^+ breath)

- ↑Ascitic protein

Cirrhosis: hepatocyte destruction; regeneration causes fibrosis
Sx: jaundice, fluid wave, asterixis
Alcoholic Sx: spider angioma, palmar erythema, Dupuytren's contractions, gynecomastia
Dx: high bilirubin

Albumin (serum – ascites):
<1.1: TB, tumor, inflammation
>1.1: portal HTN

Cirrhosis Etiology:
1) EtOH: AST > ALT *"A Scotch and Tonic"*
2) Hep B,C: ALT > AST
3) Biliary: xanthomas, xanthelasmas, anti-mitochondrial Ab
4) Hemochromatosis: bronze skin, DM, arthritis
5) Wilson's dz: copper deposits, KF rings, ↓ceruloplasmin
6) α_1-AT deficiency: emphysema
7) CHF: nutmeg liver

Ascites Tx:
1) Paracentesis (replace albumin)
2) Spironolactone

Cirrhosis Tx:

- Glucocorticoids

- Colchicine (↓inflammation)
- Spironolactone (↓ascites)

Hepatic Adenoma: due to oral contraceptives

- Tx: embolize/resection

Hepatic Encephalopathy:

- Confusion → coma, asterixis, fetor hepaticus
- Tx:

1) Protein-free diet
2) Lactulose (traps $NH_3 \rightarrow NH_4^+$ in colon) → enema
3) Neomycin (kills NH_3-producing bacteria)

Hepatitis:

Hep A: Fecal-oral (Pregnant, Asians, Shellfish)
Hep B: Needles (DNA virus, Mom → baby)
Hep C: Blood (2° Hemosiderosis)

Hepatorenal Syndrome: pt w/ liver dz → liver toxins → renal failure

- Sx: azotemia (↑BUN/Cr), Na retention, oliguria (urine Na <10), hypotension
- Precipitated by diuresis, paracentesis, dye, GI bleeding, aminoglycosides
- Kidneys are still viable for transplantation
- Tx: volume
- Ex: Furosemide to cirrhosis pts → no pee for 3 days (Tx: Stop diuretic, then volume load)

Portal Hypertension: ↑portal vascular resistance (>12mm Hg)

- Caused by cirrhosis, portal vein obstruction, hepatic vein thrombosis
- Tx: Propanolol, TIPS (hook up hepatic and portal veins)
- Tx: Vasopressin or Octreotide for bleeding

Spontaneous Bacterial Peritonitis:

- Fever, chills, rebound tenderness, altered mental status
- ↑Risk with nephrotic syndrome
- Dx: Paracentesis has >250 PMNs/μL
- Tx: Cefotaxime + Albumin (for renal perfusion pressure)

<u>AVM Complications:</u>

Tx: Coils to clot around

- Sequester platelets → bleed
- Rupture → bleed
- Sequester blood → heart failure

4) <u>Gall Bladder:</u> *Bilirubin >2 → yellow eyes*

- Bilary tract pain Tx: Demerol (does not ↑CCK)
- Diarrhea s/p cholecystectomy (due to bile salts) Tx: Cholestyramine

Ascending Cholangitis: infection of obstructed CBD

- Charcot's Triad (RUQ pain, fever, jaundice)

- Reynold's Pentad (shock, altered mental status) → 50% mortality

Biliary Colic: stone stuck in cystic duct or CBD

Cholangiocarcinoma: Clinorchis Sinensis in biliary tract

Cholecystitis: inflammation of gall bladder
- Murphy's sign (press gallbladder → pt stops breathing)
- Tx: Ampicillin + Sulbactam + Gentamycin
- Tx: Urgent transhepatic drainage tube (if acalulous)

Cholelithiasis: gallstone
- RUQ colic, US shadow sign

Choledocholithiasis: gallstone obstructs bile duct
- Tx: emergent ERCP

Cholestasis: bile can't get from liver to duodenum
- Pruritis, ↑alkaline phosphatase

Gallstone ileus: small bowel obstruction caused by gallstone erosion into duodenum

Klatskin tumor: R/L hepatic junction tumor

Mirizzi's fistula: compression of CBD by an impacted cystic duct

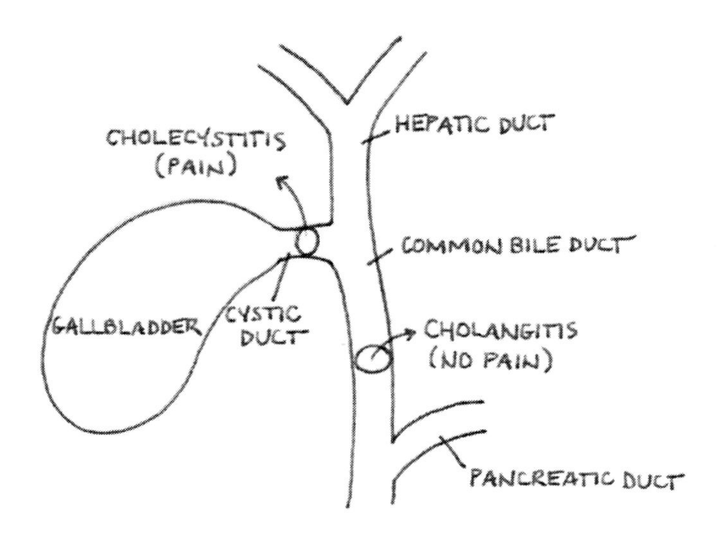

Post-Cholecystectomy Pain:
- Functional pain
- Sphincter of Oddi contract.
- Common bile duct obstruct.

The Bile-Cholesterol Relationship:

MP: eat old RBC → breakdown heme → biliverdin → bilirubin → "bile acid"

Liver: conjugates bile acid-albumin w/ glucuronyl transferase (add Gly/taurine) → "bile salts"

Small intestine: deconjugates bile salt bound to cholesterol

- 80% reabsorbed in ileum (urobilinogen => yellow pee) → kidney → recycle bile salts → liver
- 20% excreted (stercobilinogen => brown poop)

Gall bladder: stimulated by CCK to release bile

- Conjugated: "direct" (water soluble)
- Unconjugated: "indirect" (fat soluble)

Bile salt synthesis: occurs in between meals
Forms micelle: Lecithin + Fat + Bile salts
Lipase chops it all up =>

- short chain FA – have lacteals ("lymphatics" for fats), CM
- medium chain FA – transported by albumin to liver (tx steatorrhea in infants)
- long chain FA – have lacteals, CM

Gallstones: high cholesterol, low bile (most get stuck in the cystic duct)
80% gallstones: made of Cholesterol => can't see it on x-ray *"Female, Fat, 40, Fertile, Flatulent"*
20% gallstones: made of Ca-bilirubinate (w/ hemolytic anemia) => see w/ x-ray

- Brown: CBD stones (Asian, infection)
- Black: Gallbladder stones (cirrhosis, hemolysis)

Tx: Dissolve w/ Ursodeoxycholic acid or do cholecystectomy

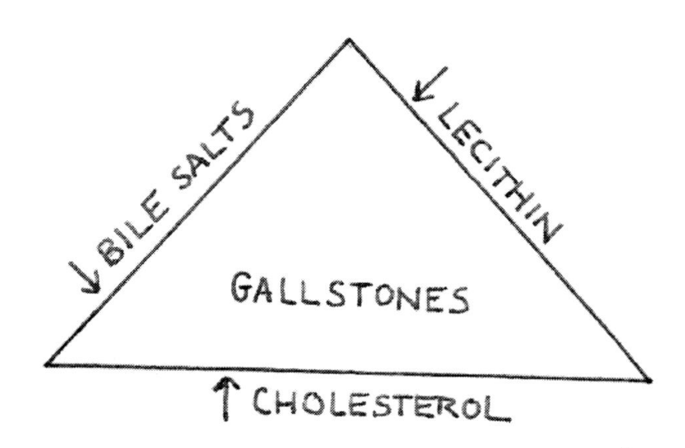

Gallstone tests:
1) Labs: ↑alk phos, ↑direct bilirubin
2) X-rays: see 20% of gallstones
3) US: see obstruction
4) HIDA scan: inject dye into veins → cystic duct → dye can't enter gall bladder → obstruction

Cholesterol Diseases:

Hyperlipidemias:
Type 1: Bad Liver LL (CM)
Type 2a: Bad LDL or B-100 receptors: trapped in ER (LDL only) ← Familial

- (+) Thompson test due to Achilles tendon rupture (can't stand on toes)

Type 2b: Less LDL/VLDL receptors (LDL/VLDL) *downregulation of receptors due to obesity*
Type 3: Bad Apo E (IDL/VLDL)
Type 4: Bad Adipose LL (VLDL only)
Type 5: Bad C2 (VLDL/CM) *b/c C2 stimulates LL*
Hypercholesterolemia: type 2 (LDL carries cholesterol)
HyperTGemia: not type 2 (they carry TG)
A-betalipoproteinemia: no B48 tags => no CM
Dysbetalipoproteinemia: type 3

Xanthoma = Cholesterol (elbow)
Xanthelasma = TG (under eye)

Driving Ms. Cholesterol:

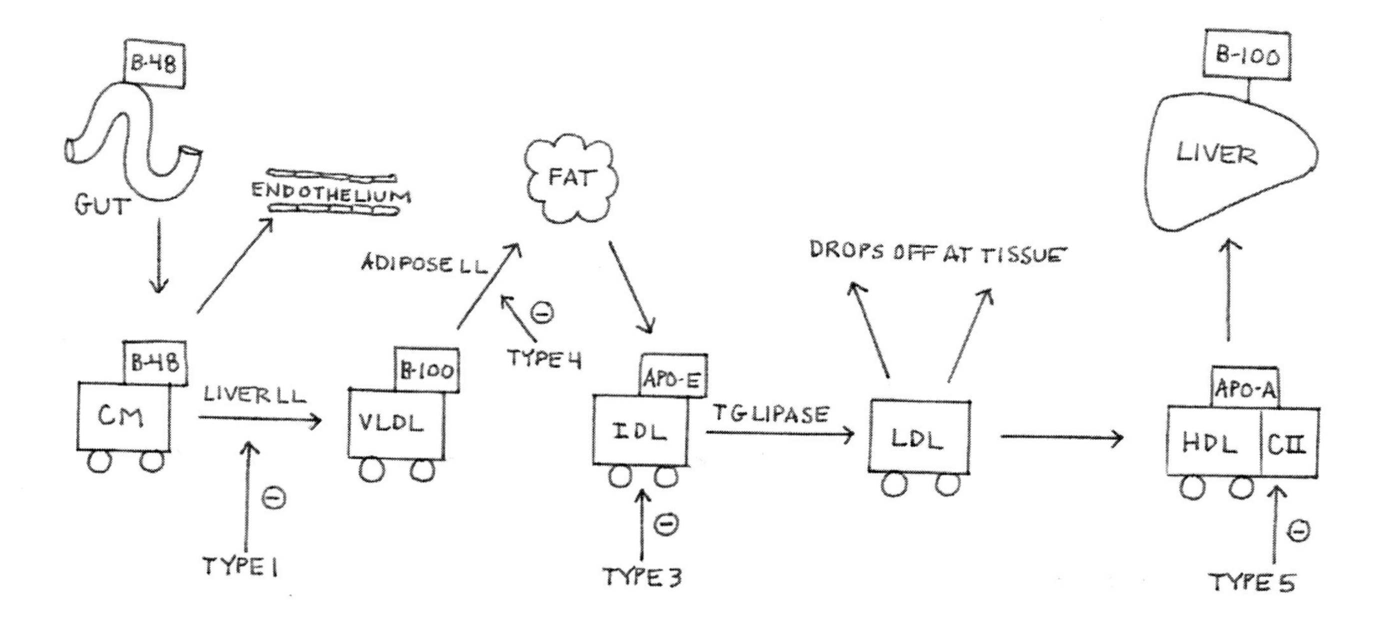

Liver-Mart: Sells B-100 tags (VLDL, IDL, LDL taxis)
Tissue-Mart: Sells B-48 tags (CM limo)
HDL: got everything in the trunk (C2 = LL and HDL activator, Apo-E, Apo-A, LCAT)
"Can Very Intelligent Livers Help?"
CM limo drives 25% TG to the *liver*, transforms into VLDL w/ B-100 tag
75% TG to *endothelium*, keeps B-48 tag
VLDL drives TG to *adipose tissue*; transforms into IDL w/ Apo-E tag
IDL – drives TG *everywhere else*; transforms into LDL by losing the Apo-E tag
LDL – drops off leftover cholesterol everywhere (B-100 binds LDL receptor in clathirin pits)
HDL – picks up the garbage to sell it later; uses Apo-A to activate LCAT → carry cholesterol to liver

Bilirubin Diseases: *"Come Get Really Drunk"*

1) Unconjugated: *(enzyme problem)*
Crigler-Najjar (AR) – unconjugated bilirubin, usually in infants *"**C**onjugation defect"*

- Type I – complete glucuronyl transferase deficiency => kernicterus
- Type II – partial glucuronyl transferase deficiency (Tx: Phenobarbital)

Gilbert's: glucuronyl transferase is saturated => stress unconjug bilirubin (Tx: hydration)

<u>2) Conjugated:</u> *(transport problem)*

Rotor's: bad bilirubin storage => conjugated bilirubin *"**R**elease defect"*

Dubin-Johnson: bad bilirubin excretion => conjugated bilirubin, black liver *"**D**eparture defect"*

Common Bile Duct obstruction => increased conjugated bilirubin, alk phos, jaundice, itching

- Newborn: choledochal cyst, biliary atresia, annular pancreas (ventral bud went wrong way)
- Kid: gallstones
- Elderly: pancreatic cancer

Gallstone ileus: gallstone eroded through gallbladder wall and fell into duodenum

5) <u>Spleen:</u> the blood's lymph node, easily crushed

White pulp: T, B cells

Red pulp: venous system

Spleen removal => ↑encapsulated infections => *pneumonia vaccine*

6) <u>Pancreas:</u> secretes enzymes

<u>Pancreatic Diseases:</u> autodigestion by proteolytic enzymes causes inflammation

Annular pancreas => ventral pancreatic bud encircles the 2nd part of duodenum, *double bubble*

Phlegmon => bowel wraps around inflamed pancreas to prevent inflamm. from spreading => ileus

Pancreatitis – severe mid-epigastric pain boring to the back, malabsorption, DM

Kids: Trauma, Infection (Coxsackie B, EBV, CMV, cystic fibrosis)

Adults: Chronic EtOH, Acute gallstones (Tx: hook pancreatic duct to jejunum)

<u>Pancreatitis Etiology:</u>

"PANCREATITIS"

- **P**eptic ulcer perforation
- **A**lcohol
- **N**eoplasm
- **C**holelithiasis
- **R**enal disease
- **E**RCP
- **A**norexia
- **T**rauma
- **I**nfection: CMV, TB
- **T**oxins: HIV drugs, ACE-I, Salicylates

<u>Default Colors:</u>

bilirubin adds normal color

Stool: clay

Urine: tea

<u>Hemorrhagic Pancreatitis:</u>

Cullen's sign: bleeding around umbilicus

Turner's sign: bleeding into flank

- Incinerations (burns)
- Scorpion bite

Pancreatitis Tests:

- Amylase – sensitive, breaks down carbs (also in mouth)
- Lipase – specific, breaks down TGs

Ranson's Critera (at presentation):
"WAGLA" => poor prognosis

- **W**BC: >16K/μL (infection)
- **A**ge: >55 (usually have multiple illnesses)
- **G**lucose: >200 mg/dL (islet cells are fried)
- **L**DH: >350 IU/L (cell death)
- **A**ST: >250 IU/L (cell death)

Splenic Flexure = "Watershed area"

little blood supply => ischemic bowel

SMA stops before watershed area

IMA starts right after watershed area

Ranson's Criteria (at 48 hrs):
"C-HOBBS"

- **C**a: <8 mg/dL (saponification)
- **H**ct: drops >10% (hemorrhage into pancreas)
- p**O**$_2$: <60mm Hg => fluid and protein leak out => ARDS (restrictive lung disease)
- **B**ase deficit >4mEq/L (diarrhea => pancreatic enzymes are dead)
- **B**UN: increase >5mg/dL (↓renal blood flow)
- **S**equester >6 L fluid => 3rd spacing (bowel swollen with water)

Ranson's Risk Factors:
3 → 15% mortality
5 → 40% mortality
7 → 100% mortality

Pancreatitis Tx:

- NPO (let the pancreas rest)
- NG tube (decompress air)
- IV fluids (NS)
- Mepiridine (pain relief, will not ↑CCK)
- Imipenum
- Ampicillin + Gentamycin + Metronidazole (if necrosis)

Appendicitis:

- Psoas sign
- Rovsing sign
- Obturator sign
- Pain at McBurney's point

Pseudocyst: fluid lined with granulation tissue
Ruptured Pseudocyst: pancreatitis + abdominal mass with bruit

VI) Lower GI Phase: *Retroperitoneal → ascending/descending colon*
Jejunum: highest absorptive capacity

Ascending colon: highest resorptive capacity

Colon: secretes K^+, all valves = α_1 receptors
1° peristalsis = haustration
2° peristalsis = mass movement

Laplace's Law:
Tension = Pr

Cecum: large pocket, no obstruction, most perforation
Ascending colon: last chance to reabsorb fluid *(most Na/K pumps controlled by Aldo)*
Sigmoid colon: stool waits to make its journey into the outside world
=> 90° angle of rectum to sigmoid kept by pubococcigious muscle → keeps feces in sigmoid

Defecation: completely parasympathetic
1) Relax PC muscle → gravity starts moving stool down into rectum from sigmoid
2) Internal anal sphincter relaxes ← pelvic nerve
3) Relax pelvic floor mm. (spread by specific angle to relax mm. = "toilet seat")
4) External anal sphincter contracts (voluntary control) ← pudendal nerve

Colon Diseases: *PAS stains fat*
Adults: usually occurs in sigmoid
Kids: usually occurs in ileum

Currant-jelly sputum: Klebsiella
Currant-jelly stool: Intussusception

Appendicitis: infection due to fecolith in appendix, periumbilical pain radiated to RLQ

Ruptured Appendicitis: pain relieved, peritonitis, pyelophlebitis (infxn of portal vein thrombosis)

Diverticulitis: hurts → do CT (colonoscopy will perforate)

Diverticulosis: bleeds (fecoliths erode into arteries)

Intussusception: 3mo -6y/o, currant-jelly stool, stacked coin enema, RLQ sausage mass

- Tx: barium enema (risk of perforation)

Volvulus: X-ray abrupt cutoff in bowel air, Barium swallow birds-beak sign, sudden pain

- Tx: sigmoidoscopy with rectal tube

Colon CA Risk:

- Villous
- Sessile
- >2.5cm

Adenomatous Polyps: pre-malignant (adults)
Tubular: stick out
Villous: flat, secrete K^+ into stool
Gardner's syndrome: familial polyposis w/ *bone tumors*
Turcot's syndrome: familial polyposis w/ *brain tumors*
Familial polyposis: 100% risk of colon cancer, dinucleotide repeats (5y/o colonoscopy q yr)
Juvenile polyposis: <10 y/o polyps + intussusception

Hyperplastic Polyps: benign (kids)
Hamartoma: hyperplastic
Peutz-Jegher syndrome: hyperpigmented mucosa → dark gums/vagina

- CA of breasts/ovaries/lymphatics, benign colon polyps (no colon CA)

Familial Colon Cancer:
1) FAP: APC gene, left-side tumors, 100% risk of colon CA
2) HNPCC: Mismatch repair gene, right-side tumors, dinucleotide repeat, microsatellites

Inflammatory Bowel Disease:

Crohn's = "regional enteritis"	**Ulcerative Colitis:** 6BM/day x 6mo
Sx: weight loss, cramps, melena	Sx: bloody diarrhea, rectal pain (tenesmus)
Transmural (all 3 layers)	Mucosal only
Non-caseating granulomas → cobblestones	Pseudopolyps (2 ulcers next to each other)
Skip lesions	Continuous lesions
Starts in ileum → distal (involves anus)	Starts in rectum → proximal (not involve anus)
Creeping Fat (due to granulomas)	Lead Pipe colon
Melena = dark stools	Hematochezia = bright red blood
3% risk for colon CA, HLA-DR1	10% risk for colon CA, HLA-B27, HLA-DR2
Fistulas: *do CT scan* • Enterocutaneous: GI to skin ▪ Enteroenteral: GI to GI ▪ Enterovesicular: GI to bladder "pneumoturia" ▪ Enteroaortic: bleed through bowel, die	1° Sclerosing Cholangitis Toxic Megacolon Pyoderma Gangrenosum (skin) Smoking cessation: leads to flare-ups
Associated w/ uveitis, amyloidosis, mouth ulcers, steatorrhea, kidney stones, oxalate stones	Associated w/ ankylosing spondylitis, cholangitis
Tx: Prednisone, Mercaptopurine, Infliximab	1) Mesalamine "5-ASA" or Sulfasalazine 2) Steroids 3) Azathioprine or 6-mercaptopurine 4) Infliximab 5) Cyclosporin A

"Creepy Mrs. Crohn Skipped down the Cobblestone lane, shaking her Fist at the Dark Stool"

Irritable Bowel Syndrome:

- Alternating diarrhea and constipation, abdominal distension
- Pain relief with bowel movement, sense of incomplete evacuation, mucus in stool
- Tx: lower stress, high fiber diet

Ischemic Colitis: sudden pain w/ bloody diarrhea

- Occurs at watershed area
- Throw clot to SMA => A Fib (ischemic tissue depolarizes)
- Can lead to mesenteric ischemia: pain out of proportion to exam, life-threatening

Ogilvie's: pseudo-obstruction
- If cecum >12cm → perforation
- No stool in rectal vault, no obstruction on colonoscopy
- Tx: Neostigmine

Pseudomembranous Colitis: overgrowth of *C. difficile*
- Sx: diarrhea, cramps, fever
- Yellow plaques on colon mucosa
- Can cause toxic megacolon = intense diarrhea, *"thumbprint" x-ray*
- Causes:
  1) Cephalosporins – prescribed more
  2) Clindamycin
  3) Ampicillin
  4) Amoxicillin
- Tx:
  - ⇨ First Episode: Metronidazole (PO)
  - ⇨ Repeat Episode: repeat Metronidazole (PO)
  - ⇨ Pregnant/Kids: Vancomycin (PO)
  - ⇨ Resistant: Vancomycin (enema) or Cholestyramine

Spastic colon: intermittent severe cramps
- Test: inject marijuana into colon => watch it spasm
- Tx: muscle relaxants

Whipple's disease: *T. whippleii* destroy GI tract, then spread
- Middle age male, arthralgia, malabsorption, PAS+ MP
- Tx: Bactrim

Upper GI Obstruction: x-ray shows air/fluid levels, high pitched bowel sounds
Newborns: Atresias
4 mo: Achalasia
6mo-2y/o: Intussusception, Hernias
>2y/o: Adhesions (from old blood)

Lower GI Obstruction:
Newborns: Hirschsprung's
Adults:
1) Adhesions

Colon CA Risk Factors:
- Low fiber diet
- High fat diet (free radicals)
- Polyps

2) Obstipation
3) Diverticulitis → *do CT (not colonoscopy; due to rupture)*
4) Cancer => *pencil thin stool, "apple core" x-ray, ↑CEA*

Upper GI bleed: coffee-ground hematemesis, above ligament of Treitz, tarry stool
Newborns: Swallowed maternal blood on the way out => *do Apt test for fetal Hb* (+ => baby's blood)
Kids: Nosepicking (nosebleed → GI → vomit up) *Tx: Phenylephrine nasal spray*
Adults: Gastritis, PUD
DDx: *"Mallory's Vices Gave An Ulcer"*

- **Mallory-Weiss tear**
- **Variceal bleeding**
- **Gastritis**
- **AV malformation**
- **Ulcer (peptic)**

GI Bleeds:
Upper => black
Lower => red

Lower GI Bleed: hematochezia
Newborns: Anal fissure from hard stool (Tx: stool softeners)
Kids: Hyperplastic polyps
>40: Angiodysplasia (varicose vein), Diverticulosis, Cancer
DDx: *"Can U Cure Aunt Di's Hemorrhoids?"*

- **C**olitis (Ulcerative)
- **U**pper GI bleed
- **C**ancer
- **A**ngiodysplasia
- **D**iverticulosis
- **H**emorrhoids

Upper+Lower GI Bleed: Duodenal-Aortic Fistula

Hemorrhoids:
Internal: no nerves fibers => *no pain* (Tx: band and allow to necrose)
External: *pain* => thrombose, ulcerate (Tx: topical anesthetic, Sitz bath, surgical excision)

Massive GI bleed:
Kids: Meckel's diverticulum – remnant of vitelline duct

- 2" long, 2ft from IC valve (on ileum side)
- 2% population
- 2 y/o (peaks)
- 2 types of mucosa = gastric and pancreatic

Adults: Peptic ulcer disease (gastrin secreting)
Test: Tagged RBC scan → where blood is pooling or ^{99}*Tc pertechnetate scan*

Abdominal Pain Management:

1) NPO
2) NG tube
3) IVF
4) Meperidine
5) X-ray (abdomen)
6) CT (abdomen)

Most Common Sources of Abdominal Pain:

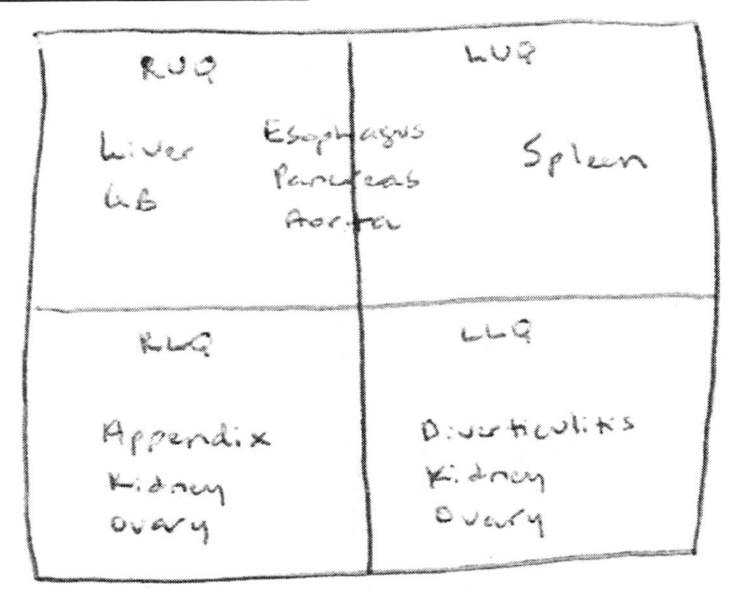

Peristalsis Review:

Esophagus:

1° peristalsis = CN10 (IP_3/DAG)

2° peristalsis = Auerbach's plexus (VIP)

Small intestine:

1° peristalsis = segmentation

2° peristalsis = MMC

Colon:

1° peristalsis = haustration

2° peristalsis = mass movement

Hyperlipidema Tx: *Screen = Total/HDL cholesterol*

Total cholesterol: **>240** (or >200 + 1 risk factor: male, HTN, etc.)

LDL: **>130** (or >100 + 1 risk factor)

HDL: <**40**

TG: >**500**

> **High Cholesterol** => atherosclerosis
> **High TG** => pancreatitis

1) Raise HDL:

- Weight loss, Exercise
- Moderate EtOH (1 glass whiskey or 1 glass wine or 2 beers/ day)

2) <u>HMG-CoA Reductase Inhibitors:</u> *most active after 8pm* (↑HDL, ↓LDL)
SE: fat soluble, rhabdomyolysis, hepatitis → check liver enzymes every 3mo

- **P**rovastatin – most water soluble *"**p**ee it out"*
- Lovastatin
- Simvastatin
- Atorvastatin
- Cerivastatin – taken off market b/c of rhabdomyolysis
- Valdestatin – taken off market b/c of rhabdomyolysis

3) <u>Bile Acid-Binding Resins</u>: force liver to take out more cholesterol to make bile (↓LDL)

- Cholestyramine – ↓absorption of lipid soluble drugs, bad if pt on multiple meds
- Colestipol

4) Niacin: inhibits VLDL prod, ↓Lipoprotein A, use ASA to avoid flushing (↓LDL)

5) <u>Fibrates</u>: enhances LL (↓TG)

- Clo<u>fibr</u>ate – associated with colon cancer
- Gem<u>fibr</u>ozil

<u>GI Drugs:</u>

Antacids:

- Tums: Ca Carbonate => acid output (gastrin), diarrhea
- Rolaids: Al-OH => constipation
- Milk of magnesia: Mg-OH => diarrhea
- Gaviscon: Al-OH and Mg carbonate

H_2 blockers: block H_2 receptors on parietal cells => bloating, cramps (need H^+ to digest food)

- Cime<u>tidine</u> "Tagamet" – inhibits p450 => gynecomastia, psychosis, ↓CrCl
- Ranitidine "Zantac"
- Famotidine "Pepcid"
- Nizatidine "Axid"

Proton Pump Inhibitors: irreversibly inhibit H/K pump in parietal cells => bloating, cramps

- Ome<u>prazole</u> "Prilosec"
- Esomeprazole "Nexium"
- Lansoprazole "Prevacid"
- Pantoprazole "Protonix"

Mucosal Protective Agents:

- Misoprostol "Cytotec" – induces abortion => do pregnancy test in all females
- Sucralfate "Carafate" – don't take w/ antacids or acid blockers (needs acid to dissolve)
- Bismuth "Pepto-Bismol" – black tongue, black stool, suffocates *H. pylori*

Anti-emetics:

- Diphenhydramine (H_1 anti-histamine)
- Odansetron (5-HT inhibitor) – used in cancer pts
- Dronabinol = Marijuana-like – increases appetite in cancer pts
- Prochlorperazine (DAr blocker in gut)

Prokinetics:

- Cisapride – 5-HT agonist (+ Erythromycin => Torsade de Pointes)
- Metoclopramide (DAr blocker in chemotrigger zone) => Parkinson's sx
- Psyllium "Metamucil" – ↑stool bulk, gas, bloating, non-absorbable sugar
- Docusate sodium "Colace" – mixes stool fat and water

Anti-diarrhea:

- Loperamide "Imodium"
- Diphenoxylate "Lomotil"

<u>Opiates:</u>

1) Muscle relaxation
2) CNS depressant
3) Analgesia (listed in order of potency)

- Fentanyl – patch, most potent
- Buprenorphine – low dependence
- Heroin – most abused on streets
- Morphine – most common in hospital for severe pain, ↑intracranial pressure
- Methadone – heroin addict recovery, crosses BBB
- Pentazocine – opioid agonist and antagonist => don't use in heroin addicts, causes nightmares
- Hydrocodone "Vicodin" – moderate pain relief
- Meperidine "Demerol"– tx abdominal pain *(does not contract Sphincter of Oddi)*
- Codeine – anti-tussive, mild pain relief
- Dextromethorphan – anti-tussive *(the "DM" in cold meds)*
- Naltrexone – opiate antagonist, oral
- Naloxone – fast opioid antagonist, IV *"the shorter name acts faster"*
- Loperamide – cough suppression

NOTES:

Pulmonary:

"Breath is the bridge which connects life to consciousness,
which unites your body to your thoughts."

–Thich Nhat Hanh

Neural Crest Cells: *"MOTEL PASS"*

- **M**elanocytes
- **O**dontocytes
- **T**racheal cartilage
- **E**nterochromaffin cells => 5-HT
- **L**aryngeal cartilage
- **P**arafollicular celss/**P**seudounipolar cells
- **A**drenal medulla/**A**ll ganglion cells
- **S**chwann cells
- **S**piral membrane

Embryogenesis:

- Notochord develops by 4 weeks, Brain by 8 wks, Lung by 12 wks
- If 90% lung doesn't develop → pulmonary aplasia → die
- Surfactant is made by 33 wks = alveoli lubricant (↓surface tension to prevent atelectasis)
- Phophatidyl glycerol = surfactant precursor, can test for this
- Lecithin:Sphingomyelin ratio is 2:1 to indicate maturity (brain sphingomyelin is done)
- Tx: Beclomethasone/Betamethasone IM to Mom => surfactant production in baby
- Tx: Blow surfactant into neonate lungs (intubation)

Premie Lung Progression: *will have to leave baby on O_2 for 18-24mo*

1) Atelectasis = collapsed alveoli
2) RDS "respiratory distress syndrome" (Tx: O_2 → free radicals)
3) Hyaline membrane dz → thicken membrane → ↓diffusion (restrictive) → ↑goblet cells
4) BPD "bronchopulmonary dysplasia" → mucus, narrow lumen (obstructive)

ARDS "Adult Respiratory Distress Syndrome": tachypnea, hypoxemia, diffuse infiltrate

- PMNs cause alveocapillary damage → increase permeability of alveolar capillaries
- Most common cause = sepsis
- *NO dilates aa => washes out surfactant, leaks proteins into interstitium*
- $pO_2/FiO_2 < 150$
- CXR: "fluffy" infiltrates although lungs sound clear
- Tx: Glucocorticoids, Ventilator (↑FiO_2, ↑pressure, ↑RR, ↑I:E ratio, ↓TV)

Predicting O_2 Saturation:

pO_2: *dissolved O_2*	O_2 sat: *bound O_2*
100 mmHg	100%
90 mmHg	98%
80 mmHg	**96%**
60 mmHg	90%
40 mmHg	75%
25 mmHg	50%

pO_2 Predictors:

pO_2:	O_2 Sat:	Action:
80	96%	Normal
55	88%	Home O_2
50	85%	Intubate

Pneumothorax: decreased breath sounds on one side, tx if covers >25% of chest
1) Spontaneous: oral contraceptives, thin male smokers, collagen vascular dz
2) Tension (can't breathe out): air in pleural space pressures lungs => tracheal shift
Tx: needle in 2nd intercostal mid-clavicle above rib on exhalation; vasoline gauze
3) Asymptomatic: observe if air occupies <25%

Lung Diseases:

The concept of restrictive versus obstructive lung diseases will help you understand 90% of the lung diseases that exist. You could never memorize the blood gas for every known pulmonary pathology out there. But if you can decide whether or not the disease is a restrictive or obstructive process, you can predict their blood gas, chest x-ray, and what they are most likely to die from. So, when you are deciding what type of process it is, ask yourself if they have trouble breathing in or out, and whether they have small stiff lungs or big mucus-filled lungs, then tell me everything you know…

Restrictive: interstitial problem (non-bacterial)

- Small stiff lungs (↓VC)
- Trouble breathing in => FEV_1/ FVC: > 0.8
- ABG: ↓pO_2 => ↑RR, ↓pCO_2, ↑pH
- CXR: reticulo-nodular pattern, ground-glass apperance or interstitial infiltrate
- Die of cor pulmonale (hypoxia leads to low energy state; heart failure due to lung disease is called cor pulmonale)
- Ex: NM diseases (breathing out is passive), drugs, autoimmune dz
- Tx: Give pressure support on ventilator, ↑O_2, ↑RR, ↑inspiratory time

Obstructive: airway problem (bacterial)

- Big mucus-filled lungs (↑RV, ↑Reid index = ↑airway thickness/ airway lumen)
- Trouble breathing out => FEV_1/ FVC: < 0.8
- ABG:↑pCO_2 => ↑RR, ↓pH
- Die of bronchiectasis
- Ex: COPD
- Tx: Manipulate rate on ventilator, ↑RR, ↑expiratory time, ↑O_2 only if needed

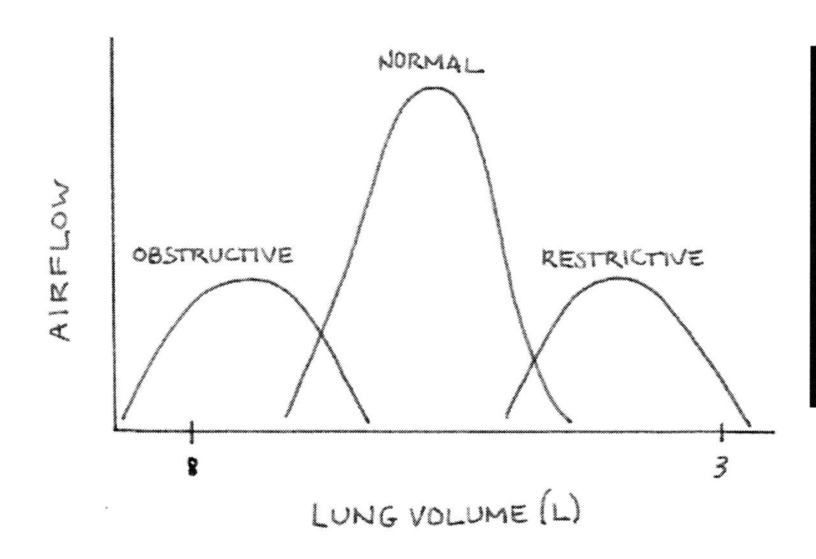

<u>Cough:</u>

each cough moves mucus 1 inch

- Postnasal drip: cold, allergies
- Pertussis: whooping cough
- COPD
- Asthma
- GERD

<u>Neck Films:</u>

Steeple sign => Croup

Thumb sign => Epiglottitis

<u>Amniotic fluid:</u> keeps pressure off baby
80% is mom's plasma
20% made by baby (must be able to swallow, absorb, filter, urinate)

Polyhydramnios: baby can't swallow (Tx: Indomethicin <34wk: stops baby's pee)

- NM problem: **Werdnig-Hoffman**
- GI problem: **Duodenal atresia**

Oligohyramnios: baby can't pee

- Abdominal muscle problem: **Prune Belly** => can't pee (Tx: catheter → UTI)
- Renal agenesis: **Potter's syndrome**: ↑atmospheric pressure => flat face

Diaphragmatic hernia – intestines are in thoracic cavity => *hypoplasia of one lung*

- **B**okdalek (90%): hole in **b**ack of diaphragm
- **M**orgagni (10%): hole in the **m**iddle of diaphragm

Dx: CXR air-fluid levels
Tx: Orogastric tube with suction (to prevent bowel distension)

Lung Anatomy:

Top: skeletal muscle (squamous cell epithelium) – *smoking increases this zone*
Bottom: smooth muscle (tall columnar ciliated epithelium)

Extrathoracic: lips to glottis (not protected by rib cage) – narrows on insp ← stridor
Intrathoracic: glottis to alveoli (protected by rib cage) – expands on insp ← wheeze

C-shape cartilage rings – compresses airway w/ swallowing to prevent aspiration
Fully encircling cartilage – where mainstem bronchus dives into lung parenchyma
Trachea divides into main stem bronchi "carina" at T_4

- Right main stem bronchus → goes straight down (bronchus intermedius)
- Aspirations → **Right Lower Lobe** *(or upper lobe if on side)*
- ***P**osterior* segment if the child is **p**laying around (upright)
- ***S**uperior* segment if patient is **s**upine

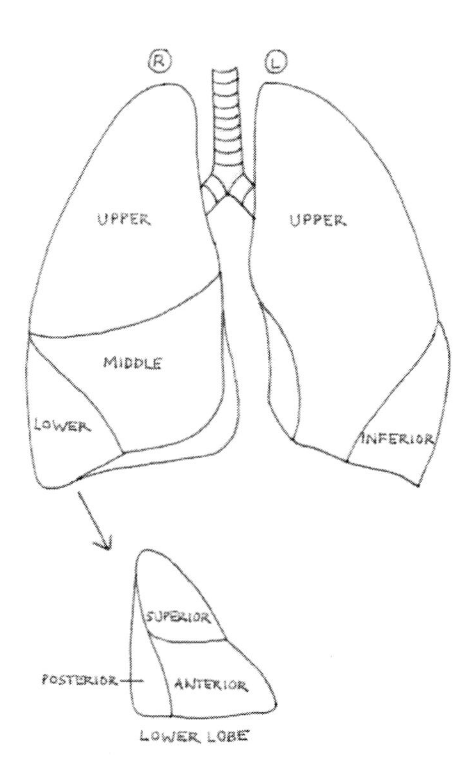

- Recurrent R upper lobe pneumonia => foreign objects (do insp/exp film)
- Most common aspirations: #1 peanut, #2 popcorn, #3 hot dog
- If foreign body makes it to the stomach, leave it alone
- Where stuff likes to get stuck: *can't talk*
- Glottis *Tx: Heimlich maneauver (adults) or back blows (kids)*
- Midway b/w glottis and carina (LA sits on it)
- Carina

Main-stem bronchus – branches in parallel => ↓ resistance, humidify and warm air

Medium size bronchioles – most dilation/ constriction *(most SM/ β_2 receptors)*
Terminal broncioles – most dependant (*small particles settle, 1° lung cancer starts*)

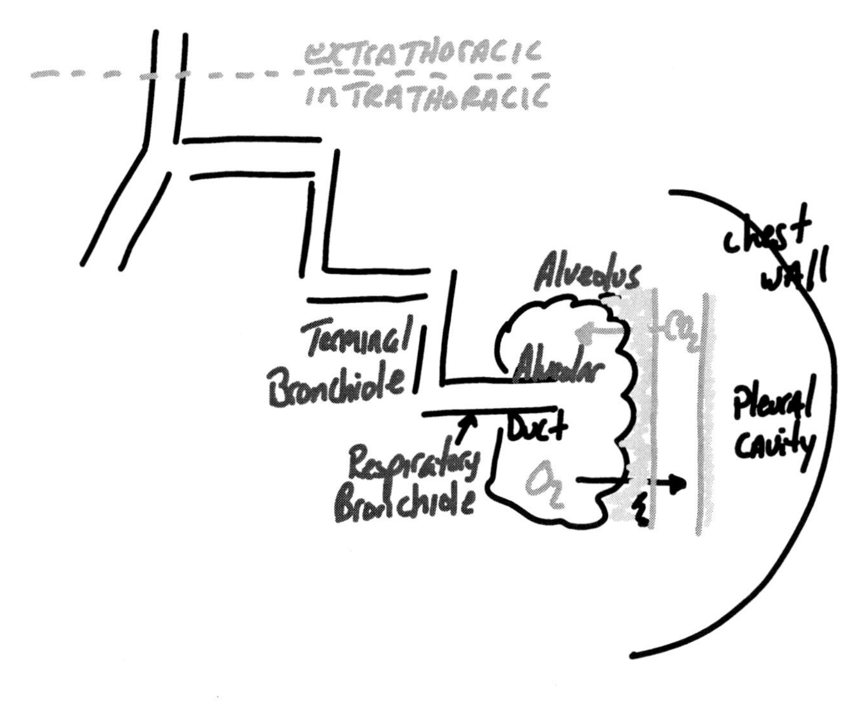

Trachea → Alveoli: velocity decreases

Respiratory unit = resp bronchiole + alveolar duct + alveolus (1 layer of epithelium)
Respiratory unit = the ONLY oxygen exchange system

Ventilation:
Dead space (V_D): lip to terminal bronchiole (everything except your respiratory unit)
Alveolar ventilation (V_A): respiratory bronchiole to alveoli
Total ventilation = $V_D + V_A$
Minute Ventilation (Vm) = **TV x RR** = how much you breathe in during 1min
Vm = 10-15cc/kg (have a Vm for V_D, V_A, V_T, etc.)

Histology:

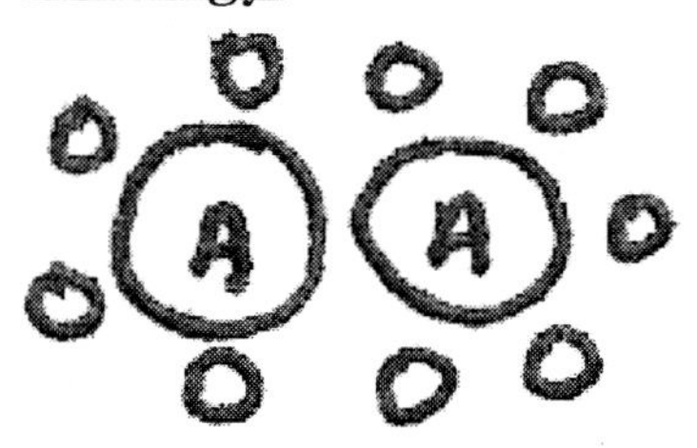

- Goblet cells: secrete mucus to trap dirt (most abundant cell type)
- Cilia: 9+2 actin configuration: *orade movement* "toward mouth" (not back and forth)
- Dyenin arm: flexibility for cilia *(virus, smoke paralyzes cilia → green sputum)*
- Type I pneumocytes (5%) = macrophage (in terminal bronchiole)
- Type II pneumocytes (95%) = surfactant producers (in alveolar bronchioles)

- Type II cells can demote themselves to type I cells after injury…
- Clara cells "dust cells": MP that eat dust (live in the terminal bronchiole)
- **Kartagener's** (broken dyenin arm): situs inversus, bronchiectasis, male infertility

Pulmonary Sounds:

Stridor: extrathoracic narrowing => narrows when breathe in => neck x-ray
Wheeze: intrathoracic narrowing => narrows when breathe out => chest x-ray
Rhonchi: mucus in airway => obstructive lung disease
Grunt: blows collapsed alveoli open => restrictive lung disease
Dull percussion: something b/w alveoli and chest wall absorbing sound (fluid, air, solid)

Hyperresonance: air
Tracheal deviation: away from pneumothorax OR toward atelectasis *"air-phobic"*
Fremitus, egophony, bronchophony: consolidation => *pathognomonic for pneumonia*
Crackles "rales": alveoli are collapsed =>

- No surfactant: washed out due to pulmonary edema or CHF
- Alveolar fibrosis: Pneumoconioses, Bleomycin, Busulfan, Amiodarone, Tocainide

Symptom:	**Disease:**	**Treatment:**
Barking cough (steeple x-ray)	Croup (Parainfluenza)	O_2, racemic Epi, Dexamethasone
Stacatto cough	Pneumonia (Chlamydia)	Fluoroquinolone
Whooping cough	Pertussis (Bordatella)	Erythromycin
Muffled voice/drool (thumb x-ray)	Epiglottitis (H. influenza B)	OR Intubation, Ceftriaxone
Expiratory wheeze – kid	Bronchiolitis (RSV)	Ribaviran, Albuterol
Inspiratory stridor – kid	Laryngomalacia	Observe
Inspiratory stridor – adult	Subglottic stenosis	Dilation

Lung Pathology: (alphabetical order)

Asthma: wheeze on expiration

- Intrinsic: Genetic (cold air, exercise, NSAIDs make it worse)
- **E**xtrinisic: **E**nvironment-induced (dust mites, roach droppings), IgE/ Eosinophils

Benign Pulmonary Nodule:

- Popcorn calcifications

Bronchiectasis: digestion of airways → hemoptysis

- Honeycomb lung = bronchiole dilation, halitosis
- Dx: high resolution CT
- Tx: Antibiotics, O_2

X-rays:

Radiolucent (black) => air

Radiopaque (white) => fluid/ solid

Bronchitis:

- Acute bronchitis = increased mucus production
- Chronic bronchitis = 3 consecutive months over 2 yrs
- Bugs: Strep pneumo (*rusty-colored sputum*), H influenza, Neisseria cattaralis
- Inflammation → dilate airways → secretion buildup → bronchial destruction

Bronchiolitis: asthma symptoms < 2y/o

- ↑AP diameter, flat diaphragm, usually due to RSV
- ↑Risk of future asthma/ear infections
- Tx: Isolation/Albuterol, Ribavirin (if resp failure)

COPD: Bronchitis, Emphysema, Asthma

- Prognosis: FEV_1
- Tx:

1) O_2 *(can alter natural history of dz)*
2) Albuterol (bronchodilator)
3) Me-Prednisolone (glucocorticoid)
4) Levofloxacin

Blue Bloater: Bronchitis
Pink Puffer: Emphysema

Croup: swelling around the glottis => *steeple sign* on x-ray

- Barking cough, stridor on inspiration
- Fluctuating course (improves/worsens within 1hr)
- Viruses:
- Parainfluenza
- RSV = most severe (Tx: Ribavirin)
- Adenovirus
- Influenza virus
- Tx: Dexamethasone, racemic Epi, O_2

Asthma:
Early Phase: IgE (Tx: Antihistamines)
Late Phase: Cytokines (Tx: Steroids)

Cystic Fibrosis (AR): Chr #7 CFTR: Cl channel broken => more Cl in secretions

- Test: Pilocarpine sweat test (Cl >60mEq/L => have CF)
- Tx: N-acetylcysteine (breaks mucus disulfides), chest percussions, future lung transplant
- Vaccinations: Influenza
- Bugs: Staph/Pseudo like to attack them (Tx: Tobramycin + Piperacillin)
- Newborn => meconium ileus (Tx: gastrografin enema)
- Lung => obstructive pulmonary disease
- Nose => obstruction
- Pancreas => malabsorption => Vit. A,D,E,K def
- Epididymis => infertility
- Urine => oxylate stones (malabsorption)
- Stool => steatorrhea

Pulmonary Eosinophilia:

- Aspergillosis
- Parasites: Strongyloides
- Drugs: Nitrofurantoin, Sulfonamide

Emphysema: **obstructive (↓pO_2, ↑pCO_2, ↓pH), pursed lip breathing**

- **Pan-acinar** (AR): α_1-AT def can't inhibit elastase, PAS(+), restrictive
- **Centro-acinar:** smoking *"comes in through the center"*
- **Distal acinar:** aging (least blood supply) → spontaneous pneumothorax
- **Bullous** "pneumatocele": elastase ⊕ bacteria = Pseudo/Staph aureus

Epiglottitis: inflammation above glottis => *thumb sign* on x-ray

- Drooling, stridor, muffled voice, high fever
- Bug: HI-B *"stick out thumb to say HI"*
- Tx: Intubate immediately in the OR, Ceftriaxone

Flash Pulmonary Edema:

- X-ray white out
- Tx: O_2, Morphine, Furosemide, Nitroglycerin (↓BP)

Laryngomalacia: epiglottis roll in from side-to-side => feed in upright position

Pneumoconioses: promote adenocarcinoma

- **Asbestosis:** shipyard workers, pipe fitters, brake mechanics, insulation installers
    - Crocodilite fibers
    - Fe coating: *"ferruginous body"* → MP take to pl cav → mesothelioma
- **Silicosis:** sandblasters, glassblowers, monument engravers → pulmonary TB
- **Beryliosis**: radio/TV welders, dental ceramics *"Berry the newscaster"* (Tx: steroids)
- **Byssinosis:** cotton workers *"Cotton blankets in bassonettes"*, chest tightness
- **Anthracosis:** coal workers => Not promote lung cancer, may get massive fibrosis

Pneumonia: consolidation of airway (dull percussion, rales, tactile fremitus, egophony)

I) Typical PNA:

- **Streptococcus pneumoniae:** most common
- **Haemophilus influenza:** 2nd most common, Gram (-) coccobacilli, kids
- **Neisseria cattarhalis:** 3rd most common
- **Staphylococcus aureus:** secondary infection after influenza virus
- **Pseudomonas:** found in cystic fibrosis
- **Klebsiella:** currant jelly sputum, bulging fissures, found in alcoholics, DM
- **Anaerobes:** gas, foul sputum, aspiration in dementia, alcoholics

Hemoptysis DDx:

- Bronchiectasis
- Bronchitis
- Pneumonia
- TB
- Lung CA

II) Atypical PNA: *dry cough (Tx: Erythromycin)*

- **Chlamydia** (0-2mo): stacatto cough, eosinophils
- **Mycoplasma** (college): reticulonodular, bullous myringitis, cold agglutinins
- **Legionella** (>40y/o): A/C ducts, CYAE, silver stain, low Na, CNS changes

- **Actinomyces:** sulfur granules

III) Fungal PNA:

- **Histoplasma**: bat droppings (Mississippi river), no true capsule, MP, oral ulcers
- **Blastomyces**: pigeon droppings (NY), **b**road-**b**ased hyphen, rotting wood in beaver dams
- **Coccidioides**: thin walled cavity (San Joaquin Valley), desert bump fever, budding yeast
- **Paracoccidioides**: looks like a ship's wheel (S. America)
- **Aspergillus**: fungal ball, moldy hay, pulmonary bleed (Tx: Prednisone)

Pulmonary Embolus: blockage of blood flow in lungs => tachypnea

- EKG: $S_1Q_3T_3$
- ↑ V/Q scan: perfusion defects – most reliable
- Venous US
- Spiral CT
- Pulmonary angiogram – gold standard
- CXR: Hampton's hump: wedge opacification

Only reason for:
Radiation: small cell CA
Surgery: V-Q mismatch (palliative)

Tx:

- Anticoagulation: Heparin, Coumadin, IVC filter
- Intervention angiography, Surgery

1° Pulmonary HTN:

- ↑PA pressure (enlarged right heart leads to cor pulmonale)
- ↑Mortality rate with pregnancy
- Pre-capillary: ↑resistance to flow in pulmonary arteries (Ex: ASD/VSD/PDA/L→R shunts)
- Post-capillary: ↑resistance to flow in pulmonary veins (Ex: LV dysfxn/constrictive pericarditis)
- ↑PCWP (LA pressure): cardiac problem
- ↓PCWP: lung problem (1° Pulm HTN, ARDS, cor pulmonale)
- Tx: Coumadin + Amlodipine

Sarcoidosis:

- Hilar lymphadenopathy
- Erythema nodosum
- Non-caseating granulomas
- Lymph node “eggshell califications”
- “Potatoe nodes” → face weakness
- Uveitis
- ↑ACE, ↑Ca, ↓T cells
- Test: Parotid gland biopsy
- Tx: Prednisone (if eye/heart involved)

Sinusitis: bacterial infection obstructing maxillary sinus

- Pain worse when bend forward
- Tx: Amoxicillin

Tonsillitis: sore throat, pooling of saliva, muffled voice

- Tx: needle drainage, Abx

Tracheitis = Diphtheria: vascularized grey pseudomembrane *(don't scrape it)*

- Look toxic, stridor w/ cyanosis, leukocytosis
- Bugs: Staph, Strep
- Toxin ADP-ribosylates EF-2 => cells die
- Tx: Ceftriaxone, cricothyroidotomy if suffocating

Tracheomalacia: soft cartilage, stridor since birth, outgrow by 1y/o

Physiologic Parts of Lung:

- Intrathoracic space: chest wall, pleural space
- Pulmonary vasculature
- Pulmonary airway

Compliance: $\Delta V/\Delta P$
Elastisticity: provides recoil

Lung Physiology:

Lung Volumes: *"LITER"*

1) **I**RV "Inspiratory Reserve Volume": air you can force in after a normal breath
2) **T**V "Tidal Volume": normal breath
3) **E**RV "Expiratory Reserve Volume": can force out after normal exp, fills dead space
4) **R**V "Residual Volume": air in lungs after forced exp, keeps alveoli open (not on PFTs)

Lung Capacities:

IC "Inspiratory Capacity": total amount of air you can breathe in = 1+2
VC "Vital Capacity": all the air you can breathe in after forced expiration = 1+2+3
TLC "Total Lung Capacity": air in lungs after deep breath = 1+2+3+4
FRC "Functional Residual Capacity": baseline (where you stop/start breathing) = 3+4
*"FIT": **F**RC + **I**C = **T**LC*

Above FRC => positive pressure
Below FRC => negative pressure
FEV_1/ FVC normal ratio = 0.8

- **FEV_1:** forced expiratory volume in 1 sec
- **FVC:** forced vital capacity

Obstructive Lung Dz: ↑RV (or FRC) first; last to change is TV
Restrictive Lung Dz: ↓VC (or TLC) first; last to change is TV

Muscles of Breathing:

Normal inspiration: => *Tidal volume*
Diaphragm – goes down
External intercostals – used during exercise *"externals breathe in"*
Innermost intercostals – *right* muscles (along sternum) move *left* chest wall

Forced inspiration: => *IRV*
Pectoralis major and minor
Head and neck muscles:

- Scalenes
- Sternocleidomastoid
- Trapezius

A-a Gradient:

A = Alveoli; a = arteriole
↑: Extracts O_2 (restrictive)
↓: Lose O_2 (polycythemia)

Normal expiration: *Recoil only (know this!)*

Forced expiration: => *ERV*
Internal intercostals *"internals breathe out"*
Abdominal muscles:

- Obliques
- Rectus abdominis
- Transversus abdominis
- Quadratus lumborum

High Altitude:

Chronic: Kidneys pee off bicarbonate
Acute: Mountain Sickness
(Tx: Acetazolamide to pee off bicarbonate)

Intrathoracic Pressure: *necessary to pull blood into thorax*
Note: Pleural space is always negative

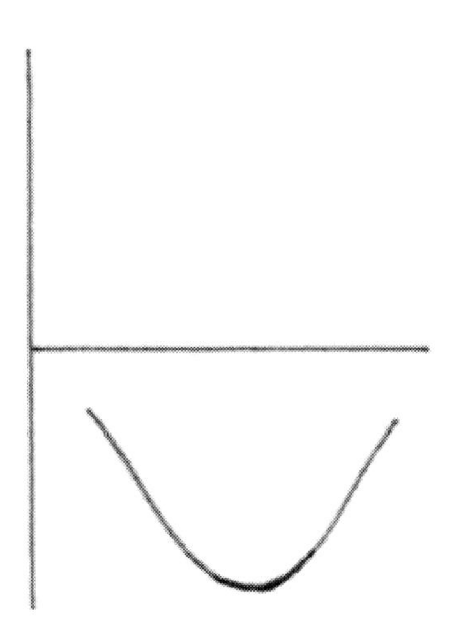

Resting: -3 to -5
Normal breath: -10 to -12
Deep breath: -20 to -24
Restrictive lung disease: -40 to -60 (negative pressure sucks in => GERD, hiatal hernia)
Pneumothorax: + pressure (Ex: oral contraceptives, thin male smokers, Staph/Pseudo)

Breathing: (P_A = airway pressure = opposite sign of intrathoracic pressure)
Highest compliance (mid-inspiration or mid-expiration) => max airflow into alveoli

Inspiration: *moves air into lungs and blood into heart*
Start: chest wall > lung expansile force, $P_A = P_{ATM}$
Mid-inspiration (50-99%): lung > chest wall expansile force, ↑compliance, $P_A << P_{ATM}$
End-inspiration: recoil of chest wall = expansile force of lung (alveoli negative pressure)

Expiration:
Start: chest wall > lung recoil, $P_A >> P_{ATM}$, effort dependant => can force out
Mid-expiration (50-99%): lung > chest wall recoil, ↑compliance, effort indep, collapse airway
End of expiration: lung recoil = chest wall expansion, $P_A = P_{ATM}$ (airway positive pressure)

Flow and Ventilation:
Top of lung: *more air* (more air flows into bottom during inspiration only)
Bottom of lung: *more blood flow* (gravity, dilated capillaries, dilated arterioles)

Every V/Q mismatch presents with a restrictive pattern, leads to hypoxia
Dead space: High V/Q => no blood flow
- Ex: PE, shock

Shunt: Low V/Q => no ventilation
- Ex: atelectasis, pneumonia

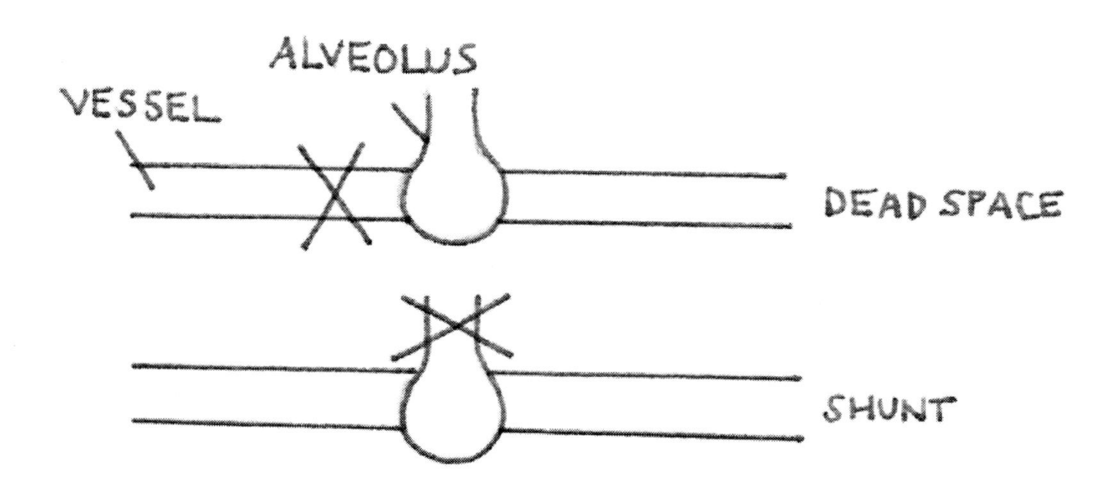

Pulmonary Airway:

At FRC: $P_A = P_{ATM}$

Inspiration: $P_A << P_{ATM}$

Expiration: $P_A >> P_{ATM}$

End of deep breath: $P_A = P_{ATM}$

Breathing Receptors:

J receptors: in interstitium => tachypnea, restrictive dz

Slow-adapting receptors: b/w ribs and muscle fibers; sense stretch, obstructive dz

Carotid Body: carotid chemoreceptor (*measures everything:* pO_2, pCO_2, $[H^+]$)

CN9 → carotid body → CN10/ phrenic nerve

Aortic Body: aortic arch chemoreceptor (*measures* pCO_2, $[H^+]$)

CN10 → aortic body → CN10/ phrenic nerve

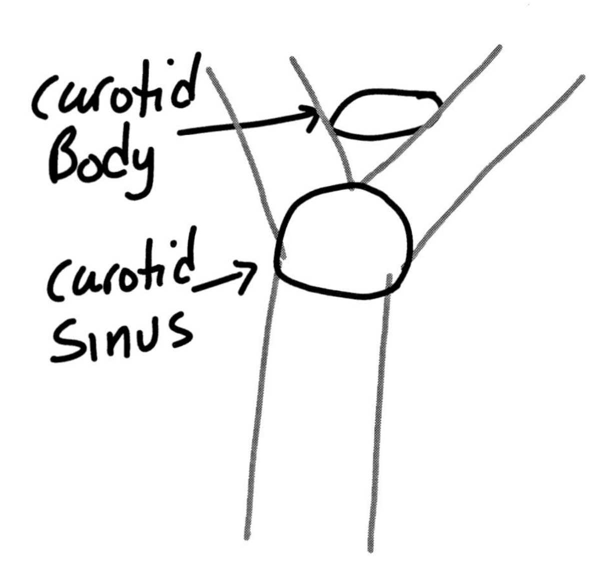

Effects of O_2 and CO_2:

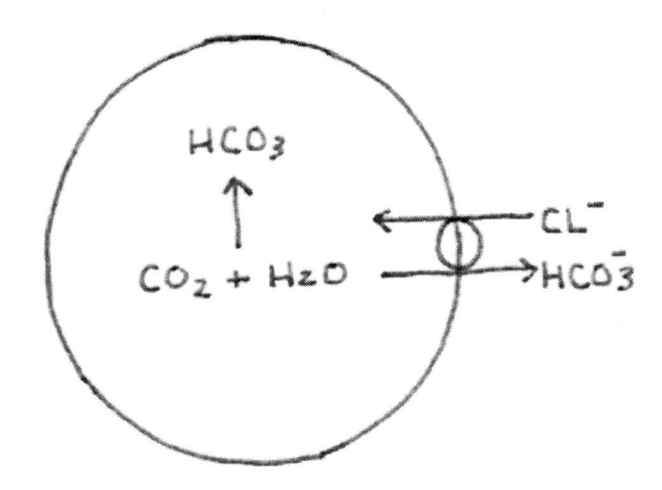

O_2: controlled by diffusion and perfusion

↓**CO_2:** controlled by ↑ventilation (i.e. airway being open)

- Don't give >1L O_2 to COPD pts b/c hypoxia is the drive for ventilation
- Tx: Bronchodilators (create more space so CO_2 can leave)

1) fills airways => CO_2 can't leave

2) knocks out apneustic center (pneumotactic center desensitized) => *coma*

- *COPD normal O_2: 55-60*

Forms of pCO_2:

90%: Bicarbonate

7%: Attached to Hb (can't measure this stuff) "carboxyhemoglobin"

3%: Dissolved (this goes to pneumotactic center) = .03 x pCO_2

CNS is affected more by high pCO_2

PNS is affected more by low pO_2 (you're almost dead if have low pO_2 and high pCO_2)

- Oxygen diffusion: alveolar endothelium → interstitium → capillary endothelium
- Oxygen is *the most potent vasodilator in the lung*

Breathing Control Centers:

Pneumotactic center (top): prevents **pneumo**thorax => breathe out (CO_2 sensitive)

Apneustic center (bottom): prevents **apnea** => makes you breathe in (O_2 sensitive)

Pons: reaction center

Medulla: sets respiratory rate (RR=8-10)

Ex: Brain death (everything above medulla is gone, can still breathe)

Ex: Central apnea of neonates = no inspiratory effort for 20 sec

- Tx: Theophylline or Caffeine to stimulate the brain

Ex: Obstructive apnea "Pickwickian" => chronic hypoxia (opposite of COPD)

- Weight loss
- Progesterone to stimulate respiration (pregnant women breathe faster)

- CPAP
- Uvulopalatopharyngoplasty (cut out soft palate)
- No BZ! (respiratory depression)

Carotid Body: Chemoreceptor
Carotid Sinus: Baroreceptor

Breathing Patterns: breathe in → hold 1 sec → breathe out → hold 1 sec
Restrictive => more time in inspiration (I)
Obstructive => more time in expiration (E)

Apneustic breathing: breathe in → hold for a long time → breathe out
Ex: Pontine hemorrhage

Cheyne-Stokes breathing: deep breathing followed by apnea (sigh)
- Lesion medulla (or low blood glucose)
- Blow to back of head cuts off blood supply to medulla via the vertebral aa.
- **Thoracic outlet syndrome:** extra rib compress subclavian, turn neck => paresthesia
- **Subclavian steel syndrome:** raised arm compress subclavian => cyanosis
- **Reversal of flow in vertebral aa → steals blood from brain**

Kussmaul breathing: rapid deep breathing (must stop talking to breathe)
- Metabolic acidosis produces GABA, which fight each other to breathe fast or slow

Medullary breathing: RR=8-10
Ex: Anencephaly (only have medulla)

Paroxysmal Nocturnal Dyspnea: wake up from sleep with air hunger
Ex: CHF

Foul Sputum:
- Bronchiectasis
- Lung abscess
- Aspiration pneumonia

Managing Ventilators:
- **O_2 amount:** Restrictive needs more, Obstructive needs less
- **Rate: 12-16** (all lung diseases have tachypnea)
- **Tidal Volume:10-15cc/kg,** peripheral hypoventilation (high pCO_2) => low E state
- **I:E ratio:** Restrictive needs more I, Obstructive needs more E (increment by 0.1)
- **CMV:** Controlled Mandatory Ventilation – total machine control (not used anymore)
- **Assist Control:** machine breathes, pt helps the least (use during sepsis)
- **SIMV/ IMV (**Synchronized Intermittent Mandatory Vent.): pt adds extra breaths
  - used to wean patient off ventilation…
  - **Pressure Support:** pt has control, machine just helps (use w/ restrictives)
  - **PEEP:** Positive End Expiratory Pressure: ↑FRC (use while intubated)
  - **ZEEP:** Zero PEEP

Steroid Side Effects:
Low Dose: thrush, dysphonia
High Dose: osteoporosis, cataracts, purpura, adrenal suppresion

- **AutoPEEP:** breath stacking
- **CPAP:** Continuous Positive Airway Pressure (use in sleep apnea, CHF)

Pulmonary Drugs:

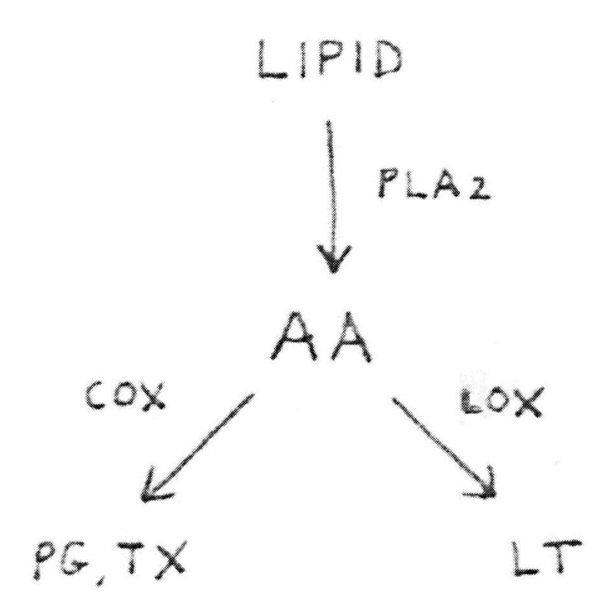

Arachadonic Acid Pathways:

1) Cyclooxygenase "COX" (forms Prostaglandins)

- PGA_2 "Thromboxane": vasoconstriction, thrombosis
- PGE_1: vasodilation (keeps PDA open) Ex: Misoprostyl for GI ulcers => induces labor
- PGE_2: vasodilation, SM relaxation, used to keep PDA open
- PGF_1: vasoconstriction
- PGF_2: vasoconstriction, menstrual cramps, abortions, found in semen
- PGI_2 "Prostacyclin": vasodilation, anti-thrombosis, made by endothelium

PG Summary:
A/F: vasoconstrict/thrombose
E/I: vasodilate/anti-thrombosis

- asa: irreversible inhibition
- NSAIDs: reversible inhibition

2) Lipooxygenase "LOX" (forms Leukotrienes)

- LT-$C_4D_4E_4$ "SRSA" = *the most potent bronchoconstrictor*
- Produced by mast cells
- asa-sensitive asthma results from closing of the COX pathway leading to LOX

Asthma Treatment:

B_2 Agonists: acute tx, bronchodilation, low K^+ levels (pushes K^+ into cells)

"RATS"

- **R**itodrine – #1 stop preterm labor
- **A**lbuterol – q4h inhalers
- **T**erbutaline – #2 stop preterm labor, q4h bronchodilator inhaler prn
- **S**almeterol – 8-10 hr inhalers

Steroids: *need adjuvant Vit D/Ca/Insulin*

- Triamcinalone – inhaled
- Prednisone
- Beclamethasone

Anti-Cholinergics: bronchodilate

- Ipratropium (↓cGMP)

PDE Inhibitors: acute tx

- Theophylline (IV)

LT receptor blockers: use if steroids fail

- Zileuton – inhibits LOX
- Zafir**luk**ast – inhibits LT_{D4} *"**Leuk**otriene inhibitor"*
- Montelukast

Prophylactic agents: stabilize mast cell membranes

- Cromolyn sodium – use before exercising (eye or nasal drops)
- Nedocromil

Steroid Anti-inflammatory Actions:

- Stabilizes: mast cells/endothelium
- Inhibits: MP migration/ PLA
- Kills: T cells/eosinophils

NOTES:

NOTES:

Renal:

"This too shall pass - just like a kidney stone."

–Hunter Madsen

Embryology:

Pronephros: forekidney
Mesonephros: midkidney → ureteric bud → collecting ducts, calyces, ureters, renal pelvis
Metanephros: hindkidney → permanent kidney
Urogenital sinus: allantois → urachus → bladder

The KIDNEY Bean:
Blood flow: Cortex → Medulla → Pyramids, Papillae, Calyces, Hilum

Kidney Anatomy: Arteries run *under* veins (higher pressure)
R kidney – sits lower at L2 (liver in the way)
R renal artery runs posterior to IVC, long course
L renal vein runs anterior to aorta, long course ← aneurysm hurts this kidney first
R gonadal vein → IVC *(so… cancer on right will spread faster)*
L gonadal vein → L renal vein

Cortex – isotonic (osmolarity is same as in plasma) ← DIC
Medulla – hypertonic (concentrates your urine) ← Clots
Nephrons are longer when it is warm (to concentrate urine)

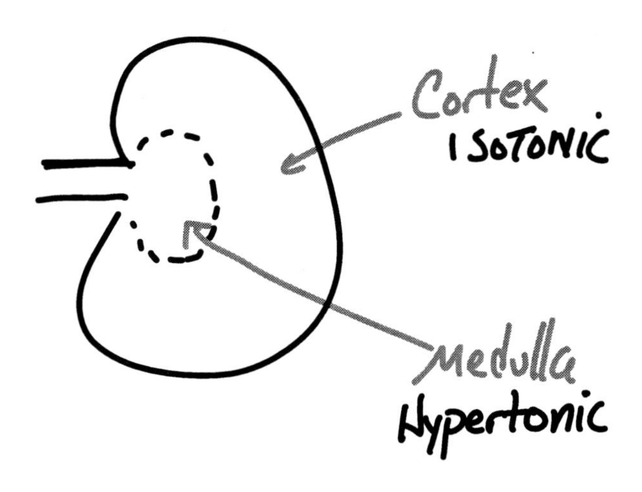

Rehydration:
1) Bolus (normal saline) = 20 cc/kg

2) Replacement (calculate based on Na):
- Measure weight loss => 1 kg = 1L
- Give ½ (minus bolus) over 8 hrs
- Give ½ over next 16 hours

Dehydration Status:
5% loss: thirsty
10% loss: tachycardia
15% loss: ↓BP/ capillary refill

3) Maintenance: What kind of fluid?

- Adults: ½-normal saline
- Kids: ¼-normal saline
- Dehydrated adult: normal saline
- Dehydrated kid: ½-normal saline
- Shock: normal saline (154mEq salt/L) or Lactated Ringer's
- Hyponatremia + Seizures: 3% NaCl

Average Weight:

Adults: 75kg

Kids: 35kg

How much do I give?

Adults: 1.5cc/kg/hr (= urine output)

Kids <8 y/o:

Per Day:

- 1st 10kg: 100cc/kg/day
- 2nd 10kg: 50cc/kg/day
- After that: 20cc/kg/day

Per Hour:

- 4/2/1 cc/kg/hr

Burn deficit: 4cc normal saline/ kg/ %burn *"Parkland Formula"*

Dialysis:

Hemodialysis: forearm AV fistula (done in hospital)

Peritoneal dialysis: peritoneum catheter (done at home)

Survival Electrolytes:

NaCl: we need 3mEq/kg/day = 225mEq/3L urine/day = 75mEq/L => use ½-normal saline

K: we need 0.75 mEq/kg/day = 56mEq/3L urine/day = 20 mEq/L => add 20mEq K/L

Burns: Rule of 9's *(don't include 1st degree burns)*

Head + neck = 9%

Chest =18%

Back =18%

Each leg = 18%

Each arm = 9%

Genitalia =1%

Burn Tx:

Most: Silver Sulfadiazine (leukopenia)

Cartilage: Sulfamylon (acidosis

Eyes: Triple Antibiotic

Burn Classifications:

1st degree = red (epidermis)

2nd degree = blister (hypodermis)

3rd degree = painless neuropathy (dermis)

RAA axis: low volume sensed by macula densa

J-G apparatus: renin release

To Liver (angiotensinogen) → ATI

To Lungs (ACE) → ATII → Aldo

1) pee out H^+/K^+, reabsorb Na

2) ADH (reabsorb water)

Aldo: controls volume (Na)

ADH: controls water

3) Vasoconstrict (↑BP)
4) ↑Thirst

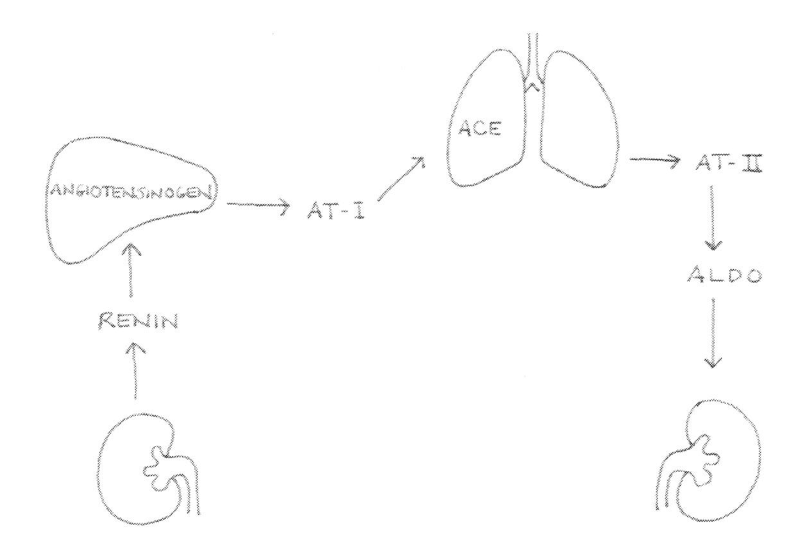

Bradykinin Pathway:

Kininogen → (Kallikrein) → Bradykinin → (ACE) → degraded Bradykinin

Bradykinin Functions:
1) Dilate veins
2) Proteinuria
3) Cough

Acid-Base Disorders:

1) pH = 7.4 — same direction (i.e. both increased or decreased) => **metabolic**
2) pCO_2 *"acid"* = 40
3) HCO_3 *"base"* = 24 — same direction => **compensated**

Note: opposite direction => respiratory or mixed, respectively

Respiratory Alkalosis: Restrictive Lung Dz (anxiety, pregnancy, Gram – sepsis, PE)
Tx: Breathe into a bag

Respiratory Acidosis: Obstructive Lung Dz (COPD, drugs)
Tx: Hyperventilation, ↑FiO_2

Metabolic Alkalosis: Low Volume State (vomiting, diuretics, GI blood loss)
Tx: Hydration

Metabolic Acidosis: Acid production ("MUDPILES", RTA II, diarrhea)
Tx: Bicarbonate if pH < 7.2

Renal Failure:

Pre-renal: low flow to kidney (BUN:Cr >20, FeNa <1%)

Oliguria: <400cc/day
Anuria: <100cc/day

1) Low volume state (ATII constricts efferent => ↑BUN, ↓Cr)
2) Vasculitis
Tx: Fluid bolus (increase flow to kidney)

Renal problem: damage glomerulus (BUN:Cr <20, FeNa >2%)
Ex: ATN, SLE, Wegener's, Renal artery stenosis
↑Cr/ ↑RPF: ↓serum BUN (filter less), then ↑ (no blood flow)
Urine osmolarity=300 (near serum osmolarity) → kidney can't concentrate urine
Tx: Fluid, Diet: low Na/K/PO_4/Protein; No renal-excreted meds
So... If you give a renal-excreted drug, make sure to give enough volume!

Post-renal: obstruction (haven't peed in last 4 days)

- Newborns: malimplantation of ureter, post-urethral valve obstruction
- Kids: Strictures (UTIs)
- Adults: Scarring (STDs, Nitrofurantoin)
- Women >40 y/o: uterine prolapse, cystocele
- Men >40y/o: BPH, Nephrolithiasis

Tx: catheter to get urine out (have a lot of urine stored in bladder)

ESRD Skin: white powder (urea in sweat), pruritis, excoriations, pallor, ecchymosis

Urinalysis:
Color: pus, blood
Specific Gravity: High => dehydration, SIADH
Low pH: Salicylate O/D
High pH: UTI, RTA Type I
Protein: High => leaky glomeruli
Glucose: High => DM
Ketones: High => DKA, starvation, isopropanol toxicity
Bilirubin: High => hemolysis
Urobilinogen: High => hemolysis or conjugated hyperbilirubinemia
Nitrite: Gram (-) bacteria
Nitrite (-): Enterococcus
Leukocyte Esterase: WBCs
RBC: stones, tumor, GN
WBC: UTI, prostatitis, vaginitis
Casts: Kidney dz
Crystals: Kidney stones, Ethylene Glycol toxicity
Squamous Epithelium: degree of contamination

Fake Sphincters:

- Ureters
- LES
- Ileocecal valve

I) Nephritic: increased size of fenestrations => *vasculitis* (HTN)
Urine: Blood, WBC casts, ↑BUN/Cr, ↓GFR, oliguria

1) Rapidly Progressive Glomerulonephritis "RPGN":

- Crescent formation => scars

2) Post-Strep Glomerulonephritis "PSGN": hematuria 2wk after sore throat

- Strain 12
- Subepithelial
- Inflamed glomerulus w/ IgG, C_3,C_4 deposition
- Anti-streptolysin Ab "ASO" (periorbital swelling)
- Tx: Furosemide (tx symptoms)

3) Interstitial Nephritis: urine eosinophils

- Caused by Fluoroquinolones

4) Vasculitidies:

- **Wegener's:** c-ANCA Ab (Tx: Cyclophosphamide + Prednisone)
- **Goodpasture's:** anti-GBM Ab, hematuria + hemoptysis (Tx: plasmapheresis)
- **Henoch-Schonlein Purpura**
- **Polyarteritis Nodosa**
- **Subacute Bacterial Endocarditis:** ↓C_3
- **Serum Sickness:** ↓C_3
- **Cryoglobulinemia:** ↓C_3

<u>II) Nephrotic:</u> lost BM charge due to deposition on heparin sulfate => proteinuria and lipiduria

- Edema (due to oncotic forces)
- Liver: makes proteins to compensate (↑LDL, hypercoagulability, ↑DVT/renal vein thrombosis)
- Urine: Protein (>3.5g), Lipid (maltese crosses)
- ↑Risk Spontaneous Bacterial Peritonitis

1) Membranous Glomerulonephritis "MGN":

- Most common in grown-ups
- Hep B (spike&dome, thick BM)
- Penicillamine, Captopril, Mercury, Gold
- Tx: Prednisone, Chlorambucil

2) Minimal Change Disease "MCD":

- Most common in children
- Fused foot processes, autoimmune
- Tx: Prednisone *(the only curable nephrotic process)*

<u>Subepithelial:</u>

- SLE
- PSGN

3) Focal Segmental Glomerulosclerosis "FSG":

- ↑Risk in African Americans and HIV patients

4) Diabetic Nephropathy:

- Glomerulosclerosis
- Kimmelstiel-Wilson nodules (PAS + ovoid hyaline masses)
- Tx: Restrict protein <0.8g/kg

<u>III) Nephritic and Nephrotic:</u>

1) Systemic Lupus Erythematosus "SLE":

- ↓C_3, anti-ds DNA
- Subepithelial
- Tx: pulsatile Cyclophosphamide

2) Membranoproliferative Glomerulonephritis "MPGN":

- ↓C_3, "tram tracks"
- Tx: ASA, Dipyramidole

3) IgA Nephropathy:

- Normal C_3
- URI, then hematuria
- Assoc w/ HIV, celiac disease, liver disease

-emia: blood
-uria: urine

<u>Total Body Water (TBW)</u> = 60% weight (measure w/ D_2O*)

- 2/3 intracellular (has K^+, Mg^{2+})
- 1/3 extracellular (measure w/ inulin*) ← decrease w/ exercise
  - ¼ plasma (measure w/ albumin*)
  - ¾ interstitial (plasma – proteins) ← isotonic saline goes here

Osmolarity (approximate) = 2 (Na) + Glucose/18 + BUN/3 = 300mOsm/kg

=> Na (or glucose, mannitol, methanol, ethylene glycol) contribute the most to osmolarity

<u>Clearance</u> = FF + secretion – reabsorption

Clearance = excretion (total of all processes)

Secrete = add to the urine (active process)

Reabsorb = subtract from urine

$FF = \frac{GFR}{RPF}$ "filtration fraction" is how much of the plasma the glomerulus filters

<u>Afferent:</u> filtration (just uses diffusion)

GFR: can measure with *inulin* (lab) or *creatinine* (physiologic) → 100% filtered

Real life: Instead of doing 24-hr CrCl, just measure serum Cr (inversely proportional)

RBF to kidney = 20% of CO = 1L/hr

20% of fluid that comes to glomerulus (RBF) is filtered =>200cc/hr

RBC are not filtered (45%) => GFR=125cc/hr => 3L urine per day => drink 3L H_2O per day to replace

GFR = UV/P *"U'v peed"*
$GFR_1 Cr_1 = GFR_2 Cr_2$

Normals:
BUN: 10-20
Cr: 0.6-1.2
GFR: 125

GFR Methods:
Cockroft-Gault: (140-age)(weight) / (72)(serum Cr)
MDRD: (age)(serum Cr)(a bunch of numbers)

GFR:
<60: Renal failure
<10: End-stage renal disease

FeNa: $(U_{Na}/P_{Na})(P_{Cr}/ U_{Cr})$

Efferent: secretion (needs transport proteins) => *Low energy state hurts efferent arteriole first*
Can measure with *PAH* (lab) or *BUN* (physiologic)
$RPF_1\ BUN_1 = RPF_2\ BUN_2$

Renal Function Summary:

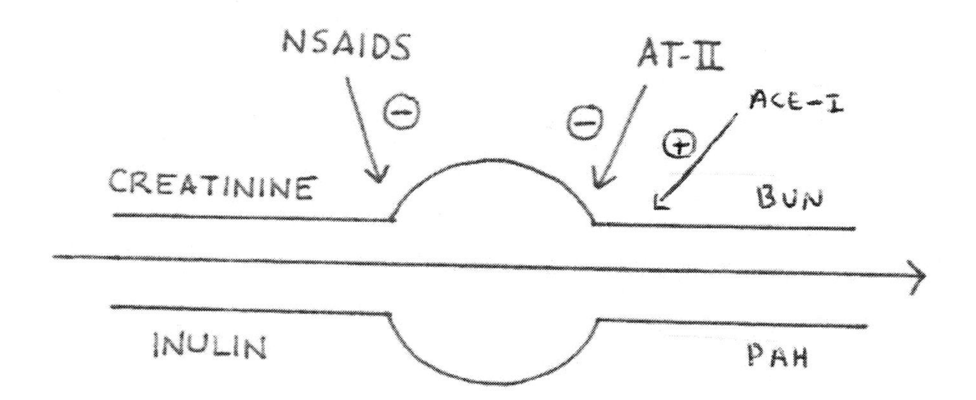

NSAIDs: constricts afferent arterioles
AT-II: contricts efferent arterioles
ACE-I: dilates efferent arterioles

	Job:	**Measurements:**	**Test:**
Afferent:	↑Filter	↓Creatinine,Inulin	GFR
Efferent:	↑Secrete	↓BUN, PAH	RPF

Small Kidneys: HTN nephropathy
1) Renal artery stenosis: think of a squeezed water hose *(most common cause of 2° HTN)*

- Increased velocity helps blood get past clot into bad kidney
- Increased pressure on contralateral kidney destroys it => malignant HTN
- Test: Captopril renal scan in both renal arteries: >1.5 difference => stenosis
- Goldblatt's Kidney: "flea bitten kidney" *(blown capillaries)*
- *<30y/o: Fibromuscular Dysplasia*
- *>30y/o: Atherosclerosis*
- Remove contralateral kidney (nephrectomy)
- Remove ipsilateral blood clot (atherectomy)
- No ACEI! (dilates efferent arterioles => ↓blood flow to kidney)

Big Kidneys:

1) PCKD = polycystic kidney disease

- Infantile type (AR): if unilateral => no problems
- Adult type (AD): if bilateral =>
    - HTN – 1st sign
    - Renal failure, azotemia, liver cysts
    - Diverticulosis, mitral prolapse, berry aneurysms (post communicating a.)

2) Medullary Cystic/Sponge Kidney:

- Polyuria, polydipsia (can't concentrate urine b/c of medulla problem)
- Low vol state => high pH => Ca ppt => kidney stones
- Sonogram => bubbles (cystic) or holes (sponge)
- *Note: Sponge kidney has more stones*

3) Amyloidosis: birefringence with Congo Red stain

4) DM: most common cause of ESRD

5) Scleroderma: tight skin, fibrosis

> **Azotemia:** ↑BUN/ Cr
> **Uremia:** azotemia + sx (bleeding, pericarditis, encephalopathy)

Proteinuria:

Benign (1+ to 2+): ↑Protein conc. gradient (stand, exercise, fever)
Malignant (3+ to 4+): Renal problem => 24-hr urine; measure protein

Ureter Constrictions: *stones get stuck here*

- Hilum (especially staghorn calculi)
- Pelvic rim
- Uretero-vesicular junction (where ureter enters bladder)

Painful Hematuria:

- UTI
- Kidney stone

- Renal infarct

Painless Hematuria:

- TB
- Kidney tumor
- Glomerulonephritis
- Prostate disease
- Sickle cell trait
- Acute Intermittent Porphyria – abdominal pain

Kidney stones: *50% recur in 10yrs*

Dehydration → painful hematuria + colic

Back pain radiating to groin

> **Crystals:**
>
> **Coffin-lid:** Triple PO_4
>
> **Rosette:** Uric acid
>
> **Hexagonal:** Cystine
>
> **Envelope:** Oxalate

1) Calcium Pyrophosphate: amorphous shape

- Cause: IBD, HyperPTH, Pseudogout
- Tx: Thiazide diuretics (↓Ca concentration in urine)

2) Triple Phosphate ($MgNH_4PO_4$)**:** coffin-lid crystals

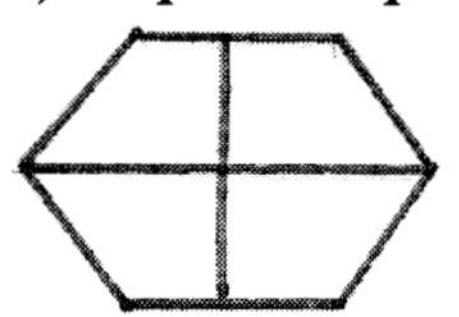

- Struvite = Staghorn calculus
- Cause: Urease (+) Bug: Proteus, Urease (+) Bugs
- Tx: lower urine pH *"lower the coffin"*

3) Uric acid: rhombic "rosette" crystals

- Cause: Gout, Chemotherapy (purines → uric acid), HCTZ, Furosemide
- Tx: raise urine pH (oral bicarbonate or citrate) *"raise: little boy peeing up"*
- Note: not seen on x-ray => do IVP *"U can't see me!"*

4) Cystine: yellow-brown hexagonal crystals

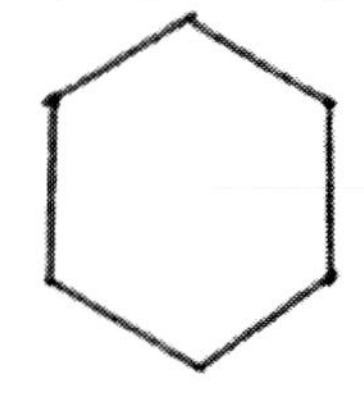

- Cause: Homocystinuria
- Tx: raise urine pH *"look up at the Sistene Chapel"*

5) Calcium Oxalate: envelope or dumbbell crystals ← most common

- Cause: Malabsorption => Ca trapped in fat => Ca can't bind oxalate to excrete it
- *"a dumb ox ate the envelope, then pooped it out"*
- Tx: Thiazide diuretics
    - Caucasian kids: **C**ystic Fibrosis
    - AA kids: Celiac sprue
    - Adults: Crohn's, Vegetarians
    - Other: Ethylene Glycol poisoning, High Vit. C (↑oxalate secretion)

<u>Kidney Stone Tx:</u>

< .4cm: Rehydration, Opiates

.4-4cm: Lithotripsy (shatter it w/ sound waves) → hematuria

> 4cm: Surgery

Septic: Stent placement to drain pus

<u>URETERS:</u>

Ureter: runs on top of psoas muscle, inferior and behind bladder

Reflux makes ureters dilate → hydronephrosis → pyelonephritis → renal failure

Bilateral Hydronephrosis: dilation of ureters due to urine reflux

Unilateral Hydronephrosis: stones => colic (pain comes in waves), radiates to groin

<u>BLADDER:</u> residual volume = 100cc

<u>Urinary Tract Infections:</u> urinary tract has podocytes

Cranberry Juice: *prevents bacteria from adhering to bladder*

Viral: Adenovirus

Bacterial:

1) E. Coli

2) Proteus

3) Klebsiella (Tx: Bactrim)

4) Enterococcus

Urethritis: urethra => dysuria alone (Chlamydia or Gonorrhea)

Cystitis: bladder => frequency, urgency (Tx: Bactrim or Nitrofurantoin)

Honeymoon cystitis: Staph saprophyticus from penis head => female UTI

Prostatitis: bacteria climb urethra to prostate, uncomfortable in sitting position

- Young: N. gonorrhea, Old: E. coli
- Tx: Bactrim IV or Norfloxacin

Pyelonephritis: ascending infxn from nephron => WBC casts, CVAT (Tx: Ceftriaxone)
Balanitis: penis head inflammation
Phimosis: foreskin scarred at penis head (foreskin stuck smooshed up)
Paraphimosis: foreskin scarred at base of penis head (retraction of foreskin strangulates penis)
Prostatic abscess: repeated UTIs that improve w/ abx; prostate fluctuance (Staph aureus)

<u>Bladder Diseases:</u>
Exstrophy of Bladder: urachus stuck outside => cancer risk (Tx: surgery at birth)
Congenital bladder obstruction: posterior urethral valves close when bladder contracts
Urinary retention: BPH drugs, Ipratroprium, Quinidine
Hypospadia: urinary opening near anus (penis fuses dorsal to ventral, zips up tip to base)

- Tx: delay circumcision so the prepuce can be used for reconstruction, repair at 6mo

Seatbelt Trauma: injures superior surface of bladder

Thoracic Aortic Dissection: tearing pain, unequal BP/pulses, CXR widened mediastinum
Type A: ascending aorta (Marfan's, syphilis)
→ emergency surgery
Type B: descending aorta (atherosclerosis in elderly, trauma in young)
→ tx HTN (Nitroprusside + Esmolol)

Abdominal Aortic Dissection: ripping pain, pulsating abdominal mass

- 90% occur below left renal artery (cause: atherosclerosis)
- Dx: US or CT (if pt is hypotensive)
- <4cm => control HTN, get CT/MRI/Angiogram
- >6cm => control HTN, surgery
- Emergency Tx: Tie aorta off, open heart massage; NO CPR!
- Types:
  - True: all 3 layers
  - Pseudo: intima/media only (Ex: femoral a. catheter injection)
- No Steriods with Aneurysms (causes stress demargination of WBCs → thinner walls)
- Repair → emboli → Ant spinal cord infarction → loss of pain/temp/DTR

<u>Urinary Incontinence:</u> *do cystometry*
Urge Incontinence: urgency → complete voiding *"Gotta go right now!"*
1) Urinate frequently to train detrusor
2) Rx:

- Oxybutinin
- Imipramine
- Glycopyrrolate (anti-cholinergic)
- Tolterodine

Stress Incontinence: weak pelvic floor muscles (estrogen effect), pee when you sneeze

Q-tip test: >30° change

1) Kegel exercise
2) Pessary (stick a plastic stopper in to plug it up…)
3) Pseudoephedrine (α_1-agonist)
4) Kelly plication

Overflow Incontinence: persistant dribble, but can't completely empty bladder

1) Obstruction
2) Detrusor hypotonia ← DM, Multiple Sclerosis
    - Tx: Bethanechol (↑detrusor contractions)
    - Tx: intermittent self-catheterization

One-way valves:

- Urethra
- Ejaculatory duct

Ectopic Ureter: continuous urine leakage in a child

Casts: take nephron's shape from PT

WBC casts: Nephritis

- WBC casts only => **Pyelonephritis** (sepsis)
- + Eosinophils => **Interstitial nephritis** (allergies) Tx: steroids
- + RBC casts => **Glomerulonephritis** (hematuria => vasculitis => HTN)

Fat casts: Nephrotic syndrome

Waxy casts: ESRD

Tubular casts: ATN

Muddy brown casts: ATN

Hyaline casts: Normal sloughing

Epithelial casts: Normal sloughing

Crescents: RPGN

Henle's Loopy Loop: Kidneys can absorb max of 2g salt/day

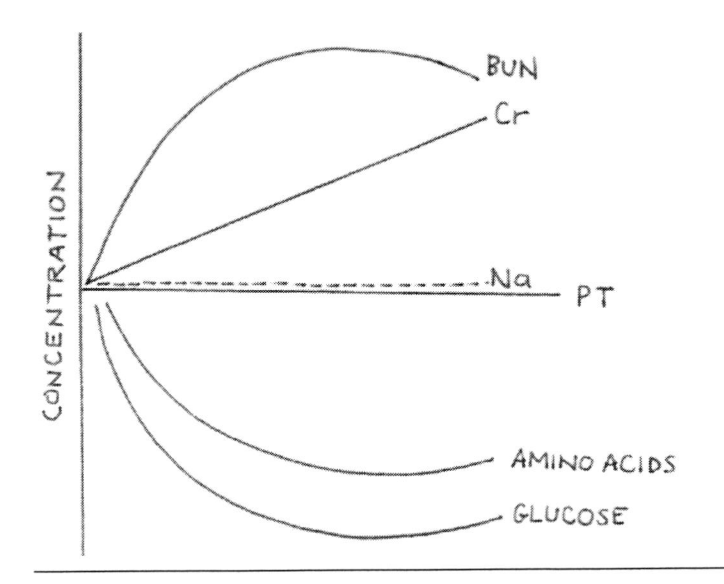

Na Measurements:

FeNa: 1-10% normal

- Pre-renal < 1%
- Renal > 2%
- Post > 4%

U_{Na} = 10-20 normal

- Pre-renal <20 mEq/L
- Renal >20 mEq/L
- Post >40 mEq/L

I) Proximal Tubule: in cortex (isotonic)
Job: reabsorb glucose, amino acids, salt, bicarb, water
Secrete: H^+ (not excrete; keeps circulating)
Reabsorb:

- 70% Na
- 70% H_2O (reabsorbed intercellularly – tiny space between cells)
- filtered HCO_3^- (into plasma) ← "contraction alkalosis"
- 99% glucose (126 = transport maximum)
- 90% aa

Main Rxns:

- H2CO3 → (CA) → CO2 + H2O
- Na/Mg/PO4/aa/glucose/lactate co-transporter
- HCO3-/Cl- antiporter

PT Diseases:
Fanconi's syndrome (old tetracycline): Urine phosphates, glucose, amino acids => low E state

Carbonic Anhydrase Inhibitors: *has sulfur*

- Acetazolamide – Tx: ↑ICP, acute glaucoma, mountain sickness, pseudotumor cerebri

II) Thin Descending Limb: reabsorbs water

III) Thick Ascending Limb: hypotonic
Job: make the concentration gradient by reabsorbing Na, K, Cl, Mg, Ca without water
Main Rxns:

- 25% Na/K/2**Cl** co-transporter
- Na/Ca or Na/Mg co-transporter
- Not water permeable

TAL Diseases:
Bartter's syndrome: baby w/ defective triple transporter (low Na, Cl, K w/ normal BP)
Psychogenic polydipsia: no concentrating ability → cerebral edema
Post-obstructive diuresis: remove obstruction → medulla feels diluted, can't conc (Tx: replace vol)

Loop Diuretics: weak acid

- blocks triple transport system in thick loop of Henle (Na/Cl/K and Ca/Mg)
- competes for uric acid excretion => gout
- have sulfur
- must replace K^+

- low Ca *"Loops lose calcium"*

"A Bum's Torso is Full of Ethanol"

- **B**umetanide
- **T**orsemide
- **F**urosemide – reversible hearing loss, Stephen Johnson syndrome, dilates lung lymphatics (use w/ renal failure and pulmonary edema)
- **E**thacrynic acid – does not have sulfur

<u>IV) Early Distal Tubule:</u>

Job: concentrate urine by reabsorbing NaCl (hypotonic)

<u>Main Rxns:</u>

- Macula Densa (MD): the policeman that measures osmolarity and effective circulating volume
- NaCl co-transporter
- Ca^{2+} reabsorption (Vit D stimulates Ca ATPase)

<u>Thiazide Diuretics:</u> weak acids, have sulfur

"Hyper GLUC":

Hyper**G**lycemia

Hyper**L**ipidemia

High **U**ric acid

Hyper**C**alcemia

- Metolazone
- Chlorthalidone
- Indapamide
- Hydrochlorothiazide – tx Ca oxalate stones

JG: measures volume
MD: measures osmolarity

<u>V) Late DT/ Collecting Duct:</u> *Hyperkalemia (in blood) → acidosis (in cell)*

Job: final concentration of urine by reabsorbing water, excretion of acid (isotonic)

- Vit D/Ca-ATPase reabsorption

<u>Principle cells:</u>

- H_2O channels ← ADH
- K (lose 90%) channel ← Aldo
- Na channel ← Aldo

<u>Intercalated cells:</u>

- H/K ATPase
- H secretion-ATPase ← Aldo

<u>CD acid sources:</u>

1) H^+ ATPases

2) Urea cycle (90% in liver, 10% in CD): $NH_4^+ \rightarrow NH_3 + H^+$

3) CA makes new bicarb in CD: H_2CO_3 → H^+ + HCO_3^- (need if you have respiratory acidosis)
4) Glutaminase: breaks Gln → Glu + NH_4^+ → NH_3 + H^+ (activated when liver fails)

CD Diseases:

Hepatorenal syndrome: high urea from liver → increase glutaminase → NH_4^+ → GABA

- GABA causes heart to stop pumping → kidney stops working

K-sparing Diuretics:

"K STAys"

- **S**pironolactone – blocks Aldo and p450, gynecomastia, galactorrhea, tx Conn's, tx hirsuitism
- **T**riamterene – blocks Na channels directly, tx Meniere's
- **A**miloride – blocks Na channels directly, sodium wasting

Renal Tubular Acidosis: serum ↑H/↓K

Type I (distal): H/K in CD is broken → *high urine pH* (inflammation, autoimmune dz, stones, Li)
Type II (proximal): bad CA → lost all bicarb → *low urine pH* (multiple myeloma, Fanconi's, metals)
Type III: I + II → *normal urine pH 5.3*
Type IV: infarct J-G → no renin → no Aldo → *high K* (DM, NSAIDs, ACE-I, Heparin, sickle cell)

Nephron Summary:

J-G: measures volume
PT: reabsorbs glucose/aa/bicarb/NaCl
Thin AL: reabsorbs water
Thick AL: reabsorbs ions only → makes concentration gradient
MD: measures osmolarity
Early DT: reabsorbs NaCl → starts concentration of urine
Late DT/CD: reabsorbs water/excretes H^+ → final concentration of urine

Anion Gap: Na – (Cl + HCO_3) = 9-14mEq/L *"Positive – Negatives"*
This gap exists b/c some Cl is trapped in RBC….

High Anion Gap: Buffer H^+ by losing HCO_3^- *"MUDPILES"*
Methanol – turns into formic acid → kills retina
Uremia
DKA
Paraldehyde
INH/ Iron
Lactic acid (Ex: bowel ischemia)
Ethanol/Ethylene Glycol (antifreeze) – turns into glyoxylate => kidney stones
Salicylates: asa

Low Anion Gap:
Multiple Myeloma

<u>Non-Anion Gap:</u> ↓HCO_3 or ↑Cl

Diarrhea

Fanconi's

RTA (type II)

Acid ingestion

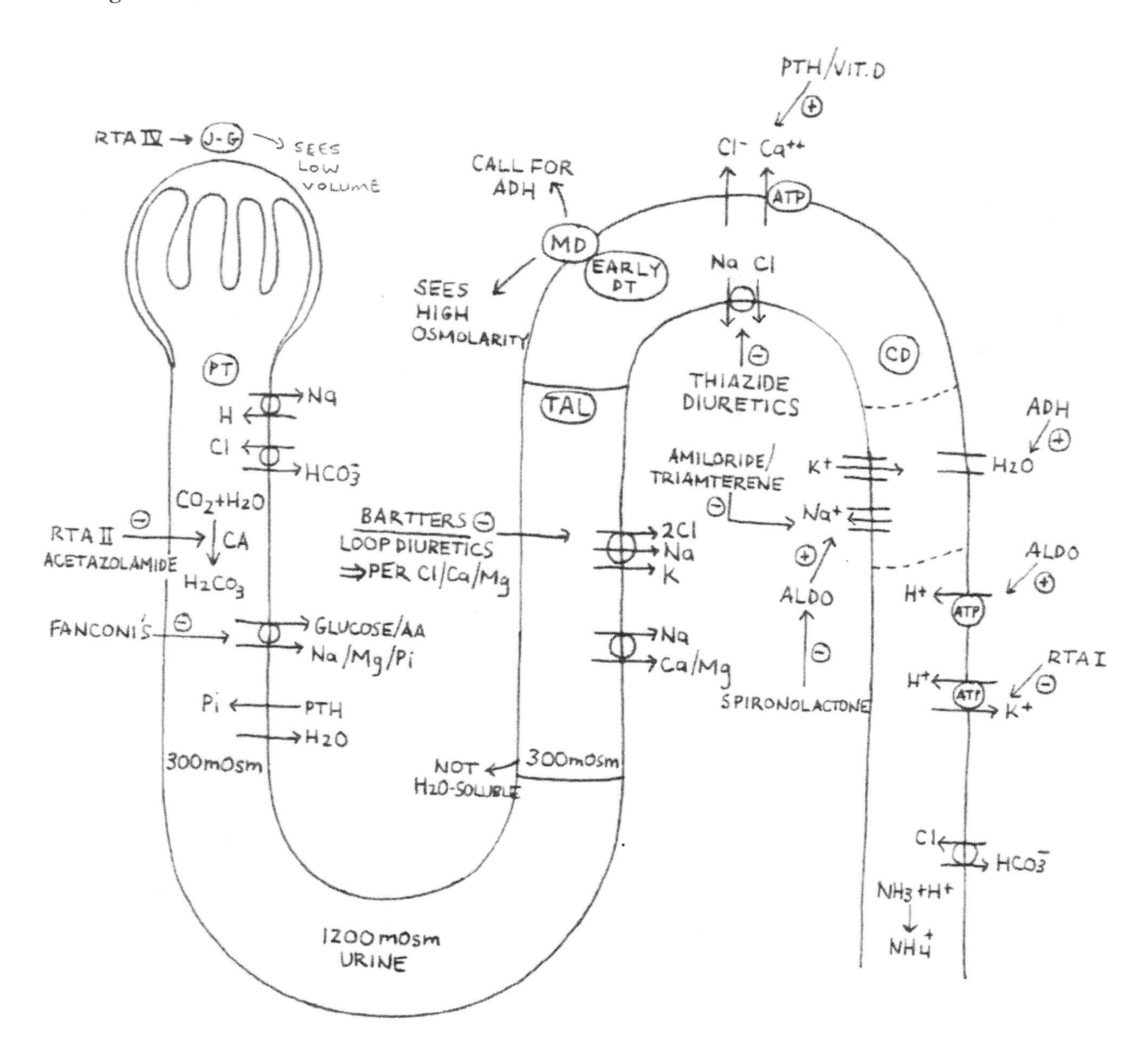

NOTES:

Neurology:

"Any man who reads too much and uses his own brain too little

falls into lazy habits of thinking"

–Albert Einstein

Embryology:

Brain has developed by 8 wks → can direct other stuff now
Primitive streak → Notochord → Spinal cord (nucleus pulposus)
Spinal cord: plates are divided by sulcus limitans

- **B**asal plate (motor) => ventral spinal cord *"**boob** side"*
- **Al**ar plate (sensory) => dorsal spinal cord *"**anus** side"*

Vertebral arch develops ventral → dorsal

- Fusion of spinal cord starts in cervical region, zips up bidirectionally

Ring-Enhancing Lesions:

- Toxo
- Lymphoma
- Abscess
- Metastasis

Brain Problems:

Anencephaly: notochord (day 17) did not make contact with brain => only have medulla => high AFP
Encephalocele: pocket at the base of the brain
Dandy Walker malformation: no cerebellum, distended 4th ventricle, big head, separated sutures
Arnold-Chiari malformation: herniation of cerebellum through foramen magnum

- Type I: cerebellar tonsils (scoliosis) => less sx
- Type II: cerbellar vermis/medulla => hydrocephalus, syringomyelia (loss of pain/temp)

Sacral Problems: *do MRI*

Spina bifida occulta: covered by skin w/ tuft of hair
Spina bifida aperta: opening *(high AFP)*
Meningocele: sacral pocket w/ meninges in it
Meningomyelocele: sacral pocket w/ meninges and nerves, problems with bowel and bladder control

CSF Production:

- Needs Vit A, carbonic anhydrase
- Drainage: subarachnoid dural sinuses → plasma
- Each ventricle has its own choroid plexus => CSF
- CO_2 can diffuse into your brain
- Bicarb and H^+ cannot diffuse into brain => resp. problems affect CSF (Not metabolic acidosis)
- Acid can make GABA

Vomiting centers: responds to increased intracranial pressure, toxic smells

- Chemotactic Trigger Zone – on floor of 4th ventricle
- Area Postrema – in BBB

Ventricle Problems:

Communicating Hydrocephalus: ↑CSF production

- Newborns: Intraventricular hemorrhage

CSF: ↓Bicarb, ↑Cl
Sweat: ↑Bicarb, ↑K

- Kids: Pseudotumor cerebri (high Vit A)
- Elderly: Normal pressure hydrocephalus – due to cortical atrophy with age

Non-communicating Hydrocephalus: obstruction

- Newborns:

#1: Aqueductal stenosis
#2: Dandy Walker Cyst (in 4th ventricle)

- Adults: Tumors (Ependymoma)

<u>**Neurocutaneous Tumors**</u> (AD) => mental retardation, cerebral calcification, seizures
Sturge-Weber: port wine stain (big purple spot) on forehead, angioma of retina

- Tx: pulsed dye laser

Osler-Weber-Rendu: AVM in lung, gut, CNS => sequester platelets => telangiectasias

Tuberous Sclerosis (AD, Chr#9): Ashen leaf spots (hypopigmentation; seen better under Wood's lamp), Brain Ependymomas, Heart Rhabdomyomas, Renal cell CA, Shagreen spots (leathery), bumpy nose, mental retardation

Von Hippel-Lindau: AVM in head, retina => renal cell CA risk *"you can only see the head and eyes of a hippo"*

Neurofibromatosis: >6 Café au lait spots (hyperpigmentation) => peripheral nerve tumors
- If café au lait spots are painful or get bigger, think malignant transformation

- Type 1 "Von Recklinghausen's": Peripheral (Chr#17), optic glioma, Lisch nodules (iris hamartoma), scoliosis
- Type 2 "Acoustic Neuroma": Central (Chr#22), cataracts, bilateral deafness, proliferation of Schwann cells

Pseudotumor Cerebri:

- HA worse when lying down, see spots when straining, tinnitus, CN6 palsy, optic disc edema
- Caused by Isotretinoin
- Tx: Acetazolamide

<u>**Embryology:**</u>
Forebrain = Prosencephalon *"Pro Die & Tell"*

- Diencephalon => Thalamus, Basal ganglia
- Telencephalon => Cerebrum

Budd-Chiari: Hepatic vein obstruction
Arnold-Chiari: Foramen magnum obst.

Midbrain = Mesencephalon (CN 3-4)

Hindbrain = Rhombencephalon *"Meet My Rhomb"*

- Metencephalon => Pons (CN 5-8), Cerebellum

- Myencephalon => Medulla (CN 9-12)

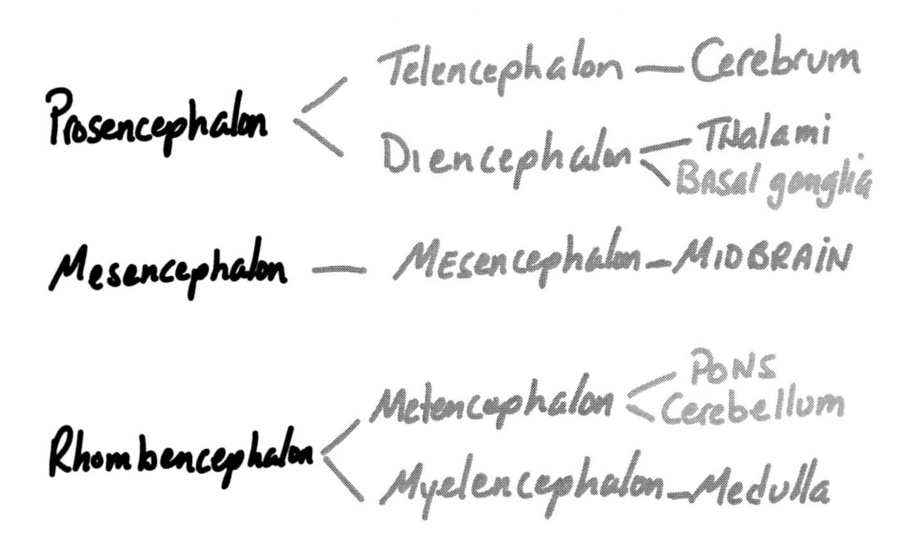

Vision Lesions: *"SO_4LR_6"*

- Nasal fibers – cross over
- Temporal fibers – go straight back

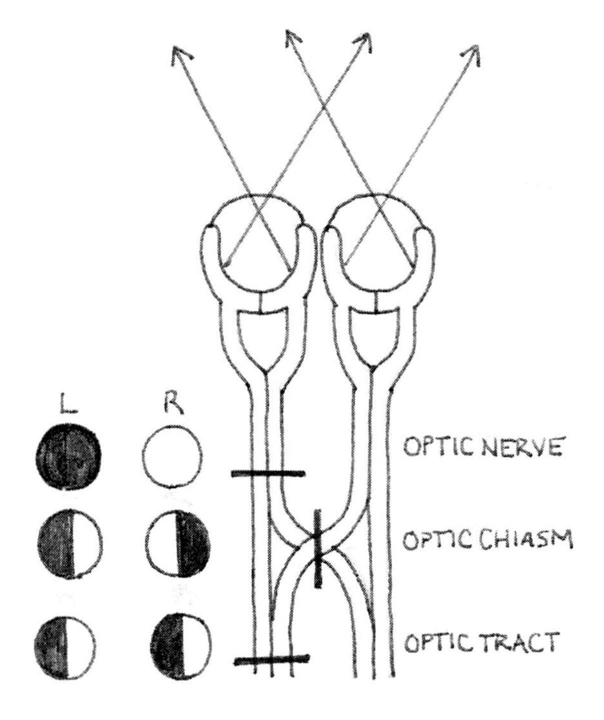

1) Ipsilateral blindness: Optic nerve lesion

Newborns: cataract or Rb (light must hit retina by 3mo or child will be blind)

Kids: optic nerve glioma

Adults: emboli

- TIA: *blind in one eye for <1 hr*
- Temporal arteritis: *HA, blind in one eye* (Tx: steroids)
- Optic neuritis: *painful, blind in one eye, change in color perception* (Tx: steroids)

- Central retinal artery occlusion => *pale retina and cherry red macula* (Tx: thrombolytics)
- Central retinal vein occlusion => *blue retina*

2) Bitemporal Hemianopsia: Optic chiasm lesion
- Pituitary tumor => high PRL
- Pineal tumor => precocious puberty

3) Homonymous Hemianopsia: Optic tract lesion
- Lose nasal same side, temporal other side

4) Quadranopsia: Calcarine fissure lesion
- Lose opposite side

5) Central Scotoma: Macula lesion

6) Acute Loss:
- Retinal detachment: flashes of light (Tx: surgery)
- Vitreous hemorrhage: floaters (Tx: photocoagulate)

7) Amaurosis Fugax: Retinal emboli => painless loss of vision, looks like a "curtain falling down"

I) <u>Cerebellum:</u> depth perception, balance
- Romberg – lose unconscious proprioception (signal doesn't go to cortex)
- EtOH: attacks vermis => ataxia
- All else: attacks hemispheres => intention tremor, dysmetria, dysdiodokinesis, pronator drift

II) <u>Brainstem:</u>

1) Midbrain
2) Pons: most sensitive to shifts in osmolarity
3) Medulla: sets stuff

<u>Frontal Lobe:</u>
- Abstract reasoning => ***Schizophrenia*** (test: interpret proverbs)
- Personality => ***Pick's disease*** (inhibition loss)
- **Broca**'s area (expressive aphasia) => ***brok****en speech "say babababa"*
- Hippocampus (short-term memory) *Ex:* ***early Alzheimer's****, drowning victims*

<u>Parietal Lobe:</u>

Dominant Lobe: (99% population: left side -- regardless of what hand you write with)
- Everything you learned in kindergarden: all long-term memory => ***late Alzheimer's***

Non-dominant lobe: right side
- Apraxia: finger function (test: trace a letter)

- Hemineglect => inability to recognize 1 side of body (usually left side)

Corpus Callosum: fibers cross from right to left side of brain (and vice versa)
- L handed people: ipsilateral connections
- R handed people: contralateral connections
- Absent corpus callosum: ipsilateral connections => ambidextrous

Temporal Lobe: *includes hippocampus, amygdala, and limbic lobes*
- Hearing
- Balance
- Hallucinations
  - Temporal lobe "partial complex" seizures: have olfactory hallucinations before seizure
  - Hypnopompic hallucinations: occur when waking up
  - Hypnogogic hallucinations: occur when **go**ing to sleep
- Wernicke's area (receptive aphasia) => can't *understand* speech or writing *"Wordy"*
- **Klüver-Bucy:** bilateral temporal lesion, hypersexual, oral fixation

Occipital Lobe: *Make sure and flip ALL the words in eye lesion problems…*
Vision: light must hit retina by 3mo or child is permanently blind => look for red reflex

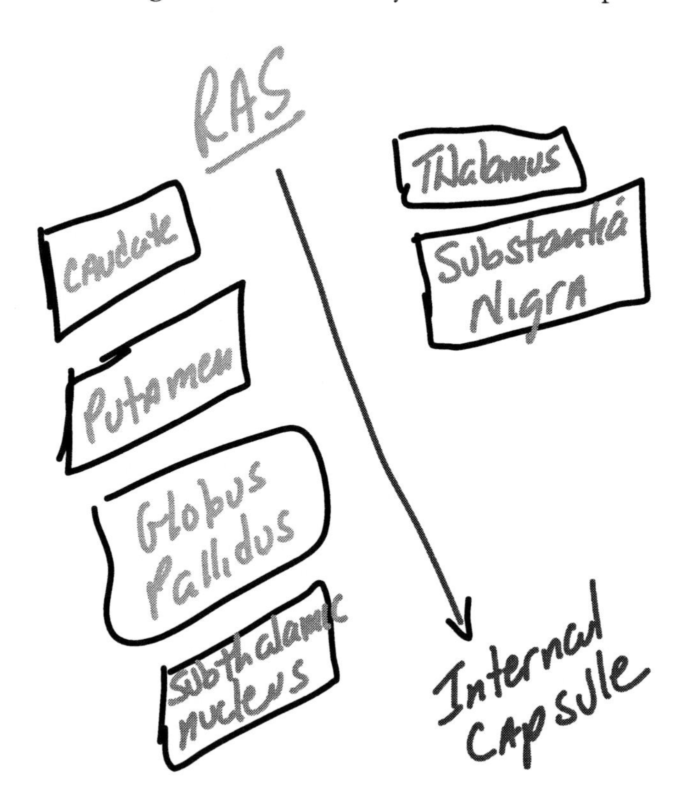

Nerve Tests:
Thumbs up: radial nerve
Finger circle: median nerve
Spread fingers: ulnar nerve

Epithalamus: no known function

Thalamus: *"The Secretary"* All sensory info must stop here
- Medial: Leg fibers

- Lateral: Arm fibers

Hypothalamus:
- Anterior nucleus: Temp regulation (dissipates heat) *"AC: Anterior Cools"* ← *Acetaminophen*
- Posterior nucleus: Temp regulation (conserves heat), parasympathetics
- Lateral nucleus: Hunger *"grab food with hands, which are lateral"*
- Medial nucleus: Satiety
- Suprachiasmatic nucleus: Circadian rhythms *"the timekeeper"*
- Supraoptic nucleus: ADH production
- Paraventricular nucleus: Thirst center

<u>**Basal Ganglia:**</u> lateral wall of internal capsule, controls gross movement
Striatum = caudate nucleus + putamen
Corpus striatum = globus pallidus + striatum
Lenticular nucleus = globus pallidus + putamen

1) Caudate nucleus
- ***Huntington's:*** triplet repeats, anticipation, no GABA, choreiform movements
- Cause of death: suicide (30 y/o), insurance will drop them upon diagnosis

2) Globus Pallidus: inhibits movement (send Ach or GABA to substantia nigra)

<u>**Movement Disorders:**</u>
Tick: involuntary movement of one muscle
Dystonia: sustained contraction of muscle groups
Tardive Dyskinesia: facial grimace, tongue thrusting
Akathesia: non-stop restless movements

<u>**Tourette's Syndrome:**</u> *Tics ↑OCD risk*
- Motor tics (Tx: Clonidine or Haloperidol)
- Vocal tics
- Echolalia: repeats everything
- Coprolalia: constant swearing

3) Putamen: *most common site of HTN hemorrhage*
- ***Wilson's:*** ceruloplasmin def. => Cu deposits in liver, eye (KF rings), brain (lenticular nucleus)

4) Substantia Nigra: med wall of internal capsule => initiates movements (sends DA to basal ganglia)
- ***Parkinson's:*** can't initiate movement (no DA)

5) Subthalamic Nucleus: "The Final Relay Station for Fine Motor Coordination" => ballismus

Internal Capsule: All info going in & out of brain must come through here
Reticular Activating System: "Gatekeeper of the Internal Capsule", maintains your focus

- ***Attention Deficit Disorder:*** can't ignore any thought

Cushing's triad:

intracranial pressure

- HTN
- Bradycardia
- Irregular respirations

Stress Response:

1st response: Parasympathetic

- Ex: Erection, diarrhea

2nd response: Sympathetic

- Ejaculation, constipation, increased GI acid output
- Stress ulcer
- Curling's ulcer (burn pts) *"Burn pts have a curling scream"*
- Cushing's ulcer (intracranial pressure ulcer)
- IBS: constipation → diarrhea → repeat

Nerve Reflexes:

L4: knee jerk, foot dorsiflex
L5: big toe dorsiflex, foot drop
S1: ankle jerk, foot eversion
S2-4: anal wink

Sleep Waves:

Increased by: ACh, 5-HT
Decreased by: DA, NE

"BATs Drink Blood"
Wide awake: β waves
Eyes closed awake: α waves
Light sleep: θ waves "Stage 1-2"/ K-complexes
Deep sleep: δ (huge) waves "Stage 3-4"

- teeth grinding, sleepwalking
- BZ, Imipramine inhibit this

REM sleep: β waves – every 90 min

- dreams, penile erections
- NE, EtOH, Barbs, Age inhibit this
- 5-HT, ACh increase this

Watershed Areas:

First to lose, last to recover

Brain: Hippocampus
GI: Splenic flexure

Sleep Disorders:

Nightmare: remember dreams, occurs in REM sleep
Sleep Terror: don't remember dreams, occurs in non-REM sleep
Dysomnia: quality of sleep
Parasomnia: sleep behavior (nightmares)
Narcolepsy: fall asleep during day, pathognomonic cataplexy, sleep paralysis, hallucinations
Sleep Apnea: fall asleep during day, wake up exhausted
Kleine-Levin syndrome: teenage boys eat and sleep a lot

Brain Tracts:

Fasciculus = few fibers

Tractus = lots of fibers
Gracilus = legs *"graceful legs"*
Cuneatus = arms

Upper vs. Lower Motor Neuron Lesions:

UMN: *"upper"*	**LMN:** *"downer"*
CS tract	non-CS tracts
Spasticity	Flaccid paralysis
Hyperreflexia	Hyporeflexia
Clonus	Atrophy
Babinski reflex	Fasciculations/ fibrillations

Glasgow Coma Scale:
<8=severe

- Eye opening (max=4)
- Verbal response (max=5)
- Motor response (max=6)

1) Descending Tracts

Corticospinal (CS) tract: *motion*
Crosses in medulla => sx are contralateral, pyramidal decussation, use CN to find level of injury

Corticorubral tract: contains red nucleus, runs right below CN3 => flexion

Hypothalamospinal tract:
Lesion => ipsilateral ***Horner's syndrome:*** miosis, ptosis, anhydrosis, enophthalmos

2) Ascending Tracts

Dorsal Column Medial Lemniscus (DCML) tract: vibration, position, 2-point discrimination
Crosses in medulla
1st synapse: Dorsal root ganglion
2nd synapse: Nucleus cuneatus and nucleus gracilis (crosses here)
3rd synapse: Thalamus
4th synapse: Post-central gyrus

- ***Pernicious anemia:*** anti-IF Ab (affects DCML and ST)
- ***Friedreich's ataxia:*** scoliosis, retinitis pigmentosa (attacks DCML and SC)
- ***3° Syphilis:*** obliterative endarteritis, lancinating stabbing pain (attacks DCML)

Spinocerebellar (SC) tract: balance, depth perception
The only pathway that crosses twice => ipsilateral symptoms

Spinothalamic (ST) tract: pain & temp
Crosses in spinal cord (ant. white commissure) => lose stuff 2 levels lower on contralateral side
All sensory fibers come in through dorsal root ganglia, go up two dermatomes, then cross

- Anterior white commissure lesion: ***Syringomyelia*** => lose pain/temp in "cape" distribution
- Spinal cord lesions: *pain & temp loss on opposite side of all other losses*

CPP = MAP - ICP
CPP = cerebral perfusion pressure (Ex: HTN + head injury => ↑CCP)

ICP = intracranial pressure (↓ w/ head injury) – No Nitroprusside! Treat ICP and BP will follow...
MAP = mean arterial pressure (↑ w/ HTN)

↑ Intracranial pressure:

- Sx: Headache
- Signs:

1) Papilledema (check CT first)
- Mass => don't do a lumbar puncture!
- No Mass => do LP for meningitis

2) Esotropia (eye moves in due to CN6 compression)
3) Dilated pupils (due to CN3 compression) => 1st sign of herniation
4) Decorticate rigidity: herniation above red nuc. => flex arms => 2nd sign of herniation
5) Decerebrate rigidity: herniation below red nucleus => extend arms => dead

Head Injury Tx: *"VMAB"*
1) **V**entilator (100% O_2 to ↓CO_2 → ↓flow → ↓ICP)
2) **M**annitol – suck fluid out of brain cells to shrink brain
3) **A**cetazolamide – inhibits CA (which makes CSF)
4) **B**urr a hole in top of head to relieve pressure, No LP!

LP Barriers: Skin → Ligaments → Dura → Arachnoid (Note: CSF is in subarachnoid space)
- Do LP at L4-L5 *"L4-L5 keep the spinal cord alive"*

CN 3-4: Midbrain → blown pupils
CN 5-8: Pons → no doll's eyes, (+) ice water calorics
CN 9-12: Medulla → tongue deviation to injured side, irregular breathing

Cranial Nerve Lesions: Put the lesion at the level of the highest CN affected...
"On Old Olympus' Towering Top, A Fin And German Viewed Some Hops"
CN 1: **O**lfactory => can't smell
CN 2: **O**ptic => blind
CN 3: **O**cculomotor => no response to light, ptosis, look down and out
CN 4: **T**rochlear => see double when they look down
CN 5: **T**rigeminal => can't chew
CN 6: **A**bducens => eye points toward nose
CN 7: **F**acial => facial paralysis
CN 8: **A**coustic => can't hear
CN 9: **G**lossopharyngeal => dry mouth, dysphagia, ↓gag
CN 10: **V**agus => hoarse voice, uvula deviation, palate does not rise with "ahh"
CN 11: **S**pinal Accessory => can't shrug shoulders or turn head
CN 12: **H**ypoglossal => tongue deviates to weak side, difficulty speaking

Eye Reflexes:
Blink reflex: CN 5 → 7 → 3
Pupillary light reflex: CN 2 → 3

Sports Concussions:
No LOC: 15 min observation

LOC <5 min: 1 week observation
LOC >5min: go to ER with C-collar

CNS Infections:

Rabies: exposure to raccoon, skunk, fox, bat, dog, cat
- Negri bodies, hydrophobia, laryngospasm
- Virus → unmyelinated nerves → CNS (hippocampus) → salivary glands, peripheral nerves

HSV-2: temporal lobe hemorrhagic encephalitis, need C/S prophylaxis

Polio: asymetric Fasciculations in a child
- 2 wks after gastroenteritis
- Trendelenburg gait (superior gluteal nerve → gluteus mm.)

JC virus: progressive multifocal leukoencephalopathy, brain demyelination

Toxoplasma: multiple ring-enhancing lesions, cat urine, parietal lobe

CMV: shallow big esophagus ulcers, yellow-white retina opacification, bloody diarrhea, tenesmus, spastic diplegia of legs, hepatosplenomegaly, blindness, central calcifications

Measles = Rubeola (paramyxovirus): *multinucleated giant cells (lymphocytes)*
Complications: otitis media, pneumonia, demyelinating disease = "*Subacute Sclerosing Panencephilitis*"
1) Cough, Coryza (thick rhinorrhea), Conjunctivitis
2) Koplik spot (white spot on buccal mucosa) – 24 hrs before rash
3) Morbilliform blotchy rash – spreads from head

Echinococcus: raw lamb/dog poop => *hydatid cyst* w/ eggshell calcifications
- Tx: Mebendazole + surgery

Taenia Solium: raw pork (*cysticercosis* = larva swims in aqueous humor)
- Tx: Praziquental + steroids

Syphilis: Rhagade's (lip fissure), Hutchison's razor teeth, saber shin legs, mulberry molars

Acanthamoeba: in contact lenses, eats through cornea

Naegleria Fowleri: swamp diving trauma
- Fulminant meningoencephalitis – eats through cribiform plate into brain => die in 48 hrs

Creutzfeldt-Jacob: prion induced, die within 1 year, post-cornea transplant
Rubella: cataracts, hearing loss, PDA, meningoencephalitis, pneumonia, "blueberry muffin" rash

Ramsay-Hunt: CN7 herpes zoster → hearing loss, vertigo, Bell's palsy (facial nerve palsy)

<u>Most Common CNS Infections:</u>
Frontal Lobe: Rubella *"Rub the Front"*
Temporal Lobe: HSV *"Wish Herpes was Temporary"*
Parietal Lobe: Toxoplasma *"The Toxic Pariah"*
Hippocampus: Rabies *"The Hippo with Rabies"*
Posterior Fossa: TB *"Posterior is **The Back**"*
DCML: Treponema Pallidum *"Don't Trip on The Columns"*

<u>Delirium:</u> loss of consciousness, lucid intervals, sundowning, *abnormal EEG (Ck UA/CMP)*

- **D**ementia
- **E**pilepsy, **E**pidural hematoma
- **L**ungs, **L**iver
- **I**nfxn
- **R**x: NPH
- **I**njury
- **U**nfamiliar environment
- **M**etabolic: vitamin deficiency

<u>Dementia/Degenerative:</u> *Ck TSH/Vit* B_{12}
Pick's disease: frontal lobe atrophy, disinhibition

Normal pressure hydrocephalus: Tr*IAD* = **I**ncontinence, **A**taxia, **D**ementia (reversible w/ LP)

<u>Alzheimer's:</u> ↓ACh in nucleus basalis of Meynert, bad ApoE, neurofibrillary tangles of tau
1) Acetyl Cholinesterase Inhibitors:

- Donepazil - best
- Tacrine – liver toxicity
- Galantamine – liver and kidney toxicity

2) NMDA-receptor antagonist: Memantine

<u>Parkinson's:</u> in substantia nigra, bradykinesia, pill-rolling tremor, shuffling gait, Lewy bodies
Pathway: ↓DA → ↑ACh → ↑GABA
1) ↑ DA: Raynaud's, VH

- L-DOPA
- Pergolide
- Bromocriptine
- Pramipexole – tx restless leg syndrome
- Ropinirole – tx restless leg syndrome

2) ↓DA Metabolism:

- Carbidopa/Levodopa
- Selegiline
- Amantidine – purple skin

3) ↓ ACh:

- Benztropine
- **T**rihexyphenidyl – tx **t**remor

4) COMT Inhibitors:

- Entacapone
- Tolcapone – hepatotoxic

Lewy body dementia: stiff, visual hallucinations, dementia within 1yr of NM dysfunction

Huntington's (AD): in caudate/putamen, triplet repeat disorder, choreiform movements

Amyotrophic Lateral Sclerosis (ALS) = Lou Gehrig disease:

- Descending paralysis, fasciculations in middle aged male
- Only motor nerves are affected
- CS tract and ventral horn
- Tx: Riluzole (↓pre-synaptic Glu)

Friedreich's ataxia: Retinitis pigmentosa, scoliosis

Vascular "multi-infarct" dementia: sudden onset, stepwise progression of deficits

<u>**Demyelinating Diseases:**</u>

Multiple Sclerosis: anti-myelin Ab, symptoms come and go

- *Middle aged woman with vision problems*
- Optic neuritis => halo vision (can't see directly)
- Internuclear ophthalmoplegia: opposite eye won't go past midline
  - MLF lesion (connects CN 3 and CN 6)
  - Bilateral trigeminal neuralgia
- LP: myelin basic proteins, MRI: plaques
- Tx: Glucocorticoids, INF-β => depression, Plasmapheresis

Central Pontine Myelinolysis: if you correct osmolarity too fast

- Increase glucose 100mL/dL => decrease Na 1.6mEq/L
- Never correct Na faster than 0.5mEq/hr

Guillain-Barre: ascending paralysis *"Ground-to-Butt"*

- 2 wks after URI or *C. jejuni* infection
- Anti-ganglioside Ab
- MP eat myelin off nerve axons → ↑CSF protein
- Polyradiculoneuropathy – many dermatomes involved
- Same presentation as tick bites, resolves spontaneously like MS
- Tx: Intubate if needed, IV Ig Plasmapheresis

Metabolic Encephalopathies:

Wilson's disease (AR): Ceruloplasmin deficiency

- Cu in basal ganglia, hepatitis, Kaiser-Fleischer eye rings
- Tx: Penicillamine (chelates Cu^{2+})

Kernicterus: sulfa exposure, Hemolytic Disease of the Newborn

Wernicke's encephalopathy: ophthalmoplegia, ataxia, psychosis

Korsakoff psychosis: mamillary bodies => anterograde amnesia, confabulation

Acute Intermittent Porphyria: **increased porphyrin production, urine δ-ALA, porphobilinogen**

- **Sx: Abdominal pain, neuropathy, red urine (hemolytic anemia)**
- **Can be set off by stress (menses, Drugs: Barbs, Sulfas)**
- **Tx: 1. Fluids – wash away porphyrin ring**
  2. Sugar – break down bilirubin
  3. Opiates – stop pain (use Meperidine for abdominal pain)
  4. Hematin – inhibits δ-ALA synthase

Headaches:

Migraines: aura, photophobia, numbness and tingling → throbbing HA, nausea

- Prophylaxis: Amitriptyline, Propanolol
- Tx: NSAIDs, Sumatriptan (5-HT agonist), Ergotamine (not if CAD)

Tension Headache: bilateral "band-like" pain, worse as day progresses, sleep disturbance

- Prophylaxis: Amitriptyline
- Tx: ASA

Cluster Headache: unilateral retro-orbital pain, suicidal, facial flushing, lacrimation, Horner's

- Acute Tx: O_2 inhalation, Glucocorticoid, Sumatriptan
- Chronic Tx: Li, Verapamil, β-blocker

Temporal (Giant cell) Arteritis: pain with chewing, blind in one eye

- Tx: Prednisone

Trigeminal Neuralgia: sharp, shooting face pain

- Tx: Carbamazepine

Chronic Daily Headache: bilateral, diffuse, pressure-like

- Prophylaxis: Nortriptyline

Brain Tumor: immediate vomiting
Migraine HA: vomit hours later

Altered Mental Status: *"MENTAL"*

Meds: BZ, Opioids (Tx: Naloxone + Flumazenil)

Electrolytes: DKA (Tx: Insulin → Fluid → K^+)

Neuro disorders: Status epilepticus (Tx: Diazepam)

Temperature

Alcohol: Vit B_{12}, folate deficiency (Tx: Thiamine → Glucose)

Liver/ kidney dz: Hepatic encephalopathy (Tx: Nitroprusside + Lactulose)

Comatose Tx:

"DONT"

Dextrose

O_2

Naloxone

Thiamine

NOTES:

Psychiatry:

"I have learned that people will forget what you said, people will forget what you did, but they will never forget how you made them feel."

–Maya Angelou

F<u>**reud's Tripartite Model:**</u> *"devil and angel on each shoulder"*

Id: Selfish instinct *"devil"*
Ego: Self
Superego: Rules *"angel"*

DSM IV:

Axis I: Clinical d/o
Axis II: Personality d/o
Axis III: Medical d/o
Axis IV: Psychosocial
Axis V: GAF

Psychological Stages of Development:

Infant (0-15 mo): attached to mom
Toddler (15 mo-2½ years): rapprochement (comes back to mom for reassurance)
Preschool (2½-6 years): "Band-Aid phase" (concerned about illness)
School Age (6-11 years): understand death

Freud's Psychoosexual Stages:

Age:	Stage:	Interpretation:
0-1	Oral	Oral gratification
1-3	Anal	Toilet training
3-4	Phallic	Penis/Clitoris fascination
4-6	Oedipal	Possess parent of opposite sex
6-12	Latency	Social skills
13+	Genital	Sex drive

Major Depression:

need 5 "SIGE CAPS" >2wks

Sleep disturbances: wake in am
Interest/ Libido loss: anhedonia
Guilt
Energy loss
Concentration loss
Appetite loss
Psychomotor agitation
Suicide: hopelessness

Childhood Disorders:

Enuresis (>5 y/o): pee whenever (Tx: buzzer pad, Imipramine, DDAVP)
Encopresis (>4 y/o): kids won't poop (Tx: laxatives)
Autism: repetitive movements, lack of verbal skills and bonding
Asperger's: good communication, impaired relationships, no mental retardation
Rett's: only in girls, decreased head growth, lose motor skills, hand-wringing
Childhood Disintegrative Disorder: kid stops walking/ talking
Selective Mutism: kid talks sometimes (Tx: Fluoxetine)
Separation Anxiety Disorder: kid screams when Mom leaves (Tx: Imipramine)
Reactive Attachment Disorder:

- Inhibited Type: decreased social skills
- Disinhibited Type: attach to everyone

Conduct Disorder: aggressive, disregard for rules, no sense of guilt *"bite"*
Oppositional Defiant Disorder: defiant, noncompliant *"bark"*
Attention Deficit Hyperactivity Disorder: overactivity, difficulty in school (Tx outcome: calmer, not euphoria)

<u>1st Line Tx:</u> Stimulants *(paradoxical effect)* → vertical nystagmus

- Methylphenidate "Ritalin, Concerta, Methylin, Metadate, Daytrana patch" – tx narcolepsy => appetite loss, tics

- Dexmethylphenidate "Focalin" – more potent
- Amphetamine "Adderall"
- Methamphetamine "Desoxyn"
- Dextroamphetamine "Dexedrine, Dextrostat" – 2x potency of Ritalin
- Lisdexamfetamine "Vyvanse" – more even effect
- Pemoline "Cylert" – taken off the market due to liver failure

2nd Line Tx: Antidepressants

- Buproprion "Wellbutrin" (NE/DA) – tx depression, tx ADD, causes seizures
- Atamoxetine "Strattera" (NE) – minimal side effects, #1 tx adult ADD

Mood Disorders: feel "down"

Sadness: situational, normal

Grief: sadness surrounding loss of someone/something

Bereavement: act of grieving, still functional, <2mo

Melancholy: deep sadness (Ex: sit in dark)

Depression: anhedonia, failure to function

- *Rule out hypothyroidism, SLE, drugs*
- **Acute Reactive Depression:** lasts <2 wk
- **Major Depression:** lasts >2wk

Dysthymia: low level sadness >2yr (Ex: Eor in Winnie the Pooh)

Cyclothymia: dysthymia w/ hypomania

Double Depression: depression followed by dysthymia

Bipolar: massive swings *(think about this if antidepressant meds cause mania/hypomania)*

- Bipolar I = Depression and Mania (psychosis for at least 7 days)
- Bipolar II = Depression and Hypomania (no psychosis)

Phases of Grief:

Tx: Empathy

"DABGA"

Denial

Anger → blame → divorce

Bargaining w/ God

Guilt

Acceptance

Catecholamine Side Effects:

Dopamine: vomiting center, basal ganglia (movement disorders, psychosis)

Norepinephrine: sympathetic side effects

Serotonin: sympathetic in brain, parasympathetic in periphery (flushing, wheezing)

Bipolar Tx: *"LiV Carefully Darling"*

- **Li:** blocks ADHr, Ebstein's anomaly, DI, hypothyroidism, polyuria, avoid NSAIDs
- **V**alproate "Depakene": blocks Na/Ca, liver necrosis, ↓bone marrow, NTD, urine incontinence
- **C**arbamazepine "Tegretol": SIADH, aplastic anemia, tx trigeminal neuralgia
- **D**ivalproex "Depakote": slowly absorbed form of valproic acid

Antidepressant Tx: takes 3-6wks to work (Li or T_3 can augment)

High rate of suicide during first 3mo of tx (have enough energy to do it)

1) SSRI: blocks 5-HT reuptake => *insomnia, anorgasmia (good for premature ejaculation)*

Uses: Depression, panic disorder, phobias, OCD, ADD, PTSD, eating disorders
"Feeling Fairly Poor Can Escalate Suicide"

- **F**lu<u>oxetine</u> "Prozac" – good for bulimics and pregnant women, long $t_{1/2}$
- <u>**F**lu**vox**amine</u> "Lu**vox**"
- **P**aroxetine "**P**axil" – tx MI depression
- **C**italo<u>pram</u> "Celexa" – less side effects
- **E**scitalopram "Lexapro" – purer form of citalopram
- <u>**S**ertraline</u> "Zoloft" – tx PT**S**D

Priapism:

- Trazodone
- Prazosin
- Chlorpromazine
- SSRIs

<u>Seratonin Agonists:</u>

- Sumatriptan – tx acute migraines
- Elatriptan
- Methysergide – die of MI
- Cisapride – off market due to Torsade

<u>2) SNRIs:</u>
"(SNRIs) Do Treat A Bad Mood Very Nicely"

- **D**uloxetine "Cymablta" – tx diabetic neuropathic pain
- **T**razodone (NE/5-HT_2) "Desyrel" – priapism (painful erection >4hr Tx=Epi)
- **A**tamoxetine (NE) "Strattera" – tx adult ADD
- **B**upropion (NE/DA) "Wellbutrin" – seizures, tx SAD, tx smoking
- **M**irtazapine (NE/5-HT) "Remeron" – weight gain
- **V**enlafaxine (NE/5-HT) "Effexor" – discontinuation syndrome, HTN
- Desvenlafaxine (NE/5-HT) – purer form
- **N**efazadone (NE/5-HT_2) "Serzone" – dizziness

<u>3) TCA:</u> *sedation, weight gain, #1 cause of child ingestion deaths, OD Tx: Bicarb*
NE/5-HT reuptake inhibitor: sympathetic side effects
Anti-cholinergic: hot, dry skin (can't sweat), impotence
α_1 blocker: dizzy, orthostatic hypotension
Na blocker: prolonged QT (slows AV conduction) => *severity of overdose*

"(TCA's) May Notoriously AAccentuate Child Ingestion Deaths"

- **M**aprotiline "Ludiomil" – seizure risk
- **N**ortriptyline "Pamelor" – good for elderly
- **A**mi<u>triptyline</u> "Elavil" – tx chronic pain/neuropathy
- **A**moxapine "Asendin" – tardive dyskinesia
- **C**lomipramine "Anafranil" – best tx for OCD
- **I**mi<u>pramine</u> "Tofranil" – blurred vision, tx child enuresis, tx separation anxiety
- **D**esipramine "Norpramin" – tx catatonia, stimulates appetite, good for elderly

<u>Psychotic Symptoms:</u>
(+): Hallucinations, Delusions
(–): Blunt affect, Apathy

4) MAOI: prevents NE/5-HT breakdown

Pre-synaptic: MAO breaks down catecholamines (MAO levels increase w/ age)

Post-synaptic: COMT breaks down catecholamines

Don't use w/ SSRIs, Pseudoephedrine, Mepiridine => Δ*mental status* (need 2wk washout)

Don't use w/ seafood (5-HT) => *serotonin syndrome* => MI

Don't use w/ wine, cheese (tyramine) → octapine → *hypertensive crisis* (Tx: Phentolamine)

"PITS"

- **P**henelzine – GABA effects
- **I**socarboxazid
- **T**ranylcypromine – amphetamine-like effects, no liver toxicity
- **S**elegiline "Eldepryl" – MAO_B selective (prevents DA breakdown): tx Parkinson's

5) Electroconvulsive Therapy: *induce 25-60sec generalized seizures*

- Good for elderly, pregnant women, active suicidality, unresponsive to meds
- Given with general anesthetic and muscle relaxant to prevent broken bones
- Can't use with ↑intracranial pressure or recent MI
- Causes retrograde amnesia, seizures

Psychotic Disorders: feel "up"

Psychosis: lost touch with reality

- **Brief reactive psychosis:** <1mo (Ex: "going postal"), change baseline personality
- **Schizophreniform:** 1-6mo
- **Schizophrenia:** >6mo
    - Paranoid: delusions + hallucinations
    - Disorganized: disinhibition, poor organization
    - Catatonic: bizarre positioning (Tx: BZ)
    - Undifferentiated: more than one of the above
    - Residual: emotional blunting, odd beliefs after previous episode
- **Schizoaffective:** schizophrenia + mood d/o (depression or mania), *sx balanced by mood d/o*
- **Shared Psychotic Disorder:** believe another's delusions

Psychotic Symptoms: *out of touch w/ reality*

1) Speech Disorders:

Loose associations: ideas switch subjects

Flight of ideas: no connections between thoughts

Tangentiality: wanders off the point and never gets back to the point

Circumstantiality: digresses, but finally gets back to the point

Clanging: words that sound alike, rhymes

Word salad: unrelated combinations of words

Perseveration: keeps repeating the same words

Minimal CYP450 Interactions:

Good for pts w/ multiple illnesses

"Meds Very Seldom Eat CYP450"

- **M**irtazapine
- **V**enlafaxine
- **S**ertraline
- **E**scitalopram
- **C**italopram

Neologisms: new words

2) Thought Disorders:
Delusion: one false belief
Illusion: misinterprets stimulus (Ex: magic shows)
Hallucination: false sensory perception

- Visual hallucination: rare
- Auditory hallucination: most common
- Tactile hallucination: EtOH withdrawal/Cocaine intoxication => *formication*

Nihilism: thinks the world has stopped
Loss of ego boundaries: not knowing where I end and you begin
Ideas of reference: believes the media is talking to you
Thought blocking: stops mid-sentence
Thought broadcasting: believes everyone can read his thoughts
Thought insertion: believes others are putting thoughts into his head
Thought withdrawal: believes others are taking thoughts out of his head
Concrete thinking: can't interpret abstract proverbs, just sees the facts

1) Typical Antipsychotics: block D_2, tx schizophrenia, tx emesis
EPS Side Effects:
Dystonia (eyes roll up, thick tongue, torticollis) *Tx: Diphenhydramine or Benztropine*
Tardive dyskinesia (facial grimace w/ **t**ongue **t**hrusting) *Tx: Clozapine "to cloze their mouth"*
Akathisia (non-stop restless movements) *Tx: Propranolol*
Parkinson's (poverty of movement) *Tx: L-DOPA/Carbidopa or Benztropine*
Neuroleptic malignant syndrome (fever >104°, muscle rigidity, ↑CPK, death) *Tx: Dantrolene*

Low Potency: blocks α_1 => orthostatic hypotension, tx mild psychosis

- **C**hlorpromazine "Thorazine": **c**ornea/lens pigmentation, tx intractable hiccups, aplastic anemia
- Thioridazine "Mellaril": retinal pigmentation, ↑QT (Tx O/D with bicarb)

High Potency: galactorrhea, amenorrhea (↓DA → ↑PRL → ↓GnRH/ ↓TRH)
"Positive Psychotics Prove Treatable, Though Hallucinate Darn Frequently"

- **P**erphenazine "Trilafon"
- **P**rochlorperazine "Compazine": tx emesis
- **P**imozide: tx body dysmorphic disorder
- **T**rifluoperazine
- **T**hiothixene
- **H**aloperidol "Haldol": can give q4wk decanoate, tx Huntington's, least seizures, ↑EPS
- **D**roperidol: tx acute psychosis
- **F**luphenazine "Prolixin": tx facial tics, q2wk decanoate

2) Atypical Antipsychotics: block D_4 and $5\text{-}HT_2$, tx negative symptoms

"Can't Quietly Regard Ziprasadone As Our (Atypical Antipsychotic)"

- **C**lozapine "Clozaril": fatal agranulocytosis (stomatitis), tx resistant-dz, tx flat affect, no BZ
- **Q**uetiapine "Seroquel": good for sleeping off withdrawal
- **R**isperidone "Risperdal": ↑prolactin
- **Z**iprasadone "Geodon": ↑QT
- **A**ripiprazol "Abilify"
- **O**lanzapine "Zyprexa": lot of weight gain, tx HIV psychosis, avoid in smokers/PKU

Anxiety Disorders:

Worry: situational, controlled

Anxiety: outward manifestation of worry

Generalized Anxiety Disorder (GAD): hyperarousal > 6mo, lacks a stressor (Tx: Buspirone)

Adjustment Disorder: abnormal excessive reaction to a life stressor (<3mo)

Social Phobia "Social Anxiety Disorder": extreme shyness, knows it is irrational

Acute Stress Disorder (ASD): PTSD < 1mo (Ex: "shell shock")

Post-traumatic Stress Disorder (PTSD): flashbacks >1mo (Tx: SSRI)

Obsessive-Compulsive Disorder (OCD): compulsive checking, counting, cleaning (Tx: SSRI)

Phobia: avoidance behavior, knows it is irrational fear

- **Acrophobia:** fear of heights *"acrobats"*
- **Agorophobia:** fear of **go**ing outside
- **Arachnophobia:** fear of spiders
- **Claustrophobia:** fear of small spaces
- **Ophidiophobia:** fear of snakes

Short-Acting BZ:

good for elderly

- Alprazolam
- Oxazepam
- Triazolam

Panic Attack: situational, think they are having a heart attack, stops instantly

Panic Disorder: panic attacks (30 min x 2/wk), *can be induced by Na lactate or inhaled* CO_2

- Tx:

1) Start BZ (immediate effect)
2) Start SSRI
3) Taper off BZ

Anxiolytics: ↑ GABA => less likely to depolarize, causes agitation in elderly

1) Benzodiazepines: bind BZr → GABAr → ↑*freq* of Cl opening => into cell => less firing

- Clonazepam "Klonopin" – wafer form, tx absence seizures, tx chronic anxiety
- Temazepam "Restoril" – tx to maintain sleep
- Flurazepam "Dalmane" – longest acting
- Lorazepam "Ativan" – tx status epilepticus (2nd line), tx anxiety disorder
- Oxazepam "Serax" – short-acting
- Diazepam "Valium" – tx status epilepticus (rectal suppository), tx muscle spasm
- Midozelam "Versed" – for short-term procedures, 45min anterograde amnesia
- Alprazolam "Xanax" – tx panic attacks, blackouts, seizures if missed dose
- Triazolam "Halcion" – tx to "try" to fall sleep, 15min anterograde amnesia, rebound insomnia

- <u>Chlor</u>diazepoxide "Librium" – tx delirium tremens
- Chlorazepate "Tranxene" – tx partial seizures
- Flumazenil – *tx acute BZ O/D*
- Charcoal – *tx chronic BZ O/D*

<u>2) Non-Benzodiazepines:</u>

- Buspirone "BuSpar" – tx generalized anxiety disorder, tx Alzheimer's anxiety
- Eszopiclone "Lunesta" – tx chronic insomnia
- Ramelteon „Rozerem" – melatonin agonist
- Zolpidem "Ambien" – tx insomnia, short-acting
- Zaleplon "Sonata" – tx insomnia, ultra short-acting

<u>3) Barbiturates:</u> (weak acids) bind Barbr → GABAr → ↑*duration* of Cl open *"Barbiduration"*
Tx: same as BZ (not for anxiety b/c too sedating)
Avoid: with alcohol (two acids) and with porphyria patients

- Pheno<u>barbital</u> – longest acting, tx generalized seizures in kids
- Primidone – parent compound of phenobarbital
- Pentobarbital – tx epilepsy
- **Seco**barbital – used on the street as a downer *"drug **seeking**"*
- <u>Thiopental</u> – anesthesia induction, distributed into tissues (i.e. skeletal muscle)
- Bicarbonate – #1 *tx Barb OD* (cause weak acid to not be absorbed → pee it out)
- Doxapram – #2 *tx Barb OD* (stimulates respiratory center → wakes you up)

<u>Personality Disorders:</u> don't think they have a problem
Cluster A: Weird => Thought Disorders
Paranoid – suspicious about everything, use projection
Schizotypal – "magical thinking", bizarre behavior
Schizoid – "recluse", don't want to fit in

Cluster B: Wild => Mood Disorders
Antisocial disorder – lie, steal, cheat, destroy property, impulsive w/o remorse
Conduct disorder – childhood (<18 y/o) antisocial disorder
Histrionic – theatrical, sexually provocative, use repression, hardest to obtain history
Borderline – "perpetual teenager", splitting (love/hate), self-mutilation, projection, acting out
Narcissistic – pompous, no empathy, need the "best" of everything, sensitive to criticism

Cluster C: Worried => Anxiety Disorders
Dependant – clingy, submissive, low self-confidence, need for reassurance, use regression
Obsessive-Compulsive – perfectionist, don't show feelings, detail-oriented, use isolation
Avoidant – socially withdrawn, afraid of rejection but want relationships

<u>Impulse Control Disorders:</u>
Kleptomania: steals for the fun of it
Pyromania: starts fires
Intermittent Explosive Disorder: loses self-control without adequate reason
Pathological Gambling: can't stop gambling, affects others
Trichotillomania: pull out their hair (eating hair "trichotillophagia" => bezoars)

<u>Somatoform Disorders:</u>
Somatization: think they have a *different* illness all the time, subconscious (Tx: frequent visits)
Hypochondriasis: thinks they have the *same* illness all the time, conscious (Tx: frequent visit)
Body dysmorphic disorder: imagined physical defect (Tx: Pimozide)
Conversion: motor and sensory neurologic manifestation of internal conflict, indifferent to disability (rule out MS and brain tumor)
Malingering: fake illness for ***monetary gain,*** avoids medical treatment
Factitious: fake illness to get *attention,* seeks medical treatment
Munchausen: Mom fakes illness to get attention *"2° factitious"*
Munchausen by proxy: Mom makes child ill for gain, move a lot
Psychogenic coma: unconscious + normal "COWS" test
Syncope: pale (conversion d/o is not pale)
Psychogenic non-epileptic seizures "pseudoseizures": normal EEG, may have tears

<u>Dissociative Disorders:</u>
Dissociative Amnesia: can't recall important facts
Dissociative Fugue: no past, create a new life, usually due to trauma, subconscious
Dissociative Identity Disorder "Multiple Personality Disorder": usually associated w/ incest
Depersonalization Disorder: "out of body" experiences, déjá vu

<u>Defense Mechanisms</u>: deal with *all* life issues this way
<u>Mature:</u>
Altruism – puts others before self
Humor – most mature type
Philanthropy – gives monetary gifts
Sublimation – substitute acceptable for unacceptable (Ex: boxer vs. fighting)
Suppression – consciously block memory

<u>Immature:</u>
Acting out – expression of impulse, "tantrums"
Countertransference – doctor identifies pt as family member, etc.
Dissociation: compartmentalization of memories (Ex: hooker by night, soccer mom by day)
Idealization – wait for "ideal spouse" while they are beating you up
Identification – modeling behavior after another person (Ex: abused becomes abuser)
Idolization – ascribe success to something else (Ex: rabbit's foot)
Imitation – replicates another's behavior
Passive Aggression – express anger indirectly

Projection – accusing others of your feelings
Regression – immature behavior
Social learning – model another's behavior
Splitting – black/white
Transference – pt identifies the doctor as a family member, etc.

Neurotic:
Compensation – doing something *different* to what you used to do
Counterphobic Behavior – mountain climb to overcome acrophobia
Denial – refuse to face the truth (MI yesterday, push-ups today)
Displacement – take action against someone b/c of something unrelated (Ex: 3 Stooges)
Fantasy – leave dead loved one's things exactly as they were
Intellectualization – act like a "know-it-all" to avoid feeling emotions
Isolation of Affect – isolate feelings to keep on functioning
Justification – make excuses after the fact
Rationalization – make excuses for all situations
Reaction Formation – act opposite of how you feel (smile when you feel sad)
Repression – subconsciously block memory
Undoing – doing the *exact opposite* of what you used to do to fix a wrong

Cognitive Therapies:
Biofeedback – use mind to control visceral changes
Cognitive therapy – replace negative thoughts with positive thoughts
Systematic desensitization – slowly introduce feared object
Flooding/Implosion – give overdose of feared object
Token economy – give "tokens" as a reward for good behavior
Aversive conditioning – tx paraphilia or addictions with unpleasant stimulus (electric shock)

Normal Sexual Cycle:
1) Excitement: penile/clitoral erection, vaginal lubrication ← Viagra ↑this; Propanolol ↓
2) Plateau: testes move up, secrete pre-ejaculate/contract outer vagina, facial flushing
3) Orgasm: release seminal fluid/ contract vagina and uterus ← SSRIs delay this
4) Resolution: refractory period (none in women)

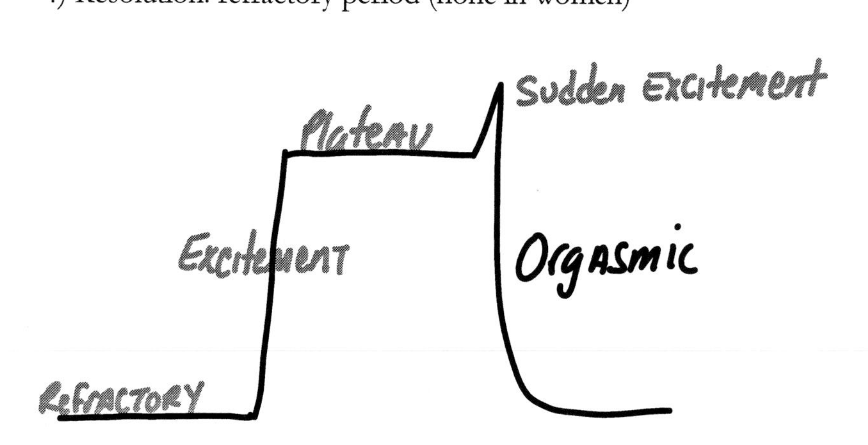

Desire Disorders: excitement phase
Hypoactive sexual desire: decreased interest (Tx: refer to sex therapist)
Sexual aversion: avoidance of sexual activity

Arousal Disorders: excitement/ plateau phase
Female Sexual Arousal Disorder: can't maintain vaginal lubrication
Male Erectile Dysfunction:
i) can't initiate - psych/neuro/endocrine problem
ii) can't fill – artery problem
iii) can't store – vein problem

Orgasmic Disorders: orgasm phase
Male/Female orgasmic disorder: inability to achieve orgasm (Tx: self stimulation)
Premature ejaculation: ejaculation before the man wants it to occur
Spinal cord injury: retrograde ejaculation, orgasmic dysfunction

Pain Disorders:
Dyspareunia: persistent pain with intercourse *(rule out endometriosis, vaginal dryness, infection)*
Vaginismus: painful spasm of vagina *(Tx: Kegel exercises, gradual dilatation)*

Sexual Paraphilias: get sexual pleasure from… (> 6 mo, interferes with their life)
Tx: Chemical castration if warranted (SSRIs, Flutamide, Estrogen)
Sadism: gives pain/humiliation
Masochism: receives pain/humiliation
Exhibitionism: exposure to others
Voyeurism: watching other people without their permission
Scatologia: phone sex
Frotteurism: rub penis against fully clothed women
Transvestite: dress up as opposite sex, no identity crisis
Transsexual: gender identity crisis "man trapped in a woman's body"
Gender Identity Disorder: strong desire to be the opposite sex
Fetish: objects (vibrators, dildos, shoes)
Pedophilia: children (watching child pornography)
Necrophilia: corpses
Bestiality: animals
Urophilia: urine
Coprophilia: feces
Klismaphilia: enemas

Medical/Psychiatric Associations:
HTN/CAD: Type A personality
Asthma: Dependant Personality Disorder
Migraines: Obsessive-Compulsive Personality Disorder

Suicide Risk Factors:

- Single
- Caucasian
- Male
- >45 y/o
- Poor health

Depression: Pancreatic CA, Multiple Myeloma, Huntington's
Mania: AIDS, Huntington's, SLE, MS

Hospitalization:

EtOH withdrawal: die after DT

Opioid withdrawal: painful, don't die

Biological Tests:

Amytal interview: diagnose conversion disorder
Na Lacate/CO_2: induces panic disorder
Hyperventilation: absence seizures

Neurotransmitters:

DA (substantia nigra): decreased in Parkinson's, increased in Schizophrenia
NE (locus ceruleus): increased in Panic, Anxiety
5-HT (raphe nucleus): decreased in Depression, Bulimia, Aggression, OCD
ACh (nucleus basalis of Meynert): decreased in Alzheimer's

Street Drugs:

Substance abuse: impairment of functioning
Substance dependence: tolerance, withdrawal

NOTES:

NOTES:

Surgery/ Trauma/ Anesthesia:

"Language is the means of getting an idea from my brain into yours without surgery."

–Mark Amidon

Pre-op Orders:

- MI history (last 6 mo): ischemia => arrhythmia – check EKG
- Heart failure – check EKG, CXR
- Uncontrolled seizures – check electrolytes
- Uncontrolled HTN – check BP
- Bleeding disorders – check PT/PTT/Platelets
- Occult infection: spreads to lungs and urine – check CBC, UA, CXR

Relative Contraindications:

- Low Mg → cripple all your kinases (need this for ATP to wake up)
- Hypothyroid – check TSH
- Hypoglycemia – check glucose
- Risk of arrhythmias – check EKG
- Abnormal K^+ levels – check electrolytes
- Anemia (↓O_2 carrying content already) – check CBC

NYHA Classifications:

1= Asymptomatic
2= Symptoms with moderate exertion
3= Symptoms with minimal exertion
4= Symptoms at rest

Informed Consent: must understand the following…

- What the procedure entails in layman's terms in the patient's language
- Risks/Benefits of procedure
- Alternatives to the procedure (including those not available at your hospital)
- Consequences of not having procedure done
- Voluntary/Competent
- Patient can change their mind at any time before receiving anesthetic
- Can't act on any intra-op unexpected findings unless it is life threatening

Pediatrics Without Parental Consent:

- Emergency situation
- Pregnancy/Abortion (most states)
- Oral contraceptives (>15y/o)
- STDs
- Psychiatric patient (ward of the state)
- Drug/EtOH detoxification
- Emancipated minor (<13 y/o living on own >1y/o, in military, married)

Confidentiality Exceptions:

- Harm to self/others
- Child/elder abuse
- Communicable diseases
- Knife/gunshot wounds

No Consent:

- Emergency situations if there is no medical personnelle at the scene
- You were trained to perform that procedure

Note: Intoxicated people have the right to refuse tx (state can enforce if they have dependants)

Jehovah's Witness:

- Can give children blood if it is an emergency (get court order if non-emergent)
- Can give children blood if one parent agrees
- Do not give blood to adults if they are dying and they have documentation

Pre-Operative Medications:

- Take normal meds with sip of water
- Stop anti-platelets 2 wks pre-op, stop NSAIDs 1 wk pre-op
- Stop warfarin, start heparin drip, stop heparin 1hr before surgery
- Stop smoking: 8 wks pre-op
- Diabetics: check glucose, start insulin drip in OR, don't give am dose/restart 2 days later
- Give stress-dose steroids if pt had Prednisone for at least 3 weeks over the past year
- Antibiotics: 1 dose 1 hour before surgery to decrease bacteremia...
- SBE prophylaxis: for heart dz except mitral prolapse (Tx: 3g Amoxicillin or Vanc)
- Pulm secretions Tx: Atropine or Glycopyrrolate
- NPO: at least 6 hrs to prevent aspiration
- Get advanced directives

Directives:

1) Advanced directives

- Living will: put in writing what you want done
- Durable power of attorney: choose someone to make that decision

2) Spouse

3) Closest kin (amount of time spent with patient)

4) Doctor (if no one else)

Fever after Surgery: *"LUIDA"*

- Day 1: **L**ungs "Wind" Pneumonia, Atelectasis (early skin infection is Clostridium or Strep)
- Day 3: **U**rine "Water" Urinary tract infection
- Day 5: **I**nfection "Wound"
- Day 7: **D**VT "Walk"
- Day 10: **A**bdominal Abscess
- **Drugs:** H_2 blockers, Pre-op antibiotics

Post-op Issues:

DVT Prophylaxis: Heparin (add Warfarin for hip/knee surgery)
Stress Ulcer Prophylaxis: H_2 blocker
Pneumonia Prophylaxis: Incentive spirometry, chest percussion therapy
Oliguria: do fluid challenge (dehydration gets better, renal failure doesn't)
Urinary retention Tx: Bethanechol, Carbachol
Adrenal insufficiency: hypotension, hypoglycemia, Δmental status (Tx: Cortisol)
Alcohol withdrawal: tremors, Δmental status (Tx: Chlordiazepoxide)
Delayed post-op healing: obesity (fat is poorly vascularized and holds stitches poorly)

Anesthesia Concepts:
Induction: the faster gas dissolves => smaller coefficient, quicker induction
Potency: how long it will keep patient knocked out => need coefficient to be high
Minimal Alveolar Concentration "MAC": concentration gradient to get into your brain

Actions of Anesthetics:
1) Anesthesia: total loss of *sensory* input (vision, taste, hearing, smell)
2) Analgesics: no pain transmission
3) ↓CNS activity: groggy, delirium, no gag reflex (don't extubate until brain function returns)
4) Muscle relaxation: atelectasis, aspiration pneumonia, can't poop/pee, DVT/PE
5) Sensitize myocardium to NE (except Isoflurane)

Topical Anesthetics:
1) Esters: *"one i"*
- Short $t_{1/2}$
- Water soluble: needs Epi to localize it (don't use Epi on fingers/toes/penis/nose)
- Metabolized by kidney
- Nephrotoxic
- Dependant on GFR (give 'em fluid to get rid of it)
- Small V_D
- ↑MAC (concentration gradient to get into your brain)
- Blood/Gas ratios
- Slow induction (has a hard time getting across lipid BBB)
- Fast recovery (wants to get out of there fast)
    - Cocaine – blocks reuptake of NE, don't need Epi
    - Benzocaine
    - Procaine – use if allergic to Lidocaine
    - Tetracaine
    - Novacaine

2) Amides: *"two i's"*
- Long $t_{1/2}$
- Fat soluble: still needs Epi to localize it

- If metabolized to NH_4 => ↑GABA
- Hepatotoxic
- Dependant on P450
- Large V_D
- ↓MAC (concentration gradient to get into your brain)
- Oil/Gas rations
- Fast induction
- Slow recovery
    - L**i**doca**in**e (fat soluble, distributes) with Epi (vasoconstricts to hold it)
    - Pr**i**loca**in**e
    - Mep**i**vaca**in**e
    - Bup**i**vaca**in**e

General Anesthesia:

1) Inhaled: *blocks Na channels from inside!* (ionized base inside neuron)

- Nitrous oxide: diffusion hypoxia
- **H**alothane: **h**epatitis, has bromine
- Iso<u>flurane</u>: used in heart surgery, has fluoride (inhibits enolase in glycolysis)

2) Intravenous:

- Barbiturates
- Benzodiazepines
- Opioids
- Propofol: lipid soluble, good for induction
- Ketamine: dissociative amnesia, stimulates heart, colorful nightmares on recovery
- Succinylcholine: malignant hyperthermia (Tx: Dantrolene)

Unresponsive: *"ABC"*

Airway: call, listen for noise:

1) Endotracheal Tube placement:

- For pO_2 <50 or pCO_2 >50
- ET Tube size: (8+age)/2
- Lift jaw, push tube straight in just past vocal cords
- Listen to breath sounds
- Blow up ET Tube balloon
- Tape in place

2) Cricothyroidotomy:

- Make cut in trachea right under "Adam's apple" and insert straw

Breathing: listen to chest

X-rays:

3: Coarctation

Boot: RVH

Banana: HCM

Egg: TGA

Snowman: TAPVR

Circulation: color and capillary refill (extremity temperature)

1) IV Access:

- Adults: femoral
- Kids: interosseus

2) IV Fluids:

- NS: 154 mEq/L NaCl → forces kidney to waste bicarb (good for hypovolemic alkalotic pts)
- LR: 130 mEq/L NaCl → causes lactate to convert into bicarb (good for bleeding acidic pts)

Wounds:

- **Clean cases:** groin vessels, open heart, prosthetics, bladder, eye (Tx: 1g Cefazolin or 1g Vanc)
- **Clean-contaminated:** abdomen, gallbladder (Tx: 1g Cefazolin or 1g Vanc)
- **Contaminated:** colorectal, appendectomy (Tx: Gentamycin + Amp or Vanc)

Pressure Ulcers: *worst position is HOB@45°*

Stage I: non-blanching erythema

Stage II: partial thickness, blisters

Stage III: full thickness, deep crater

Stage IV: muscle/bone exposed

Shock: state of hypoperfusion

No response to fluid => cardiac tamponade, internal bleed, or neurogenic shock

CVP: average volume in RA

CO: how much blood can come out of heart: 5L/min

TPR: resistance in arterioles

1) Hypovolemic Shock: pale, cold, tachycardic, bleeding (low preload)

↓CVP, ↓CO, ↑TPR, ↓PCWP (preload)

Tx: volume

2) Cardiogenic Shock: chest pain, peripheral edema, big liver, heart's not pumping

↑CVP, ↓CO, ↑TPR, ↑PCWP (preload)

Tx: Low HR: Dobutamine (β_1)

High HR: Low dose DA (D_2)

3) Neurologic Shock: warm, flushed, can't move or feel; ↓BP/↓pulse, vasodilates

↓CVP, ↓CO, ↓TPR

Tx: Phenylephrine (α_1 vasoconstrict)

4) Anaphylatic Shock: antigen exposure

↓CVP, ↑CO, ↓TPR

Tx: Phenylephrine (α_1 vasoconstrict)

5) Septic Shock: Gram (–) endotoxin dilates vessels (NO), FEVER
↓CVP, ↑CO (due to ↑HR), ↓TPR
Tx: Antibiotics

Signs of Shock:
Hypotension + Bradycardia => Neurologic or Cardiogenic shock
Hypotension +Tachycardia => Hypovolemic or Anaphylatic or Septic shock
Hypertension + Tachycardia => Severe pain
Hypertension + Bradycardia => ↑Intracranial pressure

Amyloid:

AA: **A**ny chronic disease
AB: **B**rain (Alzheimer's)
AB$_2$: β_2 microglobulinemia (renal failure)
AE: **E**ndocrine (medullary CA of thyroid)
AL: **L**ight chains (multiple myeloma)

Heart Failure: *BNP released due to change in ventricular filling pressures*
Left-side HF "systolic dysfunction": heart not squeezing, S_3, do TEE

- Fluid backup into lung: crackles, pleural edema, DOE, PND, SOB, orthopnea, ↓renal perfusion
- Ex: dilated cardiomyopathy, HTN, coarctation of aorta
- TxE:

↓Preload: Furosemide, Nitrates, ACE-I, Morphine
↓ Afterload: ACE-I
↑ Inotropy (contraction): Dopamine, Dobutamine, Digitalis

- Discharge meds:

1) Digitalis: blocks K of the Na/K pump to ↑contractility
2) ACE-I: dilates arteries/veins, blocks Aldo
3) β-blocker: ↓mortality
4) Aldosterone blocker: Spironolactone or Epleranone

Stab right chest:

- pneumothorax
- hemothorax

Stab left chest:

- pneumothorax
- hemothorax
- *tamponade*

Right-side HF "diastolic dysfunction": heart can't relax, S_4, do TTE

- Fluid backup into tissue: leg edema, JVD, hepatomegaly, pulm HTN, nocturia, ascites
- Ex: ventricular hypertrophy, amyloidosis, 1° pulm HTN
- Most common cause: Left-side HF
- Tx: Digoxin (slows AV node to ↑diastolic filling), CCB (↑preload via vasodilation)

Types of Hemorrhage:
Anterior cerebral artery => affects from waist down
Middle cerebral artery => affects from waist up
Posterior cerebral artery => affects eyes
Basilar artery => locked-in syndrome

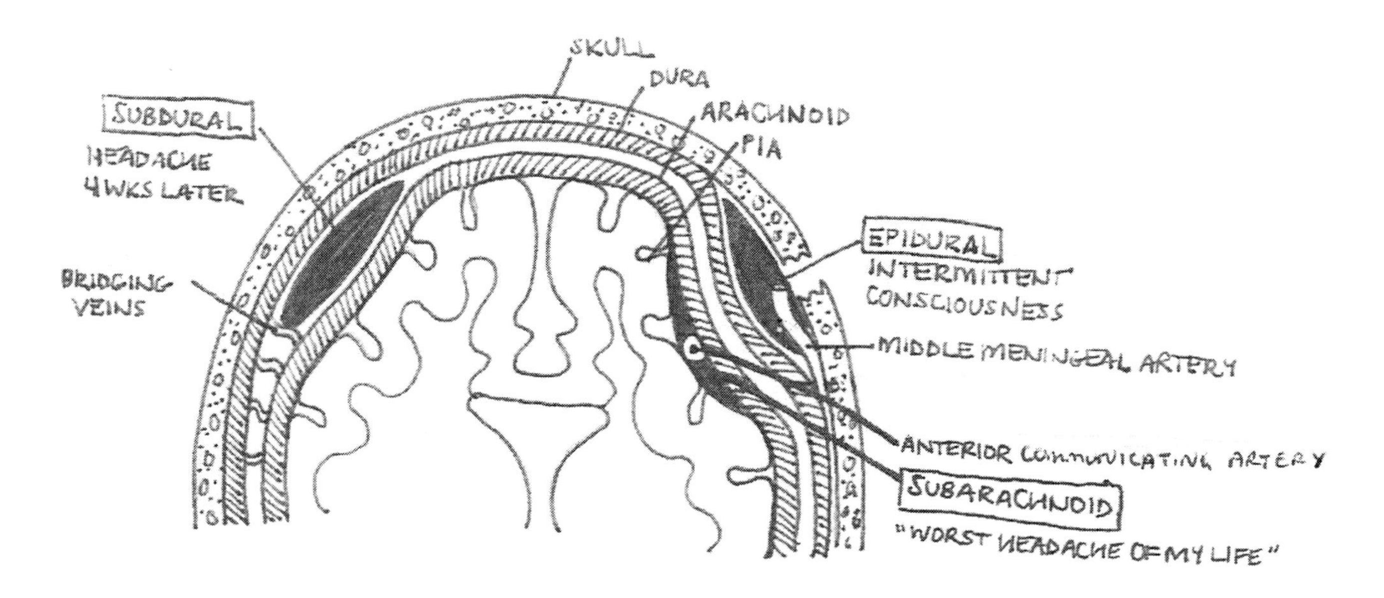

Basilar skull facture: trauma

- Bleed around eye "raccoon eyes"
- Bleed around mastoid "battle sign"
- Bleed behind eardrum "hemotympanum"
- Break cribiform plate => leak CSF into nose or ears
    - Complication: infection (meningitis)
    - Test: Put discharge on tissue, CSF will cause yellow-orange ring around blood
    - Tx: CT of head to see how big crack in skull is

<u>Circle of Willis:</u>

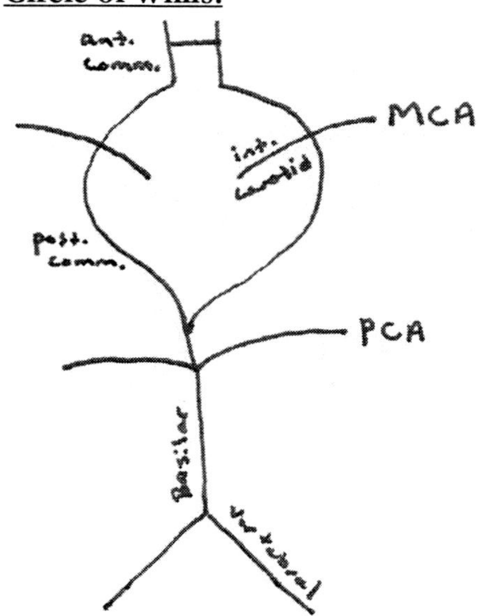

Epidural hematoma: temple trauma to MMA (baseball helmets cover MMA)

Sx: *intermittent consciousness*

CT: **e**lliptical "football" or "lenticular" shape

Tx: Immediate surgery

Intracerebral hemorrhage: blood vv. pop (at basal ganglia, internal capsule) due to HTN

Intraventricular hemorrhage: bleed into ventricles due to prematurity

- Frail medial wall of lateral ventricles
- O_2 releases free radials

Lacunar hemorrhage: lenticulostriate aa. pop (supply internal capsule) due to HTN

- "Lakes of blood"

Lobar hemorrhage: amyloid deposition in cerebral aa.

- Ex: ***Creutzfeldt-Jakob Disease*** (spongiform encephalopathy) => kills

Retinal hemorrhage: shaken baby syndrome

Subarachnoid hemorrhage: berry aneurysm emboli at anterior communicating artery

- **P**osterior communicating artery in **P**CKD
- Sx: "*worst headache of my life*" in occipital area, blood in CSF, loss of consciousness
- CT scan: white CSF, xanthochromia
- Tx: Nimodipine (for vasospasm), Phenytoin (prevent seizures)

Subdural hematoma: bridging veins

- Headache occurs 3-4 weeks after trauma
- Crescent/concave shape, crosses suture line
- Tx: Glucocorticoids

Xanthochromia:

- Subarachnoid hemorrhage
- HSV encephalitis

Subgaleal hemorrhage: trauma to scalp => prolonged jaundice

- Caput **s**uccedaneum: under **s**calp (edema crosses suture lines)
- Cephalohematoma: under bone (blood not cross suture lines)

Concussion: ↓vision, ↓multitasking ability, olfactory hallucinations, hard to find words

Brain Death:

- GCS = 3
- No brainstem reflexes (may have DTR)
- Nonreactive pupils
- No spontaneous breathing
- Heart may be beating
- Can remove life support
- Isolated limb movement = spinal reflex
- Freshly dead people stil have good reflexes

<u>**NOTES:**</u>

Reproductive/ Ob-Gyn:

"If women are supposed to be less rational and more emotional at the beginning of our menstrual cycle when the female hormone is at its lowest level, then why isn't it logical to say that, in those few days, women behave the most like the way men behave all month long?"

–Gloria Steinem

Nägele's Rule For Due Dates: based on 28-day cycle; add x days if cycle is x longer
+ 9 months from last menses → add 1 wk (inaccurate b/c not from ovulation date)

Normal Gestation: 40 wks
- Pre-term: <35 wks
- Post-term: >42 wks (Tx: Oxytocin)

Newborn Exam:
<2500g **SGA** (small for gestational age):
1) Symmetrical (baby problem): chromosomal abnormality or TORCH infection
2) Asymmetrical (mom problem): poor blood supply spares brain => small body, normal head

>4000g **LGA** (large for gestational age): DM or monochorionic twins
1st 24 hrs: Hypoglycemic (baby is used to hyperglycemia state)
2nd 24 hrs: Hypocalcemic (immature parathyroids)

Menstrual Cycle: *"FOL"*
Day 0: lining sloughs off, new follicles are starting
Day 1-10: Follicular/Proliferative stage – **high estrogen**
- proliferative endometrium, this phase is most variable

Day 10-14: Ovulatory stage – **high LH,** highest temp
- E_2 stim FSH (b/c pineal resets it), LH rises (LH higher b/c it was never inhibited)

Day 14-28: Luteal/Secretory stage – **high progesterone,** PMS
- secretory endometrium, deposition of lipids, proliferation of spiral aa

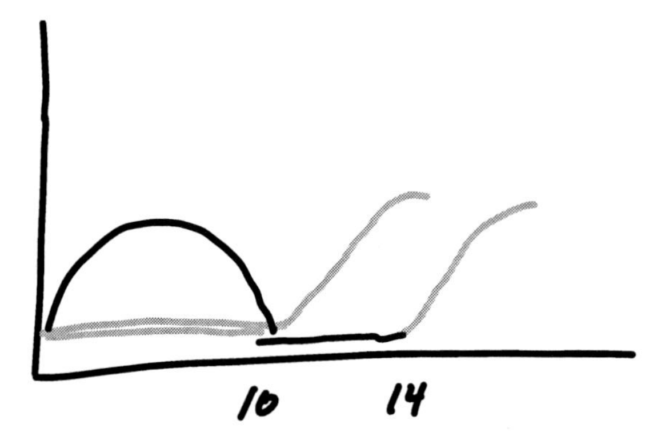

Eggs last 3 days in Fallopian tube (lose 1/month)
Sperm lasts 5 days in Fallopian tube (make 100 million/day)

Molar Pregnancy: increased risk of invasive mole and choriocarcinoma (2%)
- Increased placental villi => grape clusters
- US: *"snowstorm"* appearance
- Fundus rising more than normal, very high β-HCG, 1st trimester bleeding or HTN
  - **Incomplete Mole:** 2 sperm + 1 egg (69, XXY), has embryo parts

- **Complete Mole:** 2 sperm + no egg (46, XX – both paternal), no embryo
- Tx: D&C, use methotrexate to kill leftover tissue, follow β-HCG for 1 yr
  - Day 1: Methotrexate
  - Day 4: β-HCG should ↓15%
  - Day 7: Repeate Methotrexate if needed

<u>Oogenesis:</u>
5 mo Gestation: max # eggs
Birth: 1° oocyte (Girls have 400,000 eggs at birth), *Prophase I*
Ovulation: 2° oocyte w/ 1st polar body (8-10 eggs develop per month, 1 ovulate) *Metaphase II*
Fertilization: Ovum w/ 2nd polar body => zygote
Zona pellucida: ring around ovum

<u>Gametogenesis:</u> Mumps kills Leydig cells → no testosterone, pancreatitis, orchitis

FSH ↓
♀: LH → [Theca cells] → Progesterone/Androstenediol → [Granulosa cells] → Estrogen + Inhibin, Follicle Development
(Granulosa cells: ↑ LH receptors (at ovulation))

♂: LH → [Leydig cells] → Testosterone → [Sertoli cells] → Spermatogenesis, Blood-Testis barrier, Inhibin/MIF/ABP-T
(Leydig cells: ↑⊖ mumps; Sertoli cells: ↑ FSH)

<u>Theca externa</u> => <u>Progesterone</u> (think pregnancy state)
Uterus: Prog => proliferation of spiral aa, lipid and glycogen deposition(vacuoles)
Cervix: Prog => thickens mucus
Breast: Prog => glandular growth

<u>Progesterone → **Theca interna** => E_2</u>
Uterus: E_2 => proliferation of decidua functionalis
Cervix: E_2 => thins mucus
Breast: E_2 => ductal growth

<u>Forms of Estrogen:</u>
E_1: Estr**<u>one</u>** => menopause
E_2: Estra**<u>di</u>**ol => female
E_3: Es**<u>tri</u>**ol => pregnancy

Clomiphene: GnRH agonist => multiple births
Ovarian Hyperstimulation Syndrome: weight gain, big ovaries (can rupture)

Progesterone: ↑appetite, ↑acne, dilutional anemia, quiescent uterus, pica, violence, melasma (hyperpigmentation), dark areola, has Epo sequence in it, resorbs sodium/water (bloating), hypertension

Estrogen: muscle relaxant, constipation, ↑protein production, irritability, varicose veins, increase HDL, inhibits osteoclasts, hypotension

Amenorrhea Tests:

1) No uterus: Karyotype (Ex: testicular feminization)
2) No patent vagina: MRI (Ex: imperforate hymen/septal defects)
3) Provera challenge

Provera Challenge: 5mg x 5 days and stop *"5 for 5"*

- Bleeds => estrogen is normal (Ex: Normal/PCOS: ↑LH)
- Not bleed => estrogen is abnormal or may have scarring (Ex: Asherman's)
  - ↑FSH → ovarian failure (Ex: Turner's/Menopause)
  - ↓FSH → pituitary problem (Ex: Prolactinoma/Sheehan's/Adenoma)

Bicornate Uterus: can cerclage if previous losses

Polycystic Ovarian Syndrome: ↑LH, ↑testosterone, obese, hairy, acne, amenorrhea, DM

- ↑Cysts: no ovulation → no progesterone (↑endometrial CA) → can't inhibit LH
- Tx: Clomiphene (if pt wants to be pregnant), Metformin, Spironolactone (tx hirsuitism)

Estrogen Effects:

- Weight gain
- Breast tenderness
- Nausea, HA

Progesterone Effects:

- Acne
- Depression
- HTN

Birth Control:

I) Natural Planning: 75% effective
Periodic abstinence during ovulation

II) Barrier Contraceptives:
Male Condoms: protects against STDs (85% effective)
Diaphragms/Cervical caps: UTIs/cervicitis/TSS, leave in 8hr post-coitus (80% effective)
Spermicides: lasts 1 hr (70% effective)

III) Hormonal Contraceptives: *don't give to smokers or SLE pts (92% effective)*

- Estrogen – ↓FSH: can't select dominant follicle *"**F**ollicle **S**timulat**H**or"*
- Progestin – ↓LH: ↓ovulation, thick cervical mucus, inhospitable endometrium

Combination Pill:

- Norethindrone: most common form of progesterone
- Mestranol: most common form of estrogen

Nuvaring: Estrogen + Progesterone, placed in vagina (lasts 3 wk)
Ortho Evra Patch: Estrogen + Progesterone (lasts 1 wk)
Minipill: Progesterone only, does not block ovulation, *use w/ breast feeding*
Depo-Provera: long-acting Progesterone shot (lasts 3mo), *use in sickle cell/epilepsy/smokers*

IV) **Intrauterine Devices:** *use for smokers or bleeding disorders,* ↑PID/ectopic risk (99% effective)
Progesterone "Mirena" (lasts 5 yrs): releases 20 mcg/day Levonorgesterol (amenorrhea)
Copper T (lasts 10 yrs): inflammation kills sperm (regular menses)

V) **Sterilization:** 99% effective
Male: vasectomy – ejaculation still occurs, use caution in first 6 wks
Female: tubal ligation – ↑ectopic/pelvic pain (Electrocautery > Banding > Clipping)

VI) **Postcoital:**
Plan B: 2 pills progesterone, then 2 pills 12hr later, no prescription needed (<72hr)
Yuzpe: high dose estrogen/progesterone, repeat 12hr later (<72hr)
IUD insertion: Copper T (<5days)
RU486: blocks progesterone receptor to slough lining, abortifacient (<7wk)

Menopause test: 50 y/o → FSH >30 + 6mo amenorrhea
Hot Flash Tx: Clonidine

Hormone Replacement Therapy "HRT":
Post-menopause: E_2 + Progesterone (to protect from endometrial CA)
After 3yrs: Stop E_2 (b/c of clots), switch to Rolixifene
Post-hysterectomy: E_2 alone

HRT Risks:
↑Endometrial CA
↑Breast CA
↑DVT

HRT Benefits:
↓Osteoporosis
↑HDL

Spermatogenesis: LH → Leydig cells → Testosterone → Spermatogenesis
- Spermatogonia (near basal lamina) → 1° → 2° Spermatocyte → Spermatid → Spermatozoa
- Miss Sertoli protects the sperm and makes inhibin to inhibit FSH
- Every day guys make 20 million sperm per cc semen, have 4-5cc semen (high pH)
- Normal sperm: <50% abnormal forms, at least 50% motile after 5 days on glass slide
- Mature sperm are stored in epididymis until ejaculation, takes 76 days to mature
- Preparation for "the Battle up the Vagina" = 3 stops
- Seminal vesicle: food and clothes
- Pick up fructose as power bars
- Pick up semen for camouflage to protect against acidic environment
- Bulbourethral "Cowper's" glands: bicarb
- Secrete HCO3- to neutralize lactobacilli (get this at puberty) in vagina

- Prostate: Hyaluronidase, Acid phosphatase, Zn "The prostate HAZ it"
- Pick up hyaluronidase to eat through corona radiata to get to ovum
- Pick up acid phosphatase to eat through cervical mucus
- Pick up zinc to remove semen off sperm

Fertilization: 3 sperm reactions
Capacitation Rxn: Zn used to peel semen off
Acrosomal Rxn: sperm release enzymes to eat corona radiata
Crystalization Rxn: wall formed after 1 sperm enters (to prevent polyspermy)

Embryology:

Boy:	**Girl:**
Mesonephros = Wolffian duct (forms inside to out)	Paramesonephros = Müllerian duct (forms up to down)
Testes Vas deferens Epididymis Seminal vesicles	Ovaries Fallopian tubes Uterus (w/ cervix) Upper vagina
Labioscrotal swelling:	
Scrotum	Labia majora
Urogenital fold:	
Prostate Prostatic urethra Bulbourethral glands	Labia minora Lower vagina
Genital tubercle:	
Penis	Clitoris

Assisted Reproduction: used for inadequate spermatogenesis, 25% success rate
In Vitro Fertilization: fertilize eggs in lab → uterus
Gamete Intrafallopian Transfer: put eggs + sperm → Fallopian tube
Zygote Intrafallopian Transfer: put zygote → Fallopian tube

Premature ejaculation: stress => sympathetics, females have slow latent phase

- Tx: "squeeze technique" to cause sperm to turn around

Infertility: inability to conceive after 12 months of unprotected sex

- Risk Factors: Smoking, BMI >29
- 50% Male problem
- 30% Female problem (PID)

Impotence Tx:

No nitrates or a_1 blockers!

- Sildenafil "Viagra"
- Vardenafil "Levitra"
- Tadalafil "Cialis"

- 20% Other

Male Infertility Workup: *measure signal to see if organ is OK*
1) TSH → ↓GnRH → ↓LH
2) Testosterone
3) GnRH
4) PRL

Acute Bleeding Tx:

- Estrogen (high dose)
- Oral contraceptives (high dose)
- D&C

Female Infertility Workup:
1) Peritoneal: endometriosis, adhesions
2) Ovulatory: hypothalamus-pituitary-ovary
3) Tubo-uterine: fibroids, tubal occlusion
4) Cervix: abnormal mucus
5) Unexplained: anti-sperm Ab

Adrenal gland => Testosterone
Testicles => DHT (at puberty)

Rape: sexual contact without consent
- Erection/ejaculation does not have to occur, victim does not have to prove they resisted
- Sodomy: oral/anal penetration
- Statutory rape: <16 y/o or handicapped
- Spousal rape: sexual contact by husband w/o consent

Precocious Puberty:
- Pineal tumor
- McCune-Albright

Pathology: "male/female" is based on genotype *"It's what's on the inside that counts"*
Müllerian Inhibiting Factor "MIF" => internal genitalia
Testosterone => external genitalia
- Pseudohermaphrodite – external genitalia problem
- True Hermaphrodite – internal genitalia problem => has both sexes

Female Hermaphrodite: impossible b/c the default is female
Female Pseudohermaphrodite: XX with low 21-OHase => high testosterone
Male Hermaphrodite: XY with no MIF
Male Pseudohermaphrodite: XY that has low 17-OHase => low testosterone

Testicular Feminization = Androgen Insensitivity Syndrome (Ex: Jamie Lee Curtis)
- Bad DHT receptor → XY w/ blind pouch vagina
- No axillary/pubic hair
- Tx: Bilateral gonadectomy (to prevent cancer)

McCune-Albright: precocious sexual development
- Polyostotic fibrous dysplasia "whorls of CT"
- "Coast of Maine" pigmented skin macules
- Tx: Progesterone (anti-E_2) or Flutamide (anti-androgen)

Hirsutism: hairy
Virilization: man-like

Cryptorchidism: testes never descended => sterility after 15mo, seminomas

- Tx: GnRH to pull testes down at 6mo, Orchiopexy at 1 y/o (staple testes to scrotum)

Increased Estrogen States:

- Pregnancy
- Liver failure
- p450 inhibition

Bleeding Disorders: >80mL

Adenomyosis: growth of endometrium → myometrium, enlarged "*boggy*" uterus w/ cystic areas

- DES => increased risk in daughters
- Tx: Oral contraceptives, Leuprolide, Hysterectomy

Mom's DES → Daughter's:

- Adenomyosis → menorrhagia
- Clear cell CA of vagina
- Recurrent abortions

1° Amenorrhea: never had a period (at least 16 y/o)

Kallman's syndrome: no GnRH, anosmia (can't smell)

- Tx: pulsatile GnRH

Savage's syndrome: ovarian resistance to FSH/LH

Turner's syndrome (XO): high FSH, low E_2, ovarian dysgenesis

1° Amenorrhea Workup:

1) ↑PRL
2) ↓TSH
3) ↑E_2
4) ↑FSH → karyotype

2° Amenorrhea: stop menstruating (>6mo)

Pregnancy: #1 cause

Hypothalamic Dysfunction: exercise-induced, ↓FSH/↓LH/↓E_2

Sheehan syndrome: post-partum hemorrhage → pituitary hyperplasia infarcts → no lactation

- Tx: Synthroid, Cortisol

Asherman's syndrome: previous D&C → uterine scars

Chronic Pelvic Pain:

- Endometriosis until proven otherwise

Bleeding Most Commons:

Post-coital: cervical cancer

Post-coital pregnant: placenta previa

Post-menopause: endometrial cancer

Dysfunctional Uterine Bleeding:

- Diagnosis of exclusion, usually due to anovulation
- Tx: IV Estrogen (to stop bleeding), D&C

Dysmenorrhea: painful menstrual cramps (teenagers miss school/work)

- PG-F is responsible
- Tx: Naproxen or Oral contraceptives (need parental consent for this use)

Endometriosis: *painful* heavy menstrual bleeding, infertility, "*powder burns, chocolate cysts*"

- Blue → Brown → White lesions
- Retrograde menses through Fallopian tube into:

  - ovary
  - uterosacral ligament
  - cul-de-sac of Pouch of Douglas
- Ex: Hemoptysis with each period => endometriosis of nasopharynx or lung
- Dx: Laparoscopy => biopsy those 3 areas
- Tx: Oral contraceptives, Leuprolide, Danazol = potent androgen (↓FSH/ LH)
- Treatment-resistant Endometriosis: pelvic abscess or pelvic thrombophlebitis

Kleine regnung: scant bleeding at ovulation

Menorrhagia: heavy menstrual bleeding
Obesity: fat makes estrogen (Tx: weight loss, Oral contraceptives, Leuprolide = GnRH analog)
Fibroids = Leiomyoma: benign uterus SM tumor
- **Submucosal type** => *bleeding* (Tx: hysterectomy)
- **Subserosal type** => *pain* (Tx: myomectomy) *"serosal → think surface"*

Metrorrhagia: bleeding or spotting in between periods *"get on the Metro to travel in between places"*
- Postcoital bleeding: cervical CA (HPV 16,18,31,45)
- >40y/o: endometrial CA (Tx: TAH, BSO)

Mittelschmerz: pain at ovulation

Gynecologic Emergencies:
Adnexal Torsion: mass with no Doppler flow
Ruptured Ectopic: free fluid in cul-de-sac

Sexually Transmitted Diseases:

Disease:	Agent:	Diagnosis:	Tx:	Symptoms:
Syphilis	*Treponema pallidum* (spirochete)	FTA-ABS: + for life RPR=VRDL (>1:16) <u>False +:</u> SLE, HepC, Mono, recent vaccination	Doxycyline, Penicillin G (if pregnant)	1°: painless chancre (1-6 wks) 2°: rash, condyloma lata (6 wks) 3°: neuro, cardio, bone (6yrs)
Herpes I = oral II = genital	ds DNA virus	Tzanck test Culture PCR	Acyclovir (palliative)	1° = fulminate grouped vesicles on red base 2° = painful solitary lesion
HPV	ds DNA virus	Koilocytes Biopsy Serotype	Imiquimod	HPV 6/11: condyloma accuminata HPV 16/18: cervical cancer
Chlamydia	obligate intracellular parasite	Tissue culture Nucleic acid probe Elementary bodies	Azithromycin	Cervicitis: yellow pus Conjunctivitis 90% asymptomatic
Gonorrhea *"the drips; the clap"*	Gram – diplococcus	Gram stain Thayer Martin cult. Nucleic acid probe	Ceftriaxone or Ofloxacin (if pharyngeal)	Palmar pustules, Arthritis 90% male symptomatic 50% female symptomatic

<u>**Warts:**</u>

2° Syphilis: Condyloma **lat**a = **flat** fleshy warts, ulcerate

HPV 6/11: Condyloma accuminata = verrucous "cauliflower" warts, koilocytes

- Tx: Podophyllin, TCA, or freeze it off

<u>**Chancres:**</u> ↑*risk of HIV infection through breaks in skin barrier*

Herpes *(DNA virus)*: painful vesicles, ulcerate, intranuclear inclusion bodies

Syphilis *(T. pallidum):* painless chancre

Chancroid *(H. ducreyi)*: painful ulcer w/ irregular borders, necrotic center, Gram – rod, "school of fish" pattern (Tx: Ciprofloxacin)

Lymphogranuloma Venereum *(C. trachomatis)*: painless lymphadenopathy (Tx: Doxycycline)

Granuloma Inguinale *(C. granulomatosis):* oozing lesions, Donovan bodies (Tx: Doxycyline)

Balanitis: itchy penis papules (Tx: Fluconazole)

<u>**Epididymitis:**</u> *(Chlamydia)*

- Unilateral scrotal pain that is decreased by support

Congenital blindness: CMV

Neonatal blindness: Chlamydia

- Tx: Doxycycline (must treat partner)

Testicular Torsion:

- No cremasteric reflex, odd angle
- Pain not decreased with elevation
- Tx: immediate operation

Painless STDs:

- Syphilis
- Lymphogranuloma Venereum
- Granuloma Inguinale

Vaginitis: *"A Yoplait a day keeps the vaginitis away"*
Antibiotics: disrupt normal flora balance
Douche/ Sex: increases pH
Foreign Body: focus of infection → pediatric discharge

Candidiasis: cottage cheese discharge, itchy, pseudohyphae

- Tx: 150mg Fluconazole or 1% Terconazole cream

Trichomonas: green frothy discharge, strawberry cervix, flagellated, most common infxn s/p rape

- Tx: 2g Metronidazole (tx partners)

Bacterial (Gardnerella): *clear* discharge, *clue* cells, *fishy* odor, coccobacillus, KOH prep

- Tx: 2g Metronidazole or Clindamycin cream (if pregnant)
  "When all is Clear, you have no Clue that something Fishy is Going on"

Vulva Disorders:
Lichen simplex chronicus: raised itchy red lesions → white

- Tx: topical steroids

Lichen sclerosis: paperlike vulva, itchy white butterfly pattern from labia to anus, r/o vulva CA

- Tx: Clobetasol – steroid cream

Lichen planus: pruritic polygonal purple papules, oral lesions (white lacy streaks on buccal mucosa), assoc w/ Hep C

- Tx: Betamethasone cream

Prolapse Treatment:
Cystocele:

- Colpocleisis (obliterate vaginal canal)
- Sling (elevate urethra)

Uterine prolapse "procidentia":

- Hysterectomy

Prolapse Progression:
1st Degree: Upper 2/3 vagina
2nd Degree: Near introitus
3rd Degree: Outside vagina

TORCHS: non-bacterial fetal infections
Toxoplasma: multiple ring-enhancing lesions, cat urine, parietal lobe
Others
Rubella: cataracts, hearing loss, PDA, meningoencephalitis, "blueberry muffin" rash
CMV: spastic diplegia of legs, central calcifications, blind, most common congenital deafness
HSV-2: temporal lobe hemorrhagic encephalitis, need C/S prophylaxis

Syphilis (loves bone): Rhagade's (lip fissure), Hutchison's razor teeth, saber shin legs, mulberry molars

<u>Types of Estrogen:</u>

E_1: Estr<u>one</u> (made by fat)
E_2: Estra<u>di</u>ol (made by ovaries)
E_3: Es<u>tri</u>ol (made by placenta)

<u>Pregnancy:</u> gain ~40 lbs
Zygote: 2 cells; in Fallopian tube (ampulla)
Morula: 16 cells, enters uterus
Blastula: 256-512 cells, *blasts* into posterior wall of uterus, "decidual rxn" = "Arias Stella rxn"

- Identical twins: split into perfect halves (monochorionic)
- Fraternal twins: multiple eggs fertilized by different sperm (dizygotic)

Trophoblast: Baby => feeds off spiral aa. and lipid/glycogen
Cytotrophoblast: Mom => GnRH, CRH, TRH, Inhibin (similar to hypothalamus)
Synctiotrophoblast: Both mom and baby => HCG, HPL (similar to pituitary)

Placenta inside: baby's
Placenta outside: mom's (villi grow toward Mom) *"lacunar network"*

1 week after fertilization: Implantation
2 weeks after fertilization: β-HCG in urine

Nomenclature: GxPxxxx

- G = Gestation
- P = Pregnancies: **T**erm, **P**reterm, **A**bortions, **L**ive kids *"TPAL"*

Gestational age: from last menstrual period
Developmental age: from fertilization

(+) Pregnancy test: 8 days after conception
Teenage pregnancy risks: prematurity, perinatal mortality, cognitive disorders

Herpes Gestationalis: itchy umbilical rash (Tx: topical Triamcinolone)
Telogen Effluvium: post-partum hair loss

<u>Pregnant Endocrinology:</u>

Increased E:P ratio → labor
Decreased E:P ratio → post-partum, breast feeding

Estrogen: muscle relaxant, constipation, ↑protein production, irritability, varicose veins

- Low E_3: molar pregnancy, abortion, anencephaly, trisomies

Progesterone: ↑appetite, ↑acne, dilutional anemia, quiescent uterus, pica, violence

- Wk 1-10: Placenta → β-HCG → *corpus luteum* → progesterone
- Wk 11-40: Placenta → progesterone
- <10 ng/mL => non-viable

β-HCG: maintains corpus luteum, sensitizes TSHr => act hyperthyroid (to ↑BMR)

- Similar to LH
- Made by placenta
- Doubles every 2 days until 10 wks (when placenta is fully formed)
- False (+): proteinuria, UTI
- False (-): dilute urine
- Low: ectopic, abortion, anencephaly, Trisomy18 *"lower number"*
- High: twins, molar pregnancy, Trisomy 21 (small baby)

AFP: made by yolk sac/liver, regulates fetal intravascular volume

- Low: Trisomy 18, 21
- High: twins, openings: anencephaly, spina bifida (do US anatomy scan)

HPL: blocks insulin receptors => sugar stays high (baby's stocking up on stuff needed for the journey)

PRL: at birth, ↓progesterone => PRL acts unopposed
Inhibin: inhibits FSH => no menstruation
Oxytocin: milk ejection, baby ejection
Cortisol: decreases immune rejection of baby, lung development
Thyroid Hormones: ↑TBG = ↑bound T_4, normal free T_4 levels

Anemia of Pregnancy:

dilutional

- RBC rises 30%
- Volume rises 50%

Pregnant Physiology: *New onset A Fib → think hyperthyroid*
Brain: ↓migraines
Cardiology: ↑CO, vasodilation "glow", hypotension
Pulmonary: ↑TV, ↑V_{min}, ↓CO_2 → relative hyperventilation to remove CO_2 from baby to Mom
Renal: ↑GFR, ↓renin
GI: ↓PUD (↓H^+), constipation, GERD
Liver: ↑protein production → ↑TBG, hypercoagulable state
Endocrine: ↑total T_4, normal free T_4
Heme: ↑RBC (30%), ↑Vol (50%), telangiectasias, varicose veins
Musculoskeletal: Muscles relax
Skin: Striae, linea nigra, spider angiomas, acne, melasma hyperpigmentation "mask"
Immunology: ↓immunity => autoimmune dz gets better (except SLE), post-partum silent thyroiditis

Tests:
Fibronectin: ↑delivery by 2wk

Chadwick's sign: blue vagina (↑blood flow)
Hegar's sign: soft cervix
Fetal Heartones: >8wk (we can't detect it 'till 20wks)

Ultrasound: <20 wk

- Wk 16: can see if it's a girl/boy
- Wk 20: can feel baby move, baby fully formed/starting to grow
- Can see with vaginal US when β-HCG >1,500
- Can see with abdominal US when β-HCG > 6,000
- Crown-rump: most accurate length

Amniotic Fluid Index (AFI):

- <5: Oligohydramnios (cord compression or renal agenesis)
- >20: Polyhydramnios (DM or GI obstruction)

Fundal Height: *uterus should grow 1cm/wk (bad if >4 difference from gestational age)*

- Pubic symphysis = week 12
- Umbilicus = week 20 (can feel baby kicking here, *baby has been moving since week 8*)

Rupture of Membranes:

- Pool test: look for fluid in vagina
- Ferning: estrogen crystallizes on slide
- Nitrazine: amniotic fluid is alkaline (False +: blood, semen, infection)

Chorionic Villus Sampling: ↑risk of fetal limb defects, test for Trisomy 21 (week 10)
Amniocentesis (2% abortion rate): get fetal blood, test for Trisomies and NTD (week 16)
Percutaneous Umbilical Blood Sampling "cordocentesis": chromosomal analysis, transfusions

Radiation Levels: CXR < CT < Barium
36wk: no air travel

Biophysical Profile: *>8 = normal*
"Test the Baby, MAN!"

- Heart **T**ones
- **B**reathing
- **M**ovement: BPD, HC, AC, FL
- **A**mniotic Fluid Index
- **N**on-stress test (normal = "reactive")
    - Baby moves >32wk: ↑HR 15bpm for 15sec, need 2 over 20min

Multiple Gestations:

Twins: 37wk

Triplets: 33wk

Quadruplets: 29wk

Pelvis Types:

- Gynecoid●: circular → vaginal delivery
- Anthropoid: vertical oval → vaginal delivery

- Platypelloid: horizontal oval → C/S
- Android♥: heart → C/S

<u>Pregnant Nutrition:</u> *vitamins are for Mom's sake*

- Gain 1 lb/wk (2800 cal/day)
- Pica: urge to eat ice, clay, starch
- Folic acid: 0.4mg/day or 4mg/day if on Valproic Acid (avoid NTD)
- Ca: 1500 mg/day (bone growth, SM contraction, mitosis)
- Fe: 30 mg/day (erythropoiesis)
- Zn: 20 mg/day (sperm, taste buds, hair)
- Vegans: need protein, Vit B_{12}
- Fish: 1x/wk (no mercury: shark, swordfish, king mackerel)
- No Alcohol: inhibits nuclear division of rapidly dividing cells → mental retardation
- No Tobacco: IUGR, prematurity, SIDS
- No Cocaine, Heroin, Amphetamines
- Exercise: OK, stop if feel pain

Gestational DM: 50% will have DM type II later in life
Pregnant Glucose Control: avoid fetal anomalies

<u>Intercourse:</u> *woman on left side*

- 1st trimester: ↓ libido (β-HCG's fault)
- 2nd trimester: normal libido
- 3rd trimester: may cause uterine contractions (PGF in semen)
- 6 weeks post-partum: normal libido

<u>Doctor Visits:</u>

Month 1-7: Every month
Months 8-9: Every two weeks
> 9 months: Every week 'til delivery

<u>Prenatal Visits:</u>

1st Visit: Pap smear, GC/Rh screen
Week 16: US, Triple screen, Amniocentesis (if >35 y/o)
Week 26: DM screen, RhoGam
Week 36: GBS screen (Tx: Penicillin G during labor)

Mom Chickenpox Exposure: Ck VZV titer, give Ig within 72hr

Triple Screen:

	uE_3	B-HCG	AFP
Trisomy 18	↓	↓	↓
Trisomy 21	↓	↑	↓

<u>Requirements For Labor:</u>

1) Contractions: q5min

- Push Tx: Oxytocin

2) Cervix Dilation: >4cm is irreversible

- Incompetent Cervix Tx: bedrest, then cerclage (12-36wk)

3) Water Broke:

- Not Broken Tx: PGE_2 paper "Cervidil" if pt has good contractions
- PROM: water broke, no contractions

Leopold Maneuvers:

1) Feel fundus
2) Feel baby's back
3) Feel pelvic inlet
4) Feel baby's head

Labor Halt: Tocolytics

- **Hydration** (stop ADH=oxytocin)
- **Terbutaline** – decrease contractions
- **Ritodrine** – ↑edema

Stages of Labor: *Bishop's score >8 → delivery soon*

Come to hospital when contractions are q5 min for >1hr

Stage I: up to full dilation

1) Latent Phase (<20h): Contractions → 3cm cervical dilation
2) Active Phase (<12h): 4-10cm cervical dilation (1cm/hr)

- 7cm: place Hydromorphone Epidural

Labor Induction:

contraction → baby hypoTN

- **PGE_2** – not w/ asthma, prev C/S
- **Oxytocin**
- **Pitocin**

Monitor Baby HR: *120-160=normal*

1) Doppler
2) Scalp electrode

Monitor Uterus:

1) External: Tocodynamics: frequency/duration of contractions
2) Internal: Uterine pressure catheter: intensity of contractions

Stage II: full dilation → delivery (< 2h)

- Station 0: Baby above pelvic rim (most uteri are anteverted)
  1. Engage
  2. Descend
  3. Flex head
  4. Internal rotation
  5. Extend head
  6. Externally rotate
  7. Expulsion: LDA most common presentation
- Duration determined by the 3 P's:
  1. **P**ower of contraction
  2. **P**assenger size
  3. **P**elvis size

Vaginal Lacerations:

1st **Degree:** Skin

2nd **Degree:** Muscle

3rd **Degree:** Anus

4th **Degree:** Rectum

Stage III: delivery of placenta (due to PG-F)

- blood gush → cord lengthens → fundus firms

Braxton-Hicks contractions: irregular contractions w/ closed cervix

Stage IV: 6hrs post-partum

Post-partum hemorrhage: >500cc for vaginal delivery or >1L for C/S

1) Atony: uterus should come down to umbilicus by 24hr (Tx: bimanual massage/Pitocin)
2) Retained Placenta
3) Laceration

Episiotomy:

1°: through serosa
2°: through muscularis
3°: into perineal body

Anesthesia:

Vaginal: Epidural
Urgent C/S: Spinal anesthesia
Emergent C/S: General sedation
Forceps/Vacuum: Pudendal block

Baby Presentations: *C-section "C/S" unless noted*

Vertex: posterior fontanel (triangle shape) presents first, *normal*
Sinciput: anterior fontanel (diamond shape) presents first
Face: if mentum anterior → *forceps delivery*
Compound: arm or hand on head → *vaginal delivery*
Complete Breech: butt down, thighs and legs flexed
Frank Breech: butt down, thigh flexed, legs extended (pancake) → *deliver vaginally if >36 wks*
Footling Breech: butt down, thigh flexed, one toe is sticking out of cervical os
Double Footling Breech: two feet sticking out of cervical os
Transverse Lie: head is on one side, butt on the other → *try Leopold maneuver*
Shoulder Dystocia: head out, shoulder stuck → *try Leopold maneuver*

Shoulder Dystocia Tx:

1) Suprapubic pressure
2) McRobert's: move Mom's thighs to abdomen
3) Episiotomy
4) Wood's screw: try to rotate baby
5) Break clavicle
6) Zavanelli: push the head back in
7) C/S

Reasons for C/S:

- Arrest Disorder: adequate IUPC, no Δdilation/2hr, no Δdescent/1hr, no contractions q3min
- Fetal Bleeding: (+) Apt test
- Abruptio Placenta: painful bleeding
- Placenta Previa: painless bleeding
- Eclampsia
- Twins: unless vertex-vertex
- Breech: unless face/brow presents
- Herpes: active lesions within 2wk

C/S Types:

Classic Horizontal: must have C/S for all future pregnancies
Low Transverse: can try vaginal delivery for future pregnancies w/ Foley bulb

Fetal HR Monitor: Normal = 120-160 bpm (stress → ↓baby HR)
Fetal stress => sympathetics => meconium aspiration
Montevideo units: (increased pressure) x (contraction frequency/10 min) = 200/10min

Early Deceleration – normal, due to head compression
Late Deceleration – uteroplacental insufficiency b/c placenta can't provide O_2/nutrients (Tx: C/S)
Variable Deceleration – cord compression (Tx: O_2 + put Mom on side; amnioinfusion or C/S)

- Loss of blood => tachycardia (via carotid reflex)
- Loss of O_2 => bradycardia

Increased beat-to-beat variability – fetal hypoxemia (Tx: C/S)
Decreased beat-to-beat variability – acidemia (Tx: C/S)

Asymptomatic Bacteria: must treat it b/c ↑pyelonephritis risk → fetal mortality

Chronic HTN Tx:

1) Diet/Exercise
2) α-Me DOPA

Rule of 60's => immediate C/S

- HR below 60 bpm
- HR ↓ >60 bpm
- HR <100 for 60 sec

Most common cause of HTN:

1st trimester: Mom
2nd semester: Molar pregnancy
3rd trimester: Pre-eclampsia

Pre-Eclampsia: ischemia to placenta

- HTN (>140/85) + Proteinuria (>5g/day) + Edema (face/hands)
- If <20 wks, think hydatidiform mole
- Mom gets cerebral hemorrhage/ARDS → dies
- HELLP syndrome: *die of liver hematoma*

- **H**emolysis
- **E**levated **L**iver enzymes
- **L**ow **P**latelets

- Tx:

1) $MgSO_4$
2) Labetalol (if SBP >170)
3) Hydralazine
4) Delivery

Eclampsia: HTN + seizures (shut down pump, Na is locked in cell but K can leak out)

- Sx: Headache, blurry vision, epigastric pain
- If seizures >10min → baby will die
- Don't deliver baby while Mom is seizing
- Tx: 4g $MgSO_4$ (seizure prophylaxis) → C/S

Definitions:

Vernix = cheesy baby skin

Meconium = green baby poop

Lochia = endometrial slough

Mg Toxicity: *less likely to depolarize (Tx: Ca Gluconate)*

- 5-7 mEq/mL: ↓Uterine contractions, ΔEKG
- 8-12 mEq/mL: ↓DTR, flushing, slurred speech
- 12-24 mEq/mL: ↓Respiratory paralysis
- 25-30 mEq/mL: Cardiac arrest
-

Chorioamnionitis: fever, uterine tenderness, ↓fetal HR

- Tx: Immediate Ampicillin + Gentamycin, C/S

Amniotic fluid emboli: Mom just delivered baby and has SOB → PE, death (amniotic fluid → lungs)

Endometritis: post-partum uterine tenderness (due to E. coli)

- Tx: Clindamycin + Gentamycin at time of cord clamping

Normal Blood Loss:

Vaginal delivery: 500 mL

C-section: 1,000 mL

Ectopic Pregnancy: no nausea

- Ampulla of Fallopian tube
- Most common cause of 1st trimester maternal death → MEDICAL EMERGENCY!
- Tx: Methotrexate, Surgery

Group B Strep Tx: PCN or Ampicillin during labor (prevent meningitis)

Hepatitis B Tx: Hep B vaccine + Ig to neonate

Pseudocyesis: fake pregnancy w/ all the signs and symptoms, must consult Psych

Preterm Bleeding (<20 wks):

1st trimester abortions: Chromosomal abnormalities

2nd trimester abortions: Cervical incompetence, bicornate uterus
3rd trimester abortions: Placenta problems, incompetent cervix
Threatened abortion: cervix closed, baby intact (Tx: bed rest)
Inevitable abortion: cervix open, baby intact (Tx: cerclage = sew cervix shut until term)
Incomplete abortion: cervix open, fetal remnants (Tx: D&C to prevent placenta infxn)
Complete abortion: cervix open, no fetal remnants (Test: β-HCG)
Missed abortion: cervix closed, no fetal remnants, rule out ectopic w/ β-HCG (Tx: D&C)
Septic abortion: fever >100°F, malodorous discharge (x-ray to check for free air due to bacteria)

Term Bleeding (>36 wks):
Placenta Previa: vaginal bleeding, placenta covers cervical os; ruptures placental aa.
Vasa Previa: placenta aa. hang out of cervix
Placenta **A**ccreta: placenta **a**ttached to superficial lining (Tx: hysterectomy after delivery)
Placenta **In**creta: placenta **in**vades into myometrium (Tx: hysterectomy after delivery)
Placenta **Per**creta: placenta **per**forates through myometrium (Tx: hysterectomy after delivery)
Placenta Abruptio: severe pain, premature separation of placenta (Tx: FFP, emergency C/S)
Velamentous Cord Insertion: fetal vessels insert between chorion and amnion
Uterus Rupture: tearing sensation, halt of delivery (Tx: hysterectomy after delivery)

Post-Partum Bleeding: >500mL
Trauma: Repair
Retained Placenta: D&C
Uterine Atony (soft, boggy):

- Uterine massage
- Fluid
- Oxytocin

Baby vs. Mom Blood:
Apt test: detects HbF in vagina (brown = Mom's, pink = baby's)
Wright's stain: detects nucleated fetal RBC in Mom's vagina
Kleihauer-Betke test: detects %fetal blood in maternal circulation → RhoGam dosage

Pre-Term Babies: <36wk
1) $MgSO_4$ (tocolytic)
2) Amniotic Transfusion: flush NS continuously (max: 48hr)
3) "Window of Steroids": 28-32 wk
4) If <28wk → C/S to avoid IVH due to soft head

Post-Term Babies:
Check:

- Dating US (8-12wk) or LMP
- Landmarks
- Non-stress test: if non-reactive → do Biophysical Profile

- AFI: 5-20

Tx:

- Favorable cervix (effaced >70%, dilation >4cm): Oxytocin/Amniostomy
- Unfavorable Cervix: PGE_2 cervical ripening

Depression:

Maternity Blues: cry, irritable
Postpartum Depression: depression >2wk (Tx: antidepressant)
Postpartum Psychosis: hallucinations, suicidal, infanticidal (Tx: hospitalization)

Breast CA + Pregnancy:

Presentation: do mastectomy
2nd trimester: do chemo
Post-partum: do radiation

Gestational DM:

Goal: Glucose=60-100

A1: diet controlled

A2: insulin controlled

FDA Pregnancy Drug Categories:

A = safe in humans
B = safe in animals
C = unsafe in animals, no human studies
D = unsafe
X = very harmful

Pregnancy Medications:

Antibiotics *(cleared faster in pregnancy)*: Amoxicillin, Erythromycin
Anticoagulant: Heparin
Anti-convulsant: must continue (Phenobarbital is least teratogenic)
Anti-depressant: Fluoxetine "Paxil"
Anti-inflammatory: Acetaminophen
Asthma: Steroids, β-agonists, Theophylline, Isoproteranol, Albuterol
Bacteriuria: Nitrofurantoin
DM: Insulin
HIV: avoid Efavirenz
HTN (short-term): Hydralazine, Labetalol
HTN (long-term): α-Methyldopa
Hyperthyroid: PTU
Pyelonephritis: Ceftriaxone
TB: Rifampin/INH/Ethambutol
Toxoplasmosis: Pyrimethamine + Sulfadiazine (>2nd trimester)
Ulcerative Colitis: Sulfasalazine
Vaccines: OPV, DT, Hep B, Yellow fever, Influenza

Teratogens:

ACE-I => renal failure
Aminoglycosides => kill CN8

Amphetamines => transposition of great arteries
Carbamazepine => neural tube defects
Chloramphenicol => grey baby
Coumadin => CNS defects
DES => clear cell CA of vagina in daughter
EtOH => small stuff, mental retardation (fetal alcohol syndrome)
Fluoroquinolones => cartilage damage
Li => Ebstein's anomaly
NSAIDs => necrotizing enterocolitis
Retinoic acid => CNS defects
Sulfonamides => kernicterus
Tetracycline => ↓bone growth
Thalidomide => phocomelia (limb abnormalities)
Valproate => NTD

NOTES:

NOTES:

Pediatrics:

"The child soul is an ever-bubbling fountain in the world of humanity."

–Friedrich Froebel

Polycythemia of the Newborn:

- Hypoxia during labor → Epo (first breath will ↑pO_2 to stop Epo)

Transient Tachypnea of the Newborn:

- Compression of rib cage squeezes fluid out of lung
- C/S babies are SOB in first 3-4 hrs due to excess fluid in lungs
- >4hrs => consider septic until proven otherwise
- O_2 stimulates closing of fetal circulation => PDA, FO, DV, umbilical aa./vv.

Physiologic Jaundice:

- Spleen removes all excess blood cells → normal jaundice that peaks at day 3-4
- Leads to physiologic anemia at 2mo: Hb=6mg/dL => Epo turns back on
- 6mo: HbF → HbA

Hyperbilirubinemia:

Normal: <1mg/dL, unconjugated

Yellow eyes: >2mg/dL

Causes:

- G-6PD Deficiency
- Sepsis (bilirubin not delivered to liver)
- ABO incompatibility
- Hypothyroidism
- Breastfeeding (E_2 displaces bilirubin off albumin)

Tx:

- Phototherapy (20mg/dL): 270nm breaks down bilirubin to prevent kernicterus; toxic to retina
- Exchange transfusion: if bilirubin >25mg/dL

Premies:

- **Retinopathy of prematurity:** "retrolental fibroplasia" (↓vascularity of retina)
- **Necrotizing Enterocolitis:** GI vessels burst during feeding => ischemic bowel
- **Pneumatosis Intestinalis:** air in bowel wall => stop feedings, NG tube, TPN, tx for anaerobes

What do I do? The Baby's here!

- Suction nose/mouth before deliver 2nd shoulder (avoid meconium aspiration)
- Place under a warmer (shivering → burns sugar → hypoglycemia)
- Inject w/ Vit K – prevents bleeding (have no E coli yet to make Vit K)
- Silver nitrate in eyes – prevent Gonorrhea => "opthalmia neonatorum"
- Erythromycin in eyes – prevent Chlamydia => staccato cough, eosinophilia
- Footprints => identification (or Down's, Edward's, Patau's)
- Encourage breast feeding right away

APGAR test: (at 1 and 5 min), normal=7+, will vary if premature

A = Appearance, Color

- pink = 2
- acrocyanosis (hands and feet) = 1
- central cyanosis = 0

P = Pulse: normal=120-160

- >100 = 2
- 80-100 = 1
- <80 = 0

G = Grimace: stick something in it's nose

- strong = 2
- weak = 1
- no grimace = 0

A = Activity

- all extremities flexed = 2
- partially flexed =1
- flaccid = 0

R = Respiration

- strong = 2
- weak = 1
- none = 0

Temperature:

↑1°: ↑10bpm

Rectal > Oral > Axillary

VATER Syndrome:

- **V**ertebral abnormality
- **A**nal (imperforate)
- **TE** fistula
- **R**enal

Neural Tube Defect Risk Factors:

- Previous NTD
- Diabetes
- Valproic Acid: give 4mg Folate/day

Sepsis Workup:

- Blood cultures
- UA/ urine cultures
- CXR
- LP
- Tx: Cefotaxime

Eye Infections:

Day 1: Silver nitrate => clear discharge

Day 1-7: Gonorrhea => purulent discharge (Tx: Ceftriaxone)

> Day 7: Chlamydia (Tx: Erythromycin)

Breast Feeding:

- Bonding
- Post-partum Day 1-5: colostrum (protein) + IgA
- Post-partum Day 6: mature milk
- Immunity: Lysozyme (detergent), IgA secretion (less mucosal infxn), IL-6, memory T cells
- Breast milk has less Fe, Fluoride, Fat sol vitamins, but Fe is more absorbable (↑lactoferrin)
- Gentler proteins on GI mucosa → less bleeding
- Feed newborns q4hr
- If Mom is vegetarian, baby can get rickets
- Should stop by 1 y/o (when teeth come in)
- Avoid Breastfeeding:
  - HIV/HAART, TB, Varicella
  - Baby's Galactosemia

- Chemo/Cancer/Street Drugs/Li
- Sedatives/Stimulants
- Metronidazole: stop breastfeeding x 24hr

Pediatric Weight Gain: *need 100-120cal/kg/day = 36oz/day formula at birth*
Birth: average 5-7lbs
0-2wk: weight loss due to evaporation
6mo: double weight (gain 1 lb/mo)
1yr: triple weight (gain ¾ lb/mo)

Pediatric Nutrition: *if add new food, subtract 4oz formula*
4mo: rice cereal
6mo: fruits, yellow veggies
9mo: 2% milk, soft table foods (can get allergies/eczema if feed protein too early)
1yr: whole milk, table foods

Skin:
Milia: neonatal acne (due to progesterone in utero)
Nevus Flammeus: "stork bites" on back of neck, look like flames
Seborrheic Dermatitis: red rash w/oily skin and dry flaky hairline (Tx: baby oil/shampoo)
Hemangioma: flat blood vessels (Tx: steroid injection if growing rapidly or laser surgery if on face)
Mongolian spots: melanocytes on lower back (not child abuse), usually on Asian/Hispanics
Erythema toxicum: total body rash, eosinophils (benign), looks like flea bites
Port Wine stain: evaluate for Sturge-Weber
Acrochordon: skin tag
Vaginal bleeding: due to estrogen withdrawal from Mom
Sacral hair: spina bifida occulta

Choanale Atresia: blue w/ feed
Tetrology of Fallot: blue w/ cry

HEENT:
Microcephaly: due to Toxoplasmosis
Subgaleal hemorrhage => prolonged jaundice in newborns (trauma to scalp during birth)

- Caput succedaneum – under scalp (edema crosses suture lines)
- Cephalohematoma – under bone (blood not cross suture lines)

Epstein's pearl: white pearls on hard palate (will go away)
Persistent eye drainage since birth: blocked duct (Tx: gentle massage)
Wide sutures: Poor nutrition, hypothyroid, Down's
Midline cyst: Thyroglossal cyst (thyroid comes down from tongue)
Lateral cyst: Brachial cleft cyst
Multiple neck cysts: Cystic hygroma (Turner's)
Cleft lip: Medial nasal prominence did not fuse (reconstruct at 10wk old)
Cleft palate: Maxillary shelves not fuse => recurrent otitis media (feed w/ long curved nipple)
Saddle nose: Syphilis
Neonatal Herpes: Purulent crusted scalp blisters (do Tzanck smear)
No red reflex: Cataracts

U.S. Mental Retardation:
1) Alcohol
2) Fragile X
3) Down's

White reflex: Retinoblastoma

Kid Hernia Tx:

Inguinal: Operate

Umbilical: Observe

Hydrocele: Observe

Thorax:

Clavicle fracture (middle 1/3 L clavicle): asymmetric Moro reflex

Erb's palsy: C5-6 torn => Waiter's tip, 80% recovery (Dx: MRI)

Klumpke's: C8-T1 torn => Claw hand

Supernumerary nipple: extra nipples are always on vertical line

Abdomen/Flank:

Umbilical stump bleeding: Factor 13 deficiency

Delayed umbilical cord separation (6 wk): Leukocyte adhesion deficiency

Oomphalocele: intestines protrude out of umbilicus w/ peritoneal covering

Gastroschisis: abdominal wall defect, intestines protrude off-center

Wilm's tumor:

- Kidney tumor, aniridia, "triphasic" histology
- Hemihypertrophy: atrophy of leg on side of tumor (blood supply sucked away from leg)
- Tx: Dactinomycin

Neuroblastoma: adrenal medulla tumor, hypsarrhythmia, myoclonus, ↑VMA

Congenital Adrenal Hyperplasia:

- Females: ambiguous genetalia
- Males: premature penis development

Hip:

Congenital hip dislocation:

- **B**arlow maneuver (**B**end knee and hip, feel for clunk with middle finger) → do US
- **O**rtolani maneuver (Spread both hips **o**ut, feel for clunk)
- Tx: Triple diapers to lift hip or Spica cast (cast legs in frog position) for 3mo

Developmental Milestones: *These are estimates only…*

Age:	Language:	Fine Motor:	Gross Motor: head→neck→shoulder
Newborn	Cry	Moro, Grasp reflex	Moves head to side
2 mo	Smile, coo, goo	Swipes	Holds head up
4 mo	Listen, laugh	Reach, Parachute	Leans on arms
6 mo	Stranger anxiety (Should disappear by 2 years) Separation anxiety (Should disappear by 5 years)	Depth perception	Rolls back to belly Sits, scoots Sleeps all night
9 mo	Babble	Pincers, Waves	Crawl
12 mo	1 word "dada"	Babinksi disappears	Stand, first step
15 mo	5 words, tantrums	Feed themselves	Walk, Pick up and drop ball
18 mo	Short phrases	Scribble	Scoot upstairs, Throw ball
2 yrs	Short sentences	Draw circles	Walk upstairs, Run
3 yrs	Full sentences	Draw triangle: *"3 sides"* Sexual ID	Walk downstairs, Bend over, Kick ball, Ride tricycle
2-5 yrs	90% of language *most important time for parents to be involved*	Draw square: *"4 sides"* Draw star: *"5 sides"*	Ride bicycle with training wheels
6-12 yrs	Retains accent after this	Draw letters	Ride bicycle

Language abnormalities – due to hearing loss
Mental retardation – discrepancy between chronological age and mental age

- Approximate IQ = Mental/ Chronological x 100

Reflexes:

- Rooting: touch cheek → they turn toward it
- Moro: spread arms symmetrically when startled
- Babinski: toe extension when stroking feet
- Tonic-Labyrinthine: used to support self on a surface, "fencing" reflex
- Stepping reflex: "walking" when toes touch surface
- Parachute reflex: when held at stomach, hands will go out
- Diving reflex: when face is wet, flail arms/legs and close glottis

Sexual Development:

Puberty = pulsatile GnRH secretion
Females: Breasts "thelarche" → Growth "adrenarche" → Pubic hair "pubarche" → Menarche

Males: Testes grow → Penis grows → Growth → Pubic hair

Tanner Stage:	**Female:**	**Pubic Hair:**	**Male:**
I	nipple	None	Proportional
II	bud, growth spurt	downy, sparse	red scrotal skin
III	areola, menses	coarse, curled	growth spurt
IV	secondary mound	covers pubic symphysis	longer penis
V	separates from chest wall	spreads onto thighs	longest penis

Childhood Illnesses:

Colic: cry a lot after eating (not digesting well), will grow out of it

Fifth Disease (Parvo B-19): erythema infectiosum "slapped cheeks", red lacy body rash, arthritis in mom, aplastic anemia, keep them away from pregnant mothers for a few days, can go to school

Hand-Foot-Mouth Disease (Coxsackie A): mouth ulcers => won't eat or drink, palm/sole rash (Tx: observation)

Kawasaki's disease = Mucocutaneous Lymph Node Disease: autoimmune vasculitis
"CRASH"

- **C**onjunctivitis
- **R**ash (palm/sole)
- **A**neurysm (coronary artery) → MI in kids (Echo every year)
- **S**trawberry tongue (like scarlet fever)
- **H**ot (fever > 102°F for at least 3 days + cervical lymphadenopathy)
- Tx:
  - ASA (will decrease high platelets)
  - IV Ig (coats receptors so they don't see the body's autoimmune attack)
  - Flu vaccine (to avoid Reye's syndrome)
  - No live vaccines until 12mo after IV Ig is given

Measles = Rubeola (paramyxovirus): *multinucleated giant cells (lymphocytes)*
1) **C**ough, **C**oryza (thick rhinorrhea), **C**onjunctivitis
2) Koplik spot (white spot on buccal mucosa) – 24 hrs before rash
3) Morbilliform blotchy rash – spreads from head down "like a shower"
Complications: otitis media, demyelinating disease = *"Subacute Sclerosing Panencephilitis"*

Molluscum Contagiosum (pox virus): flesh-colored papules w/ central dimple, perianal STD in adults

Mumps (paramyxovirus): parotiditis, red Stenson's duct (behind 3rd molar), lemons hurt
Complications: pancreatitis, oophoritis/orchitis, meningoencephalomyelitis

Tx: Acetaminophen

Otitis Media (Strep pneumo): fluid in middle ear
Tx: Amoxicillin or tube placement if chronic

Pertussis: whooping cough, retinal hemorrhage, child stroke. #1 child preventable disease.

Pityriasis Rosea (HHV-7) herald patch → "C-mass tree" appearance on back, Tx: UV-B light

Reye's syndrome: uncouple ETC (↑temp → burns kid's livers)

- Causes: Pregnancy, Acetaminophen, ASA w/ influenza or varicella
- Sx: Fatty liver, brain edema, coma
- Tx=Supportive: Glucose, Albumin if fluid needed, FFP prn

Rosacea: malar rash, worse with alchohol

Rubella = German 3-day measles (togavirus): trunk rash, lymphadenopathy behind ear
Complications: bluberry muffin rash, cataracts, deafness, PDA (pre-maturity), extramedullary erythropoiesis. Don't give Rubella vaccine to pregnant women!

Sixth Disease (HHV-6): roseola, exanthema subitum (*fever disappears, then rash appears*)

Smallpox: on face, same stage of development, fever

Varicella = Chickenpox (VZV): on trunk, different stages of development
Complications: skin infections, varicella pneumonia => lethal
1) Red macule
2) Clear vesicle on red dot
3) Pustules
4) Scab => not infectious (most infectious: -2 rash +3)

Zoster: shingles, likes T4/V1, dermatome distribution, virus hiding in dorsal root

<u>When is it OK to Stay in Daycare/School?</u>

- RSV: if poop stays in diaper
- Herpes Zoster: after lesions crust
- Fifth Disease: when "slapped cheeks" appear
- Chickenpox: when all lesions are scabs

<u>Most Common Causes of Death:</u>
1st trimester: miscarriages (chromosomal abnormalities), TORCHS infections
1mo: prematurity
4-6 mo: SIDS => right ventricular hypertrophy on autopsy

6 mo-1 yr: Child abuse

- Multiple ecchymoses or cigarette burns
- Retinal hemorrhage (shaken baby syndrome), Epidural/Subdural hemorrhages
- Multiple fractures in different healing stages, Spiral fractures (twisted)
- *Rule out osteogenesis imperfecta, bleeding disorders, Fifth disease, Mongolian spots*

1-2 yrs:

- Drowning
- Accidental injestions

2-5 yrs: *Car seat until 4y/o or 40lbs (face car seat toward rear until 20 lbs or 1y/o)*

- #1: Car accidents
- #2: Accidental head injury (down stairs)

5-10 yrs:

- #1: Pedestrian injuries
- #2: Baseball => epidural hematomas (MMA)

10-19 yrs:

- Car accidents
- Homicide (Blacks/Hispanics)
- Suicide (White/Asian males/rich)

19-44 yrs:

- AIDS (got it as a teenager)
- Car accidents
- Homicide

>44 yrs:

- Heart disease
- Cancer
- Stroke
-

NOTES:

Carcinoma:

"Dream no small dreams for they have no power to move the hearts of men."

–Johann Wolfgang von Goethe

Cancer Terminology: ↑N/C ratio, *monoclonal origin* (Ames test: measures DNA damage)

Atrophy: decreased organ or tissue size (Ex: denervation)
Hypertrophy: increased cell size (Ex: bodybuilders)
Hyperplasia: increased cell number

Anaplasia: regress to mesenchymal origin (worst)
Metaplasia: change from one adult cell type to another (Ex: Barrett's esophagus)
Desmoplasia: cell wraps itself w/ dense fibrous tissue
Dysplasia "carcinoma in situ": lose contact inhibition (cells crawl on each other)
Neoplasm: new growth

Benign:

- well circumscribed
- freely mobile
- maintains capsule
- no metastasis
- obeys physiology (normal)
- hurts by compression (reason for surgery)
- slow growing

Malignant:

- not well circumscribed
- fixed/adherent
- no capsule
- can metastasize
- doesn't obey physiology (not normal)
- hurts by metastasis
- rapidly growing (outgrows blood supply → hunts for blood → secretes *angiogenin* and *endostatin*)

Cancer More Commons: MCC of death = infection (except cervical/endometrial = RF)
Men: 0-30 + >50y/o
Women: 30-50y/o (cervical, ovarian, breast)

Fast Killers:
Pancreatic cancer
Esophageal cancer

Cancer First Names: *most common cell type*
Glandular = Adeno-

Smooth muscle = Leiomyo- *"smooth liars"*
Skeletal muscle = Rhabdomyo-
Blood vessel = Hemangio-
Fat = Lipo-
Bone = Osteo-
Fibrous tissue = Fibro-

Cancer Last Names:
Tumor: -oma
Cancer: -carcinoma
Connective Tissue Cancer: -sarcoma *(the worst prognosis)*

Malignant tumors with wrong names:
- Lymphoma = sarcoma
- Melanoma = sarcoma
- Mesothelioma = sarcoma
- Seminoma = carcinoma
- Hepatoma = carcinoma
- Teratoma = carcinoma
- Retinoblastoma = carcinoma
- Neuroblastoma = carcinoma
- Nephroblastoma = carcinoma

Non-tumors with wrong names:
- Hamartoma = abnormal growth of normal tissue (Ex: keloid, polyp)
- Choristoma = normal tissue in wrong place (Ex: Meckel's, endometriosis)

Organs with most blood supply: *the most common CA here is metastasis! "BLAP"*
- **B**rain (grey-white jxn)
- **B**one (bone marrow)
- **L**ung
- **L**iver (portal vein, hepatic artery)
- **A**drenal gland (renal arteries)
- **P**ericardium (coronary arteries)
-

Cancers Caused by Bugs:
Clinorchis senesis: Gallbladder CA
Strep bovis: Colon CA
Clostridium septicum: Colon CA
Schistosoma hematobium: Bladder CA
Schistosoma mansoni: Liver CA

1-hit hypothesis:
1) Repressor damaged => childhood presentation
Ex: **Rb** mut => **E**wing's osteosarcoma, **Rb** (tumor suppressor); p53 mut => breast, ovary cancer

2-hit hypothesis: 99% of cancers
1) Initiator was damaged
2) Promoter came along

Staging:

AJC Classification:

- Stage I: localized
- Stage II: through cell wall
- Stage III: to lymph node
- Stage IV: metastasis

TNM Classification:

- T = tumor
- N = lymph node
- M = metastasis

Duke Classification:

- A: mucosa
- B1: muscularis
- B2: through serosa
- C: lymph nodes
- D: metastasis
-

Tumor Lysis Syndrome:

1) CLL
2) Non-Hodgkin's lymphoma

- ↑K (burst cells)
- ↑Phos (from inside cells)
- ↓Ca (Phos precipitated it)
- ↑Uric acid (from purines in cell)

Cancer:	**Most Common Metastasis:**
Lung	Brain
Skin	Brain
Thyroid	Lung
Liver	Lung
Kidney	Lung
Colon	Liver
Anus	Liver
Breast	Bone
Prostate	Lymphatics
Testicles	Retroperitoneum

BRAIN: cell type = astrocyte *(risk: radiation exposure, HIV)*

Intracranial tumor:

- **Meningioma:** psammoma bodies, whorling pattern, best prognosis

1° Brain tumor:

- **Astrocytoma:** Rosenthal fibers, #1 in kids w/ occipital HA (Tx: resection)
- **Glioma:** monocular blindness
- **Oligodendroglioma:** fried-egg appearance, nodular calcification *"Oligo Eggo"*
- **Schwannoma:** CN8 tumor, unilateral deafness
- **Ependymoma:** rosettes, in 4th ventricle, hydrocephalus

Brain Cancer:

- **Metastasis** (from lung, breast, skin; see at white-grey junction)

1° Brain Cancer:

- **Glioblastoma Multiforme:** pseudopalisading, necrosis, worst prognosis, butterfly

Psammoma bodies: calcified rocks due to tissue compression

Papillary (thyroid)
Serous (ovary)
Adenocarcinoma (ovary)
Meningioma
Mesothelioma

Spinal Cord Syndromes:

Spinal stenosis: back hurts when walk upstairs, relieved w/ leaning forward
Cauda equina syndrome: *"saddle anesthesia":* can't feel butt, thighs, perineum
Conus medularis (S4-S5): perianal anesthesia

Seizures:

Partial: conscious, talking (Tx: Carbamazepine)

- **Simple partial:** sensory disturbances
- **Complex partial:** incontinence, post-ictal confusion, déjà vu, lip smacking

Generalized: unconscious

- **Tonic clonic "Grand Mal":** most common (Tx: Valproate)
  - **Tonic:** freeze (contract)
  - **Clonic:** jerk
- **Absence "petit mal":** blank stare, *EEG: 3-Hz spike and wave* (Tx: Ethosuximide)
- **Status Epilepticus:** continuous seizures >20min
  - Tx: Lorazepam → Phenytoin → Phenobarbital → Midazolam/Intubation
- **Febrile:** kids, *occurs during rise in temp, not peak temp* (Tx: Acetaminophen)
- **Temporal:** have hallucinations before seizure (Tx: Carbamazepine)
- **Benign Rolandic:** kid screams in night, then eyes flutter and sleeps, outgrow
- **Myotonic:** increased muscle tone, arms fling forward (Tx: Valproic acid)

- **Infantile spasm:** looks like mytonic, but occurs <1 y/o
- **Atonic:** lose all body tone, drop to ground like a wet noodle, then writhe like a snake
- **Lennox-Gasteau:** hundreds of seizures every day (Tx: EEG, then lobectomy)

Epilepsy: recurrent idiopathic seizures, contact DMV (no driver's license)

ER Seizure Tx: 20mg/kg Phenytoin → 0.1mg/kg Lorazepam (at 30min) → Pentobarbital coma

PITUITARY:

Tumor: Adenoma

- **Functional:** Prolactinoma (5%)
- **Non-functional:** Chromophobes (95%)

Cancer: Adenocarcinoma (rare)

Mediastinum Tumors:

Anterior: Thymoma

Middle: Pericardial

Posterior: Neuro tumors

PINEAL:

Tumor: Adenoma = "Pinealoma"

- Loss of upward gaze: "Parinaud syndrome"
- Loss of circadian rhythms: Precocious puberty
- Tx: Leuprolide

Cancer: Adenocarcinoma (rare)

POSTERIOR FOSSA: early morning vomiting

Medulloblastoma: pseudorosettes, compresses brain, #1 post fossa tumor, #2 in kids

Craniopharyngioma: motor oil biopsy, tooth enamel, Rathke's pouch, bitemporal hemianopsia

POSTERIOR MEDIASTINUM:

Tumor: Neuroma

Cancer: Neuroblastoma

NEURAL CREST:

Neuroblastoma: adrenal medulla tumor in kids; highest spontaneous regression rate

- abdominal mass
- dancing eyes "hypsarrhythmia"
- dancing feet "myoclonus"

Pheochromocytoma: adrenal medulla tumor in adults => sx come/go

- 5 P's: Palpitation, Perspiration, Pallor, Pressure (HTN), Pain (HA)
- Rule of 10's: 10% are malignant/calcify, familial, found in kids, bilateral, extra-adrenal
- Tests: Phentolamine (short acting α_{ns}-blocker => drop in BP), urinary VMA
- Tx: Phenoxybenzamine (irreversible α_{ns} blocker), then remove tumor
- Tx: No β-blockers! (unopposed α stimulation causes rapid ↑BP)

THYMUS:

Tumor: Thymoma (found in all autoimmune diseases except Grave's)
Cancer: Adenocarcinoma (rare)

THYROID: cell type = glandular
Mass: Thyroglossal cyst (do US, then FNA)
Tumor: Follicular adenoma
Cancer:

- **P**apillary carcinoma – **p**sammoma bodies, "Orphan Annie eyes", previous neck radiation
- Follicular carcinoma – 2nd most common
- Anaplastic carcinoma – 0% survival rate at 5yrs
- Medullary carcinoma – amyloid deposition, ↑calcitonin, parafollicular C-cells

Thyroid Nodule Management:

- ↑TSH: check free T_4
- Normal TSH: do FNA
- ↓TSH: do iodine scan (cold → malignant)

PARATHYROID:
Tumor: Adenoma (MCC of isolated hypercalcemia in adults)
Cancer: Adenocarcinoma (rare)

MEN Tumors:
MEN I = "Wermer's": **P**ancreas, **P**ituitary, **P**arathyroid adenoma (high gastrin) *"PPP"*
MEN II = "Sipple's": Pheo, Medullary thyroid cancer, PTH (high calcitonin)
MEN III = "MEN IIb": Pheo, Medullary thyroid cancer, Oral/GI neuromas (high calcitonin)

PARAFOLLICULAR:
Tumor: Adenoma (rare)
Cancer: Medullary carcinoma of thyroid (high calcitonin)

ENDOCARDIUM:
Tumor: Atrial myxoma (female who faints, diastolic plop, ball-like valve)
Cancer: Angiosarcoma (rare)

MYOCARDIUM: skeletal muscle
Tumor: Rhabdomyoma
Cancer: Rhabdomyosarcoma (age < 3)

PERICARDIUM:
Tumor: Fibroma
Cancer: Metastasis

LUNG: cell type = glandular
Mass (kids): Hamartoma
Mass (adults): RL lobe Granuloma
Tumor: Adenoma
Cancer: Metastasis

Lung Cancer Risk Factors:

5yr changes, 15yr regress to normal risk

1) Smoking
2) Radon
3) 2nd-hand smoke (sidestream worse)
4) Pneumoconioses (except anthracosis)

Lung Most Commons: major cell type = glandular
Lung mass (kids): hamartoma
Lung mass (adults): RL lobe granuloma
Lung tumor: adenoma
Lung cancer: metastasis

Optimal Pre-op FEV_1: >800mL (can estimate with V/Q scan)

Central cancers:
1) Squamous cell: (smoking) => PTH => ↑Ca^{2+}, ↓Pi (Tx: resect + chemo)
- **Pancoast tumor:** apical tumor squishes inferior cervical ganglion → Horner's
- Dx: apical lordotic CXR

2) Small cell "oat cell": at carina, most malignant (Tx: Etoposide + Cisplatin)
- ACTH (Cushing's dz), ADH (SIADH), Lambert-Eaton syndrome
- **SVC Obstruction:** facial edema, dyspnea, JVD (Tx: emergent radiation)

Peripheral cancers:
1) Bronchogenic: asbestos, PTHrp, SVC syndrome, ↑Ca^{2+}
2) Bronchoalveolar: pneumoconioses (not smoking), looks like pneumonia
3) Bronchial Carcinoid: 5-HIAA => flushing, wheezing, diarrhea
- from pancreas/ileum → liver (or stays in appendix) "fungating mass"
- only symptomatic in lung/ileum
- Tx: Odansetron, then remove

4) Large cell: large stuff, gynecomastia
5) Adenocarcinoma: non-smokers, looks like pneumonia, mets to brain, ↑CEA
Smoking Benefits: *weird, huh? Of course, you may still die of squamous cell lung CA…*
- ↓Fibroids (↓estrogen release from ovary)
- ↓Brochoalveolar CA or Adenocarcinoma
- ↓Ulcerative Colitis flare-ups
- ↓Hypersensitivity Pneumonitis (↓Ab response)

PLEURAL CAVITY:
Tumor: Mesothelioma
Cancer: Mesothelioma (Fe coating: *"ferruginous body"* → MP take to pleural cavity)

NASOPHARYNX:
Mass: Nasal polyp (rule out Cystic Fibrosis and asa-sensitive Asthma)

Tumor: Fibroma
Cancer: Nasopharyngeal cancer (Chinese women, EBV)

<u>ORAL CAVITY:</u>
Tumor: Fibroma
Cancer: Squamous cell carcinoma (inside the mouth)

<u>SALIVARY GLAND:</u> *Sjogren's increases risk*
Tumor: Pleiomorphic "mixed" adenoma, *recurrent after excision*
Bilateral Tumor: Warthin's
Cancer: Mucoepidermoid adenocarcinoma

<u>ESOPHAGUS:</u> cell type = SM
Tumor: Leiomyoma
Upper 2/3 Cancer: Squamous cell carcinoma (floor of mouth, tip of tongue, lower lip)
Lower 1/3 Cancer: Adenocarcinoma (odynophagia)

<u>STOMACH:</u> cell type = SM
Tumor: Leiomyoma
Cancer: Adenocarcinoma => early satiety (due to stomach distension)

- Signet ring cells
- Linitis plastica → leather bottle appearance
- Virchow's nodes → metastasis to left superclavicular lymph nodes
- Krukenberg tumor → seeding of ovaries

<u>APPENDIX:</u>
Tumor: Leiomyoma
Cancer: Carcinoid

<u>COLON:</u>
Tumor: Leiomyoma
Cancer: Adenocarcinoma

- Apple-core lesion on x-ray
- Pencil-thin stool
- Risk factors: Low fiber/High fat diet, Polyps, Ulcerative Colitis

<u>ANUS:</u>
Tumor: Fibroma
Cancer: Squamous cell CA (Tx: Chemo-rad)

<u>LIVER:</u> cell type = glandular
Mass: Cyst
Tumor: Hepatic Adenoma

- Risk Factors: Oral contraceptives, injected steroids, AVM

Cancer: Metastasis

1° Cancer: Hepatocellular Adenocarcinoma

- Risk Factors: Smoking, EtOH, HepB/C/D, Aniline dye, Benzene, Aflatoxin

Vinyl Chloride (CCl_4) Exposure: Angiosarcoma of the liver

<u>GALL BLADDER:</u> cell type = SM

Tumor: Leiomyoma

Cancer: Adenocarcinoma (steatorrhea)

Gall bladder CA: thin wall calcified "porcelin" gallbladder w/ no stones → do CT

Ampulla of Vater CA: ↑Alk phos + hematochezia → do duodenum endoscopy

<u>BILIARY TRACT:</u>

Mass: Choledochal cyst

Tumor: Cholangioma

Cancer: Cholangiosarcoma

- Risk Factors: Chronic scarring (1° Biliary Cirrhosis, 1° Sclerosing Cholangitis)

<u>PANCREAS:</u> cell type = glandular

Mass: Cyst

Tumor: Adenoma

- **Insulinoma:** ↑insulin, ↑C-peptide
- **Gastrinoma "Zollinger-Ellison syndrome"**
    - Drug-resistant GI ulcers
    - ↑Stomach acid → kills pancreas
    - Enzymes → steatorrhea
    - Test: Secretin injection => ↑gastrin (>1,000pg/mL)

<u>Sporadic Zollinger-Ellison Syndrome:</u>

- Small multifocal tumors
- Found in duodenum
- High survival rate (>90%)

<u>MEN1-Associated Zollinger-Ellison Syndrome:</u>

- Large solitary tumor
- Found in pancreas
- Low survival rate (<70%)
- High metastatic potential
- **Glucagonoma:** ↑glucose, rash (weight loss, diarrhea, necrotizing dermatitis → mets to liver)
- **SSoma:** steatorrhea
- **VIPoma:** watery diarrhea (irritates bowel wall)
- **Carcinoid syndrome:** diarrhea, flushing, wheezing

Cancer: Adenocarcinoma (head of pancreas => bile duct obst) => *painless jaundice*

- Acute depression; 90% die within 6 mo of diagnosis
- Risk factors: Smoking, chronic pancreatitis, DM, old AA males
- Tx: Whipple surgery = pancreaticoduodenectomy

Trousseau's sign: migratory thrombophlebitis (clot that moves to other leg → red streaks)

- Males: pancreatic cancer
- Females: ovarian cancer

OVARY: *cell type = glandular* (CA-125 = marker to follow dz progression)
Mass: Follicular Cyst (decreased pain as cycle goes on)
Tumor: Serous Cystadenoma (cyst picked up glandular tissue and fluid)
Cancer: Serous Cystadenocarcinoma (widening abdominal girth)
Others:

- **Sister Mary Joseph Nodule:** ovarian CA spread to umbilicus
- **Mucinous Cystadenocarcinoma:**
  - Pseudomyxoma peritonei => bubble bursts
  - Krukenberg tumor = mets from stomach
- **Brenner tumor:** benign "nests" of transitional cells
- **Fibroma:**
  - **Meig's syndrome** => pleural effusion, ovarian fibroma, ascites
- **Granulosa-Theca cell tumor:** ↑estrogen, precocious puberty, monitor progression w/Inhibin
- **Sertoli-Leydig cell tumor:** ↑testosterone, masculinization
- **Teratoma:**
  - Ectoderm: hair, skin, teeth
  - Endoderm: thyroid tissue in ovary = "struma-ovarii"
- **Seminoma** = Dysgerminoma: ↑placenta alk phos, ↑LDH
- **Choriocarcinoma:** ↑β-HCG
- **Yolk sac cancer** = Endodermal sinus tumor: ↑AFP, α_1AT
- **Paraneoplastic syndrome:** cerebellum degeneration

Most Common Gyn Cancers:

1) Endometrial CA – due to E_2
2) Ovarian CA = deadliest
3) Cervical CA = most curable

↑Risk Endometrial CA:

- Estrogen
- DM
- HTN

UTERUS: cell type = smooth muscle
Note: *Endometrium = lining of uterus*
Mass: Uterine polyp (pass blood clots)
Tumor: Leiomyoma => menorrhagia
Cancer: Adenocarcinoma (>40y/o w/ metrorrhagia, Tx=SERMs)

Ovary Anatomy:

R ovary → IVC
L ovary → renal vein → IVC

CERVIX: clinical diagnosis, mets to vagina, associated w/ HPV
Mass: Wart
Tumor: Fibroma
Cancer: Squamous cell carcinoma (post-coital bleeding, die of pyelonephritis)

- Risk Factors: HPV, sex at young age, smoking, oral contraceptives
- Vaccine "Gardasil": ↓cervical CA by 70%, ↓genital warts by 90%

- Tx: Hysterectomy

VAGINA: lower cell type = skeletal muscle (upper vagina = mucosa), clinical diagnosis
Mass: Wart
Tumor: Fibroma
Cancer: bloody discharge

- **Rhabdomyosarcoma** (lower vagina, spreads via femoral nodes)
- **Squamous cell carcinoma:** growing downwards (upper vagina, spreads via iliac nodes)
- **Clear cell carcinoma:** DES exposure, starts as white ridge
- **Sarcoma Botyroides:** ball of grapes

VULVA: *pruritic*
Tumor: Bartholin's cyst

- Tx: Ward catheter, Marsupialization if recurrent (sew it open)

Cancer: Squamous cell carcinoma (Risk: Paget's disease of the vulva)

Female Cancer Risk Overview:
Estrogen: ↑endometrial CA
Progesterone: ↓endometrial CA/↓ovary CA
IUD: ↓endometrial CA
Tubal Ligation: ↓ovary CA
Lesbians: ↓cervical CA (HPV 16/18 lives on penis)
Smoking: ↑cervical CA
HIV: Highest risk of cervical CA
Tamoxifen: Highest risk of endometrial CA
Nulliparity: Highest risk of ovary CA

Breast CA Risk Factors:
#1: Having breasts (large)
#2: Female
#3: Age (old)
#4: Unopposed estrogen
#5: Previous breast cancer
#6: Fam Hx
#7: Nulliparity

KIDNEY: cell type = glandular
Mass: Renal cyst
Tumor: Adenoma
Cancer:
Kids = **Wilm's tumor:** abdominal mass → painless hematuria

- Aniridia → no iris
- Hemihypertrophy → 1 leg thinner than the other

Adults = **Renal cell Adenocarcinoma:** hematuria, flank pain, palpable mass

- Usually in upper pole of kidney
- "Cannonball mets" to lungs, 20% risk of contralateral kindey cancer
- Risk factors: Smoking, VHL, Tuberous sclerosis, Aflatoxin, Analine dye, Cyclophosphamide
- Tests:
    - High Epo (polycythemia)
    - Angiogenin => very vascular => erodes into retroperitoneal fat (check arteriogram)
    - US: blood

- Ateriogram: shows where to cut
- IVP: check kidney function
- US: mass
- Abdominal x-ray: dead cells calcify
- CT: metastasis

- Tx: partial nephrectomy (if in upper pole of kidney only) OR total nephrectomy

ADRENAL GLAND:

Mass: Cyst
Tumor: Adenoma
Cancer: Adenocarcinoma

BLADDER: cell type = transitional cell
Mass: Diverticulum (pocket) => infxn or stones
Tumor: Leiomyoma
Cancer:
Transitional cell CA: painless hematuria, multiple primaries

- Risk Factors: Smoking, Benzene, Aflatoxin, Cyclophosphamide
- Tx: Cystectomy + radiation (cecum can be used to make new bladder)

Squamous cell bladder CA: **S**chistosoma Haematobium

PROSTATE:

Tumor: Benign Prostatic Hypertrophy: (begins in center)

- **α_1-blockers:**
  - Terazosin
  - Doxazosin
  - Tamsulosin: loosen sphincters => tx HTN and BPH
- **5-α reductase inhibitors:**
  - Finasteride: can decrease prostate size
  - Dutasteride
  - Ketoconazole
- **TURP:** transurethral resection of prostate (worry about pudendal nerve impotence)

Cancer: Adenocarcinoma (begins in periphery - posterior lobe → osteoblastic CA, PSA >10)

- **GnRH analog:**
  - Leuprolide
- **DHT receptor inhibitors:**
  - Flutamide: hepatotoxic, medical castration
  - Spironolactone

> **Flutamide:** medical castration
>
> **Leuprolide:** medical menopause

TESTICLES:

Note: *Testes Mass → US (no biopsy) → Orchiectomy w/ inguinal incision*
Mass (newborns): Hydrocele
Mass (children): Hematoma

Mass (older adults): Varicocele
Tumor: Adenoma
Cancer: Seminoma or Yolk sac cancer (1y/o)

SKIN:
Mass: Skin tags or Hemangiomas (Tx: observe or steroid injection)
Tumor: Dermatofibroma
Cancer: Basal cell carcinoma
Malignancy: Squamous cell carcinoma (ulcerates)

UV-A => **A**ging
UV-B => **B**urns + Cancer
SPF-15: blocks 94% UV-B

Skin Cancer:
"ABCD": Asymmetry, irregular **B**orders, variegated **C**olor, *4mm **D**iameter*

- Clark level: level of invasion in dermis
- Breslow's classification: tumor thickness (from epidermis down), determines tx/prognosis

I) Malignant Melanoma: male back or female leg (*most prognostic factor: Sentinal LN)*
1) Superficial spreading melanoma: most common, flat brown
2) Nodular: worst prognosis, black, dome-shape, radial growth
3) Lentigo maligna melanoma: elderly pts, fair-skin, vertical growth
4) **A**cral lentigous: **AA**, Hispanic, on nailbeds
5) Japenese: occurs in skin, eyes, brain

BCC → mets
SCC → kills

II) Squamous Cell Carcinoma: flat flaky stuff on lower face, keratin pearls, ulceration

- Precursor: Actinic keratosis (red scaly plaque) ← arsenic poisoning
- Most common skin cancer in organ recepients

1) Bowen's disease: SCC in situ on uncircumcised penis dorsum (HPV 16,18)
2) Verrucous carcinoma: SCC wart on anus

III) Basal Cell Carcinoma: pearly papules on upper face, most common skin CA, good prognosis
1) Nodular: pearly white + teleangiectasias
2) Superficial: red scaly plaques w/ white border, looks like cigarette paper
3) Pigmented: brown w/ white border
4) Sclerosing: yellow waxy plaques

Cancer-Associated Rashes:
Acanthosis nigricans: dark leathery neck/axilla => Lung or GI CA, DM, obesity
Amyloidosis: non-specific red rash, protein deposition => big kidneys and liver

- Stains Congo red, Apple-green birefringence (due to β-sheets)

1°: Congenital (AD)
2°: Acquired – any chronic inflammatory disease: CA, endocarditis
Heliotropic rash: purple eyelid and knuckles "Gottron's sign" => Dermatomyositis
Kaposi's sarcoma: red-purple plaques (HHV-8)
Paget's disease: ulcer or rash around nipple => Breast CA

BONE:

Epidural Spinal Cord Compression: back pain radiating to front

- Tx: immediate Dexamethasone → MRI spine

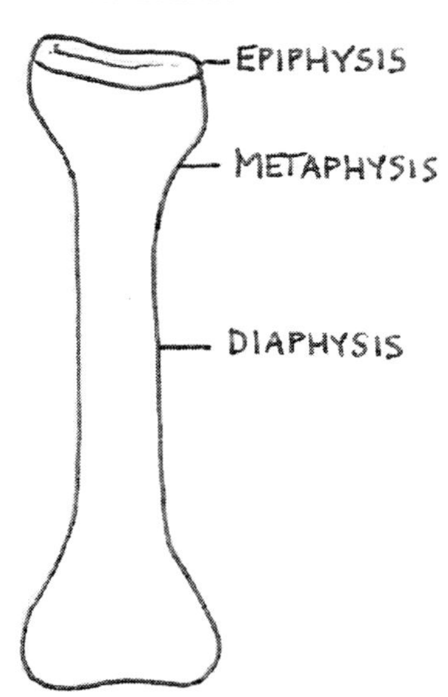

1°: Multiple myeloma: multiple lesions, >50y/o, flat bones and spine, IgG, κ light chain

Bone Metastasis:

Female: from breast

Male: from lung

Epiphysis (cartilage): cell type = chondroblasts

Tumor: Chondroma

Cancer:

- Chrondrosarcoma: cartilage tumor that destroys bone
- Giant cell tumor: moth-eaten area, "soap bubble" x-ray

Metaphysis (bone): cell type = osteoblasts

Tumor: Osteoma

1° Cancer: Osteosarcoma (Codman's triangle, "sunburst" x-ray - cancer explodes out of bone)

Diaphysis (bone): cell type = osteocytes

Tumor: Osteoma

Cancer: Metastasis

1° Cancer: painful at night

- Ewing's osteosarcoma t(11,22): onion skinning, round blue cells, pseudorosettes
- Plasmacytoma: 1 lytic lesion
- Multiple myeloma: multiple lytic lesions, IgG, Kappa chain, Roleaux

BREAST: *cell type = glandular,* lytic and blastic bone lesions, BRCA-1,2

Histologic grade is the most important prognostic factor…

Soft Mass: Cyst (Tx: US, FNA)

Firm Mass: Microcalcification (Tx: Surgery, follow with chemo or radiation if post-menopausal)

Tumor <25 y/o: Fibroadenoma, E_2-dependant, painless, mobile (pain: 1st two weeks of cycle)

Tumor >25 y/o: Fibrocystic change, Progesterone-dependant (pain: 2wk before menses)

Cancers:

- **Intraductal papilloma:** nipple bleeding
- **Intraductal adenocarcinoma:** dimple, only one with osteoblastic metastasis
- **Lobular carcinoma:** cells line up single file, contralateral primary
  - **LCIS Tx:** Tamoxifen/Observe
- **Ductal carcinoma:** worst prognosis, linear calcifications, 50% become invasive
  - **DCIS Tx:** Lumpectomy/LND/Radiation
- **Inflammatory carcinoma:** infiltrates lymphatics, "peau d'orange", the most malignant
- **Comedocarcinoma:** multiple focal areas of necrosis, "blackheads"
- **Cystosarcoma phylloides:** "exploding mushroom", rubbery, moveable, sarcoma
- **Paget's:** rash and ulcer around nipple

Osteoblastic CA:

- Intraductal Breast CA
- Prostate CA

Female Preventative Medicine:

Breast Exam: 13 y/o → annually (+ self exam at end of each menses)
Mammogram: 40 y/o → annually
Pap Smear: 21 y/o (or 3 years after sexually active) → annually until 65 y/o (or 60 if all neg Paps)
Repeat every 3-5yrs after three negative results (unless CIN/DES/HIV)

Low-grade findings: Infection/CIN I/ASCUS (repeat Pap in 6mo)
1) Tests: Syphilis RPR/HepB/HIV/GC culture
2) Tx for GC

High grade findings: CIN II/III (repeat Pap in 3mo)
1) Colposcopy: acetic acid turns dysplasia white (if pregnant, just repeat q trimester)
2) LEEP: electrocautery excision of tissue (can lead to cervical stenosis)
3) Cone biopsy: laser excision; not if inflamm/pregnant (can lead to incompetent cervix)
4) Hysterectomy: if invades cervix (removes cervix; no more Paps, only q yr pelvic exams)
5) Radiation: if invades beyond cervix

SERMs: Selective Estrogen Response Modulators => *hot flashes, vaginal dryness*
Tamoxifen:

- Good: Anti-E_2 at breast
- Bad: Pro-E_2 at endometrium (cancer)

Raloxifene:

- Good: Anti-E_2 at breast, Pro-E_2 at bone (protects)
- Bad: Pro-E_2 at liver (DVT)

Aromatase Inhibitors: *myalgias, arthralgias (Androgens –(aromatase)→ Estrogens)*

- **Anastrozole**
- **Letrozole**
- **Exemestane:** irreversible

Her2 (+) Tx: Herceptin "Trastuzumab"

<u>Chemo Exceptions:</u>

- Post-menopause without invasion
- Lymph node negative
- ER+/PR+
- <1cm size

<u>Tamoxifen Exceptions:</u>

- Pre-menopause
- ER-/PR-

<u>Cancer Antigens</u> (non-specific and non-sensitive): use to follow progression of CA
CA-125: Ovarian
CA-19: Pancreatic
S-100: Melanoma
BRCA: Breast
PSA: Prostate
CEA: Colon, Pancreatic
AFP: Liver, Yolk sac
Rb: Ewing's sarcoma, Retinoblastoma
Ret: Medullary thyroid cancer
Ras: Colon
bcl-2: Follicular lymphoma
c-myc: Burkitt's lymphoma
L-myc: Small cell lung carcinoma *"L for lung"*
N-myc: Neuroblastoma => pseudorosettes *"N for neuro"*
Bombesin: Neuroblastoma
β-HCG: Choriocarcinoma
5-HIAA: Carcinoid syndrome
p53: The Guardian of the Genome => stops cell cycle, starts apoptosis
Ki-67: Neoplasm growth rate

<u>Chromosome Abnormalities:</u>
#3: von Hippel Lindau
#4: Huntington's *"hunt = 4 letters"*
#5: Cri-du-chat (cat-like cry), HNPCC *"HNPCC = 5 letters"*
#7: Cystic Fibrosis
#11: β-thalassemia, Wilm's *"Wil, the 11y/o boy w/ an abdominal mass"*
#13: Rb
#15: Prader Willi *"Willi, the 15 y/o hungry boy"*
#16: α-thalassemia, APKD
#17: NF I, p53, BRCA

#19: Myotonic Dystrophy
#22: NF II, DiGeorge

Translocations:

t(9,22) = CML (bcr-abl gene)
t(14,18) = Follicular lymphoma (bcl-2 gene)
t(8,14) = Burkitt's lymphoma (c-myc gene)
t(15,17) = AML M3
t(11,14) = Mantle cell lymphoma
t(11,22) = Ewing's sarcoma

HLA Types:

A3, A6: Hemochromatosis
B5: Behcet's
B13: Psoriasis
B27: Psoriasis w/ arthritis, Ankylosing spondylitis, Reiter's
DR2: Goodpasture's, MS
DR3: Celiac sprue
DR4: Pemphigus vulgaris
DR5: Pernicious anemia

Malignant Lymph Nodes:

#1: Supraclavicular

#2: Epitrochlear – above elbow

#3: Inguinal

Multiple Genes Dz:

DR2,3: SLE

DR3,4: DM

DR4,5: RA

NOTES:

NOTES:

Amino Acids:

"Organic chemistry is the chemistry of carbon compounds. Biochemistry is the study of carbon compounds that crawl."

–Mike Adam

Sources of Energy:

- Proteins
- Fats
- Sugars: easiest to mobilize

Amino Group *(-NH_3^+): Charged, more soluble, attracted to water*
Acid Group (-COOH): Uncharged, crosses membranes, reflection coefficient closer to zero, more bioavailable

Buffers:

- Intracellular: Protein (RBC's=hemoglobin)
- Extracellular: Bicarbonate
- Best Buffer: Histidine (pK = 6.0, sidechain closest to pH=7.4)

Amino acids: have an NH_3^+ and a COOH
Proline: imino acid → think kinks, twists, bends, turns (Ex: GI, hair, blood vessels)

- Dissociate => take H+ away!
- Soluble => charged or polar, water soluble (Sulphur, Oxygen, Nitrogen)
- Bioavailable => neutral, can cross membranes, fat soluble

Dissociation curve:

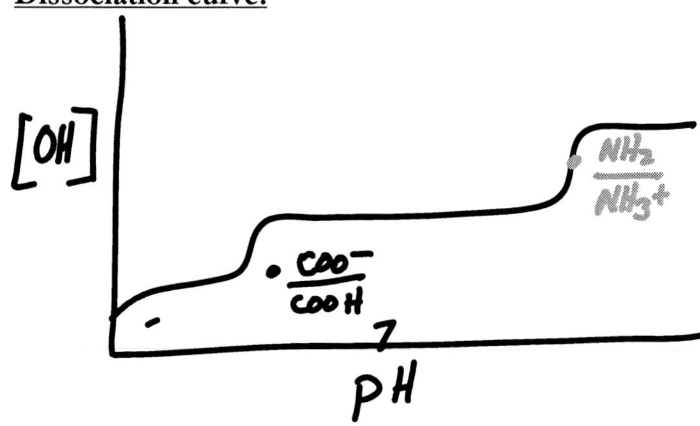

Acids: pKa<7, likes to give up hydrogen ions, dissociates early
Base: pKa>7, likes to accept hydrogen ions, dissociates late

- Strong acid: pKa 1-3
- Weak acid: pKa 4-7
- Weak base: pKa 7-9
- Strong base: pKa 10-14
- Note: pKa of 4-9 can be a weak acid or weak base
- When acids dissociate: gains negative charge, more soluble, less bioavailable
- When bases dissociate: loses a positive charge, less soluble, more bioavailable

<u>Henderson-Hasselbach Equation:</u>
pH= pK - log Base/Acid

<u>When pH= pK:</u>
- Half acid and half base
- 50% dissociated
- 50% soluble
- 50% bioavailable
- 50% reflection coefficient close to 1
- 50% reflection coefficient close to 0
- 50% crosses BBB (bioavailable)
- 50% metabolized by liver (bioavailable)

Dissociation Relationship For ACIDS: *reverse for bases*

pH:	Dissociated = Water Soluble:	Bioavailable = Fat Soluble:
pK + 2	99%	1%
pK + 1	90%	10%
pK	50% *"Best Buffer"*	50%
pK - 1	10%	90%
pK - 2	1%	99%

<u>Rules to keep molecules neutral (bioavailable):</u>
- To absorb more acid, place in a stronger acid
- To absorb more base, place in a stronger base
- Common Acids: NH4Cl, ASA, Barbiturates, Myoglobin, Juice, Coke
- Common Bases: Bicarb, Amphetamines, Baking soda, Activated charcoal, Milk

<u>Anti-Inflammatory Drugs:</u>
1) Acetaminophen: works at hypothalamus to decrease temp
- Liver toxicity: microsteatosis (highest levels in first 4 hours)

- Tx: N-acetylcystine (absorbs free radicals)

2) NSAIDs: COX inhibitors => *GI bleeding, interstitial nephritis*
COX-1: on gut tissue
COX-2: on endothelial tissue

- Indomethacin – most potent => RTA, closes PDA, tx gout
- Phenylbutazone – 2nd most potent
- Ketorolac – morphine-like, use w/ drug addicts
- **Bac**lofen – tx **back** pain (GABA-ergic effect)
- **C**yclobenzaprine – anti-**ch**olinergic effects (hot, dry skin)
- Naproxen – tx dysmenorrhea
- Ibuprofen – OTC
- asa – acetylates COX-1/2 irreversibly => most effective platelet inhibitor
    - ↓urate excretion → don't use w/ gout pts
    - displaces T_4 from TBG → don't use w/ hyperthyroid pts
    - induces asthma sx → don't use w/ asthma pts (↑LT by blocking COX)

Low LAP: CML/ PNH
High LAP: Leukemoid rxn

3) COX-2 Inhibitors: decreased GI irritation

- Celecoxib – has sulfur
- Rofecoxib – taken off market due to stroke increase
- Valdecoxib – no sulfur
- Meloxicam

Platelet Haptens:

- Quinidine
- ASA (>81mg)
- Heparin

4) Sulfur Side Effects: *don't give to pregnant women*

- Anaphylaxis
- Rash
- Interstitial nephritis
- Hemolytic anemia
- Met-Hemoglobinemia
- Displaces stuff off of albumin
- Kidney stones

ASA Toxicity:
1) Resp Alkalosis (↑RR)
2) Metab Acidosis (uncouples OP)
3) Mixed Acidosis (↑GABA)

Street Drugs:
Uppers: dilated pupils

- **PCP** "angel dust": blocks NMDAr => **p**owerfully violent, nystagmus, ↑CPK, ↑SGOT
- **LSD** "acid, trips": 5-HT agonist => **l**aid back => colorful hallucinations
- Amphetamines "speed, ecstasy, ice=smoked": hyperactivity, vertical nystagmus
- Nicotine: ↓DA reuptake => agitation, cotinine metabolites in urine (Tx: Bupropion)
- Ecstasy: very thirsty
- Cocaine "coke, snow, crack=smoked": ↓DA reuptake
    - Condescending, grandiose, big pupils, formication, HTN, ↑TPR

- Fetus: ↓limbs, VSD, brain bleed, placenta abruptio

Cocaine Chest Pressure Tx:

1) BZ (↓HR/BP)
2) Nitrate (↓vasospasm)
3) Aspirin (↓thrombi)

Cocaine HTN Tx:

1) Lorazepam (anxiolytic)
2) Phentolamine (↓BP)
3) Avoid β-blockers

Opiod Withdrawal Tx:

Neonates: Paregoric

Adults: Clonidine/Methadone

Cocaine Rhabdomyolysis Tx:

1) IVF
2) Bicarbonate (keep Mb soluble)

Downers:

- **M**arijuana "dope, grass, weed, pot": Δ-9 THC => **m**unchies, red eyes, impaired time orientation
- Heroin "smack, shit": inhibits AC => constipation, pinpoint pupils (Tx: Naloxone, Methadone)
- Alcohol => ataxia, euphoria, slurred speech, ↑GGT, AST>ALT (Tx: thiamine/glucose + BZ)

Isoelectric Point (pI): pH at which there is no *net* charge (pI= (pK1 + pK2)/2)
Zwitterions: have a negative carboxy and positive amino end
Cathode: where cations go, the negative electrode
Anode: where anions go, the positive electrode

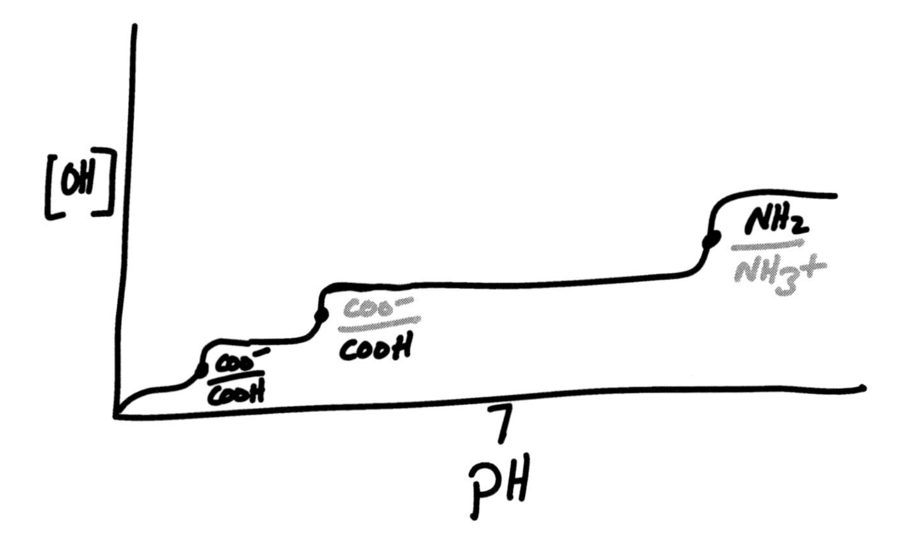

Amino Acid Functions:

Groups:	Amino Acids:	Notes:
Acidic	Asp, Glu	
Basic	Lys, Arg	Trypsin cuts *"Trip to LA"*
Sulfur	**Cys**, Met	β-ME cuts
O-Bonds	**Ser**, Thr, Tyr	Tyr makes catecholamines, melanin
N-Bonds	**Asn**, Gln	Acid hydrolysis denatures
Branched	Leu, Ile, Val	Maple syrup urine disease
Aromatic	Phe, Tyr, Trp	Chymotrypsin cuts *"bulky word/rings"*
Small	Gly	Spinal cord inhibitor
NMDA pathway	Asp	Brain excitatory
Kinky (Imino)	Pro	Yellow on Nurhydrin reaction
Active sites	Ser	
Ketogenic	Lys, Leu	Broken down to Acetyl CoA
Glucogenic + Ketogenic (Bulky)	**P**he, **I**so, **T**hr, **T**rp	*"Mr. PITT never Tyrs"*

Diabetes I: avoid Ketogenic diet
Diabetes II: avoid Glucogenic diet

Disulfide Bonds:
"PIGI"

- **P**RL
- **I**nsulin
- **G**H
- **I**nhibin

Reasons for Dialysis:

- Symptomatic Uremia
- Symptomatic Acidosis
- Hyperkalemia

The GABA Connection:
So…what happens when you decide to spend your evening at the local bar? You drink some alcohol, and drink some more, and then who knows what happens after that. Why do you have trouble remembering what happened? It's all because of GABA. Increased GABA levels lead to bradycardia, lethargy, constipation, impotence, and memory loss. This means that anything or any disease that increases acidosis, urea, or ammonia will lead to an increase in GABA which will slow everything down.
$NH_3 + H^+ \rightarrow NH_4^+ + \alpha\text{-KG} \rightarrow \text{Glu} \rightarrow \text{GABA}$

Essential Amino Acids: you gotta eat 'em: *"PVT TIM HALL"*

- **Phe → Tyr**
- **Val**
- **Trp → 5-HT**
- **Thr**

Essential FA (must eat 'em):

- Linolenic
- Linoleic → AA → PG

- **Ile**
- **Met → Cys**
- **His**
- **Arg**
- **Leu**
- **Lys**

Newborn screening:

"Please Check Before Going Home"

- **P**KU (Guthrie test)
- **C**ongenital adrenal hyperplasia
- **B**iotinidase deficiency
- **G**alactosemia
- **H**ypothyroidism

Energy Utilization:

1) Plasma Glucose: 2-4 hr
2) Liver Glycogen: 24-48 hr
3) Proteolysis for Gluconeogenesis
4) Fats for Lipolysis
5) Ketones for Ketogenesis

AA Disorders (AD):

Met + ATP → SAM → Homocysteine → Cys + propionylCoA → MMCoA

PKU

Phe → Tyr → DOPA → DA → NE → Epi

↙ ↘ ***Albinism***

T_4 Melanin

Phenylketonuria: No Phe → Tyr (via Phe-OHase)

- Mental retardation (can't make DA, NE, Epi)
- Pale, blond, blue eyes (no melanin)
- Musty odor (phenylacetate + phenylpyruvate)
- Nutrasweet sensitivity (Phe)
- Sclerodermatous plaques
- Test: Guthrie test (bacterial inhibition assay)

Albinism: No Tyr → Melanin (via Tyrosinase)

- Paleness, predisposed to skin cancer

Vitiligo:

- Anti-melanocyte Ab
- Pale, predisposed to skin cancer
- Ex: Michael Jackson

Alkaptonuria "Ochronosis":

- Kids w/ osteoarthritis (black tendons), black urine
- Homogentisic acid oxidase deficiency

Maple Syrup Urine Disease: "*Life is sweet, so LIV it up!*"
Defective renal transport of branced aa (Leu, Iso, Val) => aa leak out

Homocysteinuria (AD): No Homocys → Cys (via cystathione synthase)
Caused by high fat diet: increased Met levels
Dislocated lens from top *"always looking down toward stones in urine"*
Defective amino acid transport
Recurrent kidney stones
4 Amino Acids show up in Urine: *"Basic COLA"*

- Cystine stones ← + urine CN Nitroprusside
- Ornithine
- Lysine – basic aa
- Arginine – basic aa

NOTES:

NOTES:

Proteins:

"Time is but the stream I go a-fishing in."

–Henry David Thoreau

PROTEINS:

1°: aa sequence (peptide bonds C-N-C, planar, flat, restrictive motion trans configuration)
2°: α-helix (twisted organs: GI, vv) vs. β- pleated sheet (flat stuff: skin, flat bones)
3°: 3-D shape, consider hydrophobic/hydrophilic interactions
4°: ≥2 proteins interact (Ex: Hb)

Allosterism: rate-limiting (slowest) enzymes

Sequencing: *find the first clue, then look at your answer choices!*
Acid Hydrolysis: dip protein in acid => denatures (Gln → Glu, Asn → Asp)
Gel Electrophoresis: separates protein based on size and charge

- cations(+) => cathode (-)

Ninhydrin Reaction: all aa = purple (except Pro = yellow)
Edmund Degradation: uses *PITC* to remove 1 aa at a time (100 aa limit)
Restriction peptidases: cut on the right

- Carboxypeptidase – cuts to *left* of any amino acid on carboxy terminal
- Aminopeptidase – N-terminus
- CNBr – Met
- Mercaptoethanol – Cys, Met (disulfide bonds)
- Elastase – Gly, Ala, Ser (small) – *"GAS"*
- Trypsin – Arg, Lys (basic groups) – *"take a basic Tryp to LA"*
- Chymotrypsin – Phe, Tyr, Trp (bulky)

α_1-AT: inhibits trypsin from getting loose (b/c he can activate everything)

High Plasma Proteins:

- **↑ESR or CRP → inflammation**
- **Falsely ↑ESR: anemia**
- **Falsely ↓ESR: sickle cell anemia, polycythemia**
- **↑Acute phase reactants (IL-6): amyloidosis**

Amyloidosis: *stains Congo red, Echo Apple-green birefringence*

- 1° Amyloidosis (AD): big organs, ↑protein causes intracranial hemorrhage
- 2° Amyloidosis (chronic disease): Scleroderma, asthma, Wegener's

Amyloid: ↑*ESR*
AA: **A**ny chronic disease
AB: **B**rain (Alzheimer's)
AB_2: β_2 microglobulinemia (renal failure)

AE: **E**ndocrine (medullary CA of thyroid)
AF: **F**amilial (MEN2)
AL: **L**ight chains (multiple myeloma)

Drug Metabolism:
Zero-order Kinetics – *metabolism independent of concentration*
Ex: Phenytoin, Chemo drugs
Ex: EtOH (100mg/dL/hr): 1 glass wine, 1 shot whiskey, 2 cans of beer
Ex: high dose asa

Alcohol Limit:
0.08: legal limit
0.1: zero-order kinetics
0.3: coma
0.4: "embalming"

1st-order Kinetics – *constant drug percentage metabolism over time*

- Ex: 10% of drug (conc=100mg/dL) eliminated every 2 hours:
    - T=0 hrs: [D]=100mg/dL
    - T=2 hrs: [D]=90 mg/dL
    - T=4 hrs: [D]=81 mg/dL

Drug Dosage: **the link b/w kinetics and dynamics**

- $t_{1/2}$ = (.693)(V_d) ÷ clearance
- V_d: total drug ÷ plasma conc (large V_d => most of drug is sequestered)
- Loading dose: (desired plasma conc)(V_d)
- Maintance dose: (desired plasma conc)(clearance)
- Steady-state plasma conc (C_{ss}): availability rate = elimination rate, *takes 4.5 half-lives*
- Clearance: volume of plasma cleared of drug
- Excretion rate: (clearance)(plasma conc) – rate of elimination
- TI: toxic dose ÷ therapeutic dose (high TI => safe drug) = LD_{50}, ED_{50}
- Peak level: 4 hrs after dose (too high => decrease dose)
- Trough level: 2 hrs before dose (too high => give less often)

Dose-response relationships:

- Efficacy – *max effect regardless of dose* (lower w/ non-competitive antagonist)
- Potency – amount of drug needed to produce effect (lower w/ comp antagonist)
- K_d – [D] that binds 50% of receptors
- EC_{50} – [D] that produces 50% of maximal response
- Competitive Inhibition: potency decreases

Competitive Inhibition: fights for active site, no ΔV_{max}

$$\uparrow K_m = \frac{1}{affinity \downarrow}$$

K_m = [S] at ½ V_{max}

Non-competitive Inhibition: binds a regulatory site, no ΔK_m

Non-competitive Inhibition: efficacy decreases

Thus, K_m affects *potency* and V_{max} affects *efficacy*

Collagen: **every 3rd aa is Gly**

4 types of collagen*: "SCAB"*

- **Type I:** **Skin, bone**
- **Type II:** **Connective tissue (tendons, ligaments cartilage), aqueous humor, blood → fasciitis, cellulitis**
- **Type III:** **Arteries (coronary aa. affected first) → vasculitis**
- **Type IV:** **Basement membrane → GN**

Collagen Requirements:

- **Glycine (Every 3rd aa is Gly, smallest amino acid)**
- **Lysine**
- **Proline**
- **OH-Proline (requires Vit C)**
- **OH-Lysine (requires Cu2+)**

Collagen Synthesis: **(Pro-OHase, Lys-OHase → needs Vit C, Cu)**

- **PreProCollagen → ER →**
- **ProCollagen → Golgi →**
- **Tropocollagen → Plasma →**
- **Plasma peptidases tighten it up at the site of action**

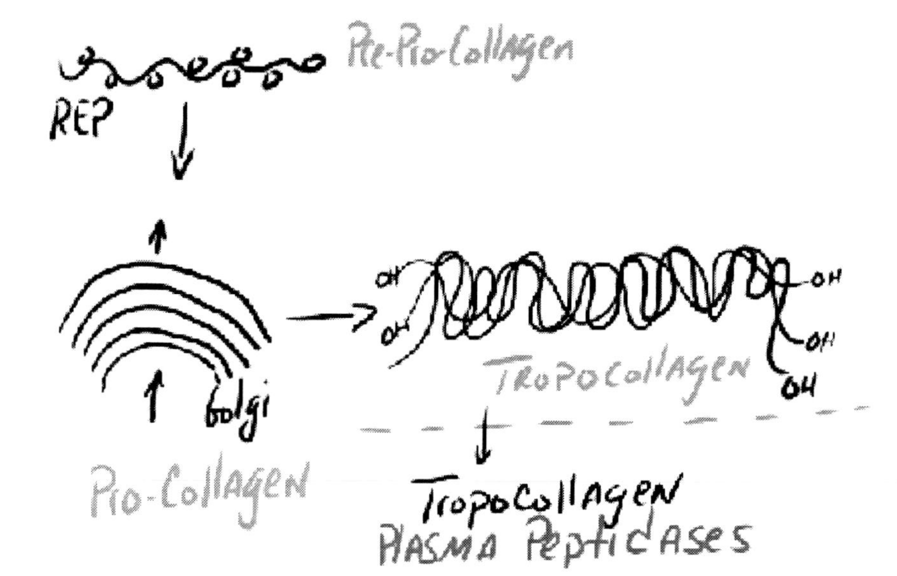

<u>Who makes collagen?</u>
Fibroblasts: **scars**
Myofibroblasts: **wound contraction (contracture means it happened too much)**

<u>Collagen Diseases:</u> ***BLEED!***

1) Keloids = hamartoma
- Fibroblasts release too much collagen

2) Scleroderma:
- Collagen types I (skin, bone) and type III (aa.)
- Tightened skin, blood vessel problems

3) Ehlers Danlos:
- Collagen types I (skin, bone) and type III (aa.)
- Hyperstretchable skin "velvety"

4) Marfan's (AD): fibrillin problem
- Collagen type II (CT) and type III (aa.)
- Hyperextensible joints, arachnodactyly, wing span longer than height
- Aortic root dilatation, aortic aneurysm, mitral valve prolapse
- Dislocated lens from bottom of eye "always looking up to Mars"

<u>5) Homocysteinuria:</u>
- LysOHase not inhibited by homocysteine
- Increased Met levels
- Marfanoid features
- Childhood strokes
- Dislocated lens from top → *"always looking down toward "COLA" urine stones"*

<u>6) Kinky Hair Disease:</u>
- Cu deficiency (Lys-OHase affected)
- Hair looks like copper wire

<u>7) Scurvy:</u>
- Collagen type III (aa.)
- Vit. C deficiency
- Bleeding gums, hair follicles

<u>8) Takayasu Arteritis:</u>
- Collagen type III (aa.)
- Asian female with very weak pulse = *"pulseless aortitis"*

9) 3° Syphilis:

- Collagen type III (aa.)
- Obliterative endarteritis => *"tree bark" appearance*

10) Osteogenesis Imperfecta:

- All 4 types of collagen involved; pleiotropy
- Shattered bones, looks like child abuse
- Blue sclera
- 11) Desmoplasia:
- Collagenous reaction surrounding a tumor

Elastin: Pro-OHase (no OH-Lysines)

- Desmosine (Lys box) => ⊕ repel each other, making it elastic
- Elastase breaks up elastin => lose recoil => Emphysema

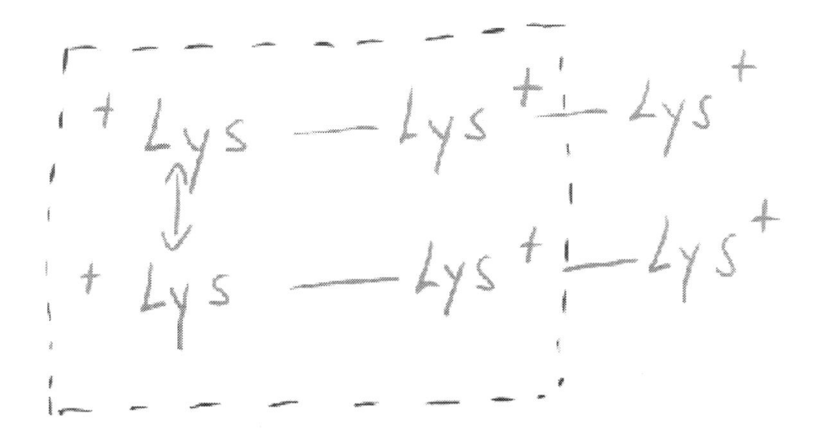

Emphysema: elastin destroyed, loss of recoil

Pan-acinar (AR): α_1-AT def can't inhibit elastase, PAS(+), neonatal hepatitis
Centro-acinar: smoking *"comes in through the center"*
Distal acinar: aging (least blood supply) → spontaneous pneumothorax
Bullous "pneumatocele": elastase ⊕ bacteria = Pseudo/Staph aureus

Keratin: **Cys → disulfide bonds → tensile strength**
Curly hair -> more bonds -> broken down by heat/chemicals to straighten hair

NOTES:

NOTES:

Enzymes:

"Always the beautiful answer who asks a more beautiful question."

–E. E. Cummings

Enzymes:

- Lower the Free Energy of Activation
- Bring substrates together in space and time
- Stabilize high energy intermediate
- Is never consumed in the reaction
- Allosteric enzyme = the slowest one

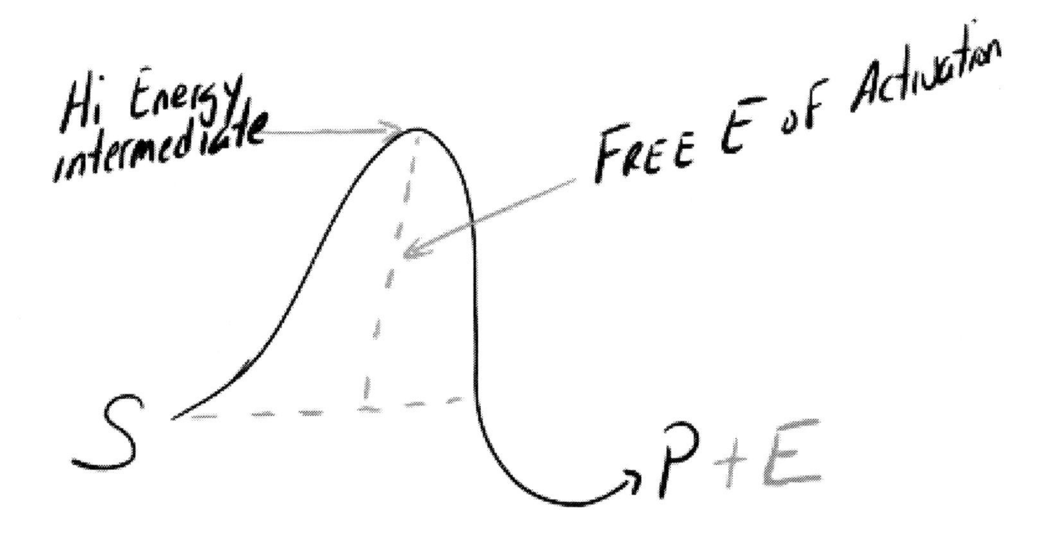

Dose-Response Relationships:

- Efficacy – *max effect regardless of dose* (lower w/ non-competitive antagonist)
- Potency – amount of drug needed to produce effect (lower w/ comp antagonist)
- K_d – [D] that binds 50% of receptors
- EC_{50} – [D] that produces 50% of maximal response

- Competitive Inhibition: potency decreases, $\uparrow K_m$

Competitive Inhibition: fights for active site, reversible, no ΔV_{max}

$$\uparrow K_m = \frac{1}{affinity \downarrow}$$

K_m = [S] at ½ V_{max}

Non-competitive Inhibition: binds a regulatory site, no ΔK_m

Non-competitive Inhibition: irreversible, efficacy decreases

Thus, K_m affects *potency* and V_{max} affects *efficacy*

Free Energy of Reaction:

$\Delta G = \Delta H - T\Delta S$

$-\Delta G \rightarrow$ drives reaction forward

ΔH = enthalpy (heat)

Endothermic = endergonic → add heat
Exothermic = exergonic → gives off heat => spontaneous, favorable

- ***Ex:*** *Silver Sulfadiazine* ***(for burn victims) creates endothermic reaction!***

+ΔS = entropy (randomness)

- **High energy/randomness to low energy/randomness**

T = temperature

- **T proportional to V_{max}**
- **If T increases too much → proteins denature → V_{max} will drop**
- **Most common cause of death is heart failure**

+ΔE = redox potential
-ΔE => has too many electrons

- **If it is being oxidized ⇒ reducing agent**
- ***"OIL RIG":*** *O****xidized*** *I**s* *L****osing electrons,*** *R****educed*** *I**s* *G****aining electrons***

Electron Transport **– cash in your electrons and get some ATP (in mitochondria)**

COQ / ubiquinone matrix e^- Cytochrome oxidase ATP Synthetase

NADH FADH$_2$ Fe Fe O_2 → H_2O

Comp I II III IV V

Complex I: **NADH feeds in electrons**
Complex II: **$FADH_2$ feeds in electrons**
Complex III: **requires iron for heme**
Complex IV: **Oxidation portion, Cytochrome oxidase (oxygen used to form water and requires iron for heme and copper)**
Complex V: **Phorphorylation portion (forms ATP)**
Coenzyme Q: **Shunts electrons to Complex III (ΔE gets progressively positive as it passes through the ETC)**

ETC Inhibitors:
Complex I: Amytal, Rotenone
Complex II: Malonate
Complex III: Antimycin D
Complex IV: CN, CO, **Chloramphenicol** ← use Cu/Fe *"4 C's affect Complex IV"*
Complex V: Oligomycin

- **NADH → 3 ATP**
- **$FADH_2$ → 2 ATP (already passed complex I, only have 2 more places to do it)**
- **Complex IV "cytochrome oxidase" – has the most ⊕ΔE => e- are driven toward it**
- **Need O_2, Cu, and Fe => low E state without 'em**

Uncouplers: ***can't make ATP now***
- **DNP: insecticide**
- **asa (Ex: Reye's syndrome)**
- **Free fatty acids**

Aplastic Crisis: low retics
Sequestration Crisis: high retics
(RBCs trapped in big spleen)

Hb Poisoning:
CO = **Competitive inhibitor of O_2 on Hb => nl O_2 sat, low pO_2 (Tx: O_2)**
- **Cherry-red lips, pinkish skin**

CN = **Non-competitive inhibitor of O_2 on Hb => normal O_2 sat, normal pO_2**
- **Almond breath**
- **Drug induced (Sulfas, Antimalarials, Metronidazole, Nitroprusside)**
- **Tx: *"A Tortured Man Breathes"***
  1. Amyl **Nitrite – converts Hb to MetHb => CN can't act**
  2. Thiosulfate **– binds CN => pee out thiocyanate**
  3. Methylene **blue – converts Fe^{3+} to Fe^{2+}**
  4. Blood **transfusion**

MetHb **(Fe^{3+}) => low O_2 sat, normal pO_2 (can't bind O_2)**

Nephrotoxicity: water-soluble (charged)

Hepatotoxicity: fat-soluble (bioavailable)

Naming 90% of Enzymes:

1st name: **Substrate**

Last name: **What is done to the substrate**

1) Move around:

- **Isomerase – creates an isomer (Ex: glucose → fructose)**
- **Epimerase – creates an epimer, differs around 1 chiral carbon (glucose → galactose)**
- **Mutase – moves sidechain from one carbon to another (intrachain)**
- **Transferase – moves sidechain from one substrate to another (interchain)**

2) Add stuff:

- **Kinase – phosphorylates using ATP (-P makes it stay inside cell)**
- **Phosphorylase – phosphorylates using Pi**
- **Carboxylase – forms C-C (w/ ATP and biotin)**
- **Synthase – two substrates are consumed (name after product)**
- **Synthetase – two substrates are consumed, uses ATP**

3) Remove stuff:

- **Phosphatase – breaks phosphate bond**
- **Hydrolase – break a bond (w/ H_2O)**
- **Lyase – cut C-C bonds (w/ ATP)**
- **Dehydrogenase – removes H (w/ cofactor)**
- **Thio – break S bond**

Body Fuels:

2-4 hrs: Glucose

24 hrs: Glycogen

48 hrs: Protein ← Muscle

>48 hrs: Fat

Rate Limiting Enzyme Overview:

Glycolysis: PFK-1

Gluconeogenesis: Pyruvate carboxylase

HMP shunt: G-6PD

Glycogenesis: Glycogen synthase

Glycogenolysis: Glycogen phosphorylase

FA synthesis: AcCoA carboxylase

β-oxidation: CAT-1

Cholesterol synthesis: HMG CoA reductase

Ketogenosis: HMG CoA synthase

Purine synthesis: **PR**PP synthase

Pyrimidine synthesis: Asp transcarbamoylase (also uses CPS-II)

T**C**A cycle: Iso**citrate** dehydrogenase

Urea cycle: CPS-I

Heme synthesis: δ-ALA synthase

NOTES:

Anabolic/ Catabolic Overview:

"Uttering a word is like striking a note on the keyboard of the imagination."

–Ludwig Wittgenstein

THE BIG PICTURE

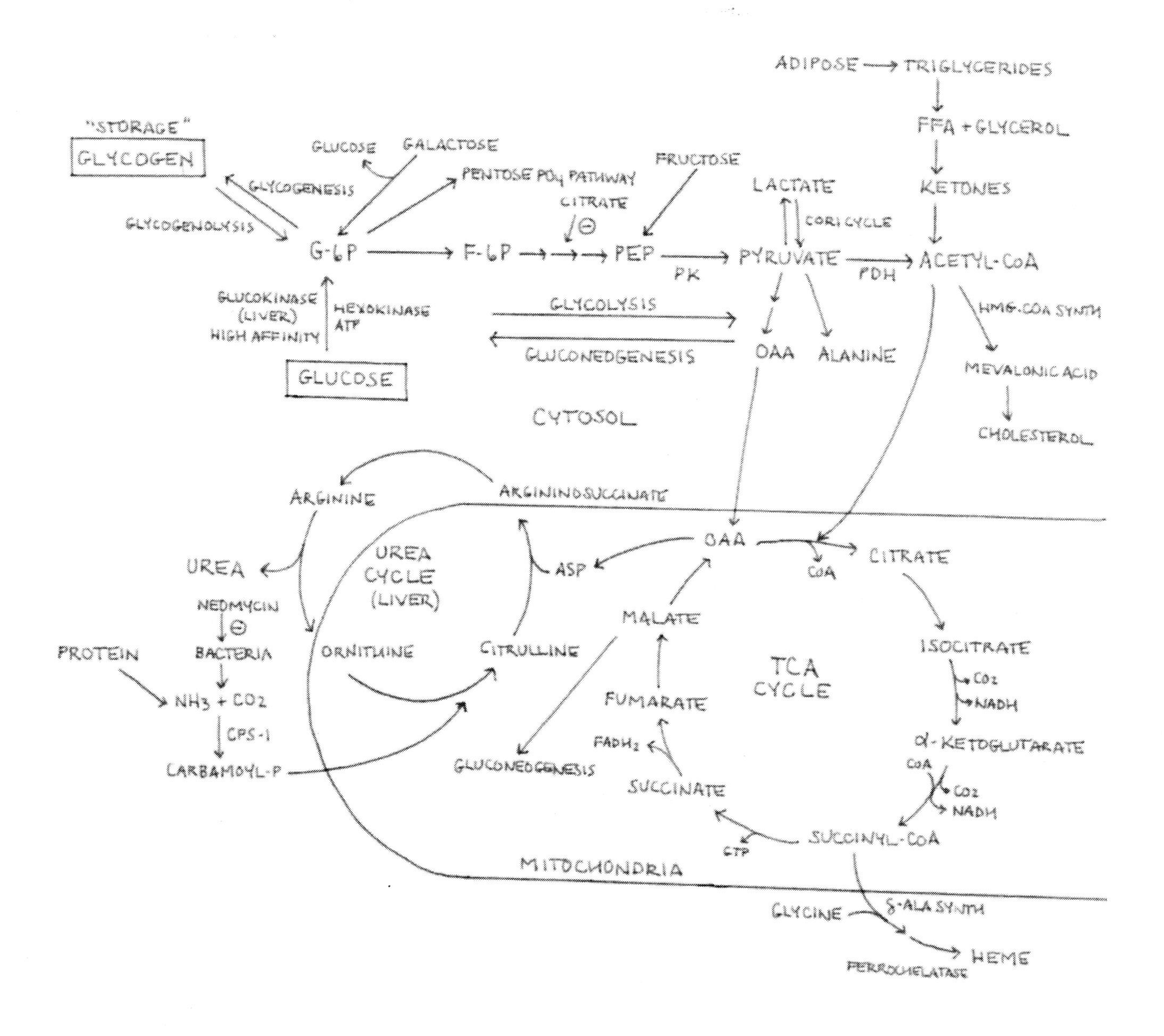

Catabolic Pathways: **create energy (mitochondria)** ***"ABC"***

- **AcetylCoA production**
- **β-oxidation**
- **Citric acid cycle**

Anabolic Pathways: **store energy (cytosol)** ***"EFGH"***

- **Endoplasmic Reticulum**
- **Fatty acid synthesis**
- **Glycolysis**
- **HMP shunt**

Anabolic + Catabolic Pathways: ***"HUG"***

- **H**eme synthesis
- **U**rea cycle
- **G**luconeogenesis

NOTES:

Glycolysis / Gluconeogenesis:

"A man does not seek to see himself in running water, but in still water.

For only what is itself still can impart stillness into others."

–Chuang-tzu

Energy Sources:

	Normal:	Stress:	Extreme Stress:
Brain:	Glucose	Glucose	Ketones
Heart:	Free FA	Glucose	Glucose
Muscles:	Glucose	Free FA	Free FA
RBCs:	Glucose	Glucose	Glucose

RBC Connection:

- RBC's use only glucose for energy.
- Hypoglycemia will always affect RBC's first, causing a hemolytic anemia.
- The only other pathway RBC's have is the Pentose Phosphate Pathway for making NADPH to maintain the membrane.

Glycolysis:

- The most active pathway in your body.
- Catabolic in all cells except the liver, where it is anabolic.

Catabolic State:

- Controlled by the sympathetic system
- Second messenger is cyclic AMP
- Controlled hormonally by Epinephrine and Glucagon

Energy Use:

Plasma Glucose:	Liver Glycogen:	Proteolysis:	Lipolysis:	Ketogenesis:
2-4 hrs	24-28 hrs	> 36 hrs	> 36hrs	> 36 hrs

Glucose → G6P:

Glucokinase:

- Found in hepatocytes and β cells of pancreas
- High Km
- ATP → ADP + Pi

Hexokinase:

- Found everywhere else
- Low Km
- ATP → ADP + Pi

G6P → F6P

- Phosphoglucose Isomerase

F6P → F1,6DP

- **PFK -1:**
- Activators: AMP and F2,6DP
- Inhibitors: ATP and Citrate
- Uses ATP
- **PFK-2:**
- Forms F2,6DP (the allosteric activator of PFK-1)
- Uses ATP

F1,6DP → DHAP and G3P:

- Aldolase A
- Triose phosphate isomerase (med conversion of DHAP to G3P and v.v.)
- DHAP
- Glycerol 3 phospohate shuttle
- Triglyceride synthesis

G3P → G1,3DP:

- Glyceraldehyde 3 phosphate dehydrogenase

1. Sulphur in active site
2. Blocked by mercury poisoning

- NAD^+ → NADH

Mercury Toxicity:

- MCC: (1) Tuna and (2) Biting into thermometer
- Blocks Glyceraldehyde 3 phosphate dehydrogenase
- RBCs affected first; brain affected the most

G1,3DP → G2,3DP:

- Bisphosphoglycerate mutase
- Shifts curve to the right decreasing affinity of hemoglobin to oxygen

G1,3DP → 3PG:

- Phosphoglycerate Kinase
- ADP + Pi → ATP (substrate level phosphorylation)

3PG → 2PG:

- Phosphoglycerate mutase

2PG → PEP:

- Enolase
- Enol group (phosphate group next to a double bond)

PEP → Pyruvate:

- Pyruvate Kinase
- ADP + Pi → ATP (substrate level phosphorylation)

Fluoride Poisoning:

- Blocks the enzyme Enolase
- Caused in the past by eating rocks of Fluoride
- Rare today since fluoride added to water and toothpaste
- Clue: extra white teeth and bones

Gluconeogenesis:

- Controlled by Epinephrine and Glucagon
- Second messenger is cyclic AMP
- Occurs only in the liver (90%) and adrenal cortex (10%)
- Occurs while other tissues are running glycolysis
- Occurs in the mitochondria and cytoplasm

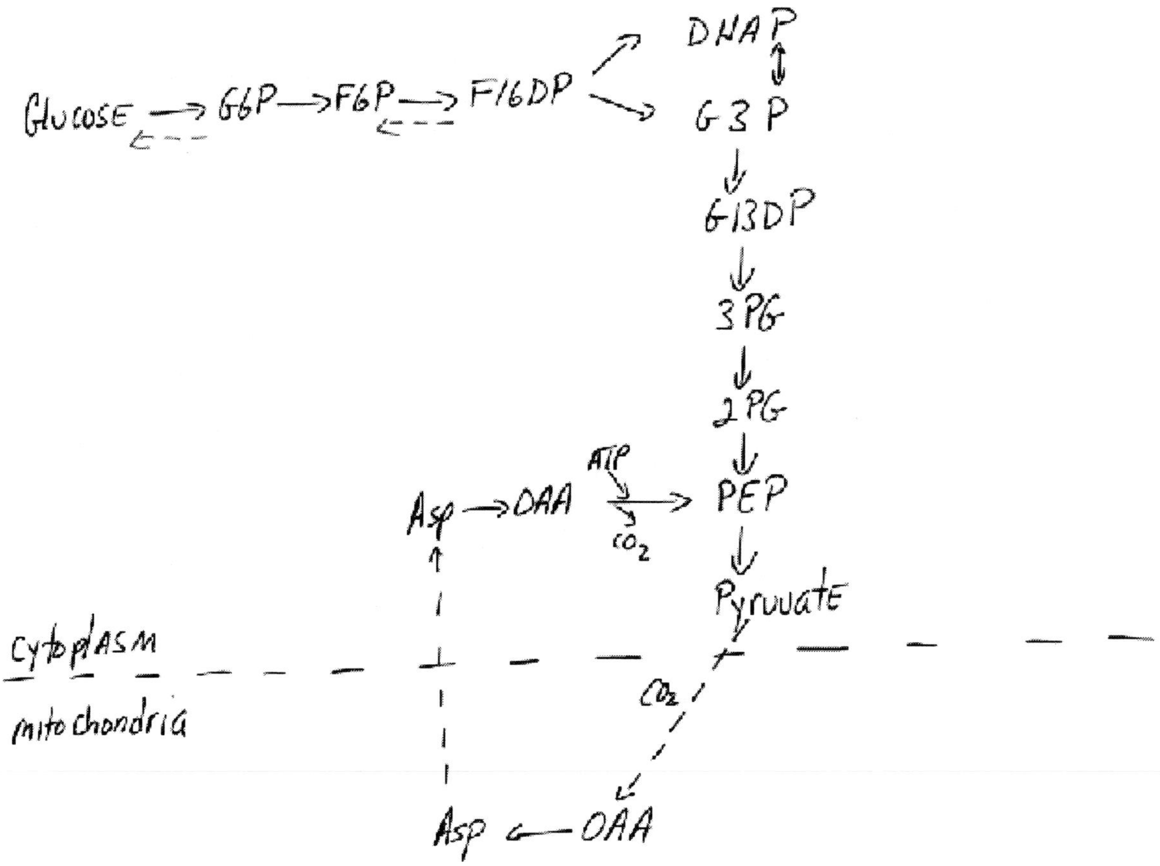

DHAP
Glucose → G6P → F6P → F16DP → G3P
G13DP
3PG
2PG
ATP
Asp → OAA → PEP
CO_2
Pyruvate
cytoplasm
mitochondria
CO_2
Asp ← OAA

Gluconeogenesis Enzymes:

- Pyruvate carboxylase (rate limiting)
- PEP Carboxykinase
- F16DPase
- G6Pase

Pyruvate + CO2 → OAA:

Pyruvate Carboxylase:

- Anapleuritic
- Cofactor: Biotin
- Activator: Acetyl CoA
- Inhibitor: Glucose, ADP

OAA → PEP:

- PEP carboxykinase
- Bypasses Pyruvate kinase
- 1 molecule of GTP is required
- Decarboxylated

F1,6BP → F6P:

- F1,6DPase

G6P → Glucose:

G6Pase:

- Missing in muscles, so unable to raise blood sugar
- Only in liver and adrenal cortex

Starvation State:

- Liver → Gluconeogenesis
- Tissues → Glycolysis

Well-fed State:

- Liver → Glycolysis
- Tissues → Gluconeogenesis

Summary:

Lactate $\xrightarrow{\text{Cori cycle}}$ **Glucose**

Pyruvate + NH_4^+ $\xrightarrow{\text{Ala cycle, PC}}$ **Glucose** (Activator = AcCoA)

Glycerol $\xrightarrow{\text{DHAP}}$ **Glucose**

Glucose $\xrightarrow{\text{PFK-1}}$ 2 pyruvate + 2 ATP + 2 NADH (Activator: F2,6BP/PFK-2)
(Inhibitors: Citrate, ATP)

Glycolysis Irreversible Enzymes:

- Hexokinase (only hexokinase is feedback inhibited by G6P)
- PFK-1
- PK

Gluconeogenesis Irreversible Enzymes:

- G-6Pase (liver,adrenals) = *"reverse glucokinase"*
- F-1,6DPase = *"reverse PFK-1"*
- Pyruvate carboxylase + biotion (pyruvate → OAA)
- PEP carboxykinase + ATP (OAA → PEP)

Glucose transporters:

GLUT 1: all tissues

GLUT 2: Liver, Pancreatic β cells

GLUT 3: all tissues

GLUT 4: Fat, Sk. muscle, Heart

NOTES:

<u>**NOTES:**</u>

Fructose/ Galactose:

"Whenever two people meet there are really six people present. There is each man as he sees himself, each man as the other person sees him, and each man as he really is."

–William James

Fructose Metabolism:

Glucose —ATP→ G6P — F6P —ATP→ F16DP
Hexokinase, Glucokinase (Glucose → G6P); PFK-1 (F6P → F16DP)

F16DP → DHAP, G3P
G3P ↓ NADH
G13DP ↓ ATP
3PG ↓
2PG ↓
PEP ↓ ATP (Pyruvate Kinase)
Pyruvate

Fructose —ATP→ Fru 1 Phos → DHAP + Glycerol

Detecting Sugars:

- Urine: Clinitest
- Stool: Reducing substances

Fructose:

- Enters Glycolysis after PFK-1
- Diabetics: least likely to raise blood sugar
- Dieters: fruits eaten at late night can still be metabolized

Fructose → F1P:

- Fructokinase
- Requires ATP

Fructosuria:

- Fructokinase is missing; Hexokinase fills in
- Fructose in the urine (positive Clinitest)
- Polyuria, polydypsia

F1P → DHAP + Glycerol:

- Aldolase B

Summary:

Sucrose → Fructose + Glucose $\xrightarrow{\text{Fructokinase}}$ F-1P $\xrightarrow{\text{Aldolase B}}$ DHAP or Glyceraldehyde

Essential Fructosuria:

- Fructokinase deficiency
- Causes excreted fructose (still have hexokinase)

Fructosemia "Fructose Intolerance":

- Aldolase B deficiency
- Fructose 1 phosphate is trapped within cells causing cellular swelling and lysis

Galactose Metabolism:

Glucose $\xrightarrow{\text{ATP}}$ Glu-6-Phos

↑

Glu 1 Phos

↑

UDP-Gal-1 Phos

↑ UDP-Glu

Gal 1 Phos

ATP ↑

GALACTOSE

Galactose → Gal-1-P:

- Galactokinase
- Requires ATP

Galactosuria:

- Galactokinase deficiency
- Hexokinase fills in for galactokinase
- Galactose in the urine (Clinitest positive)
- Polyuria; polydypsia

Gal-1-P + UDP-Glu → UDP-Gal-1-P

- The glucose will enter glycolysis

UDP-GAL-1-P → Glu-1-P

- UDPgalactose-4-epimerase

Glu-1-P → Glu-6-P

- Phosphoglucomutase

Summary:

Lactose → Galactose + Glucose —Galactokinase→ Gal-1P —Uridyl transferase→ G-6P

Galactosemia:

- Uridyltransferase deficiency
- Galactose 1 phosphate builds up in the cells, causing cellular swelling/lysis
- Screened for early, due to galactose in lactose
- Cataracts, mental retardation, liver damage

Galactose Deficiency:

- Cataracts (still have hexokinase)

NOTES:

NOTES:

Pyruvate:

"Everything should be as simple as it is, but not simpler."

–Albert Einstein

Pyruvate:

Three complexes require 5 factors:

- α-KG dehydrogenase
- Branched chain dehydrogenase
- PDH Complex

Vitamin Order:

(evens closest to middle)

1+4=5

3+2=5

Cofactors:	**Action:**	**Vitamin Derivatives:**
TP**P**	Decarboxylation	Thiamine (B_1)
Lipoic acid	Accepts Acetyl group	Lipoic acid (B_4)
Coenzyme **A**	Final acceptor → AcCoA	Panthenoic acid (B_5)
NAD	Oxidizes $FADH_2$ → NADH	Niacin (B_3)
FAD	Oxidizes Lipoic Acid → $FADH_2$	Riboflavin(B_2)

5 Fates of Pyruvate:

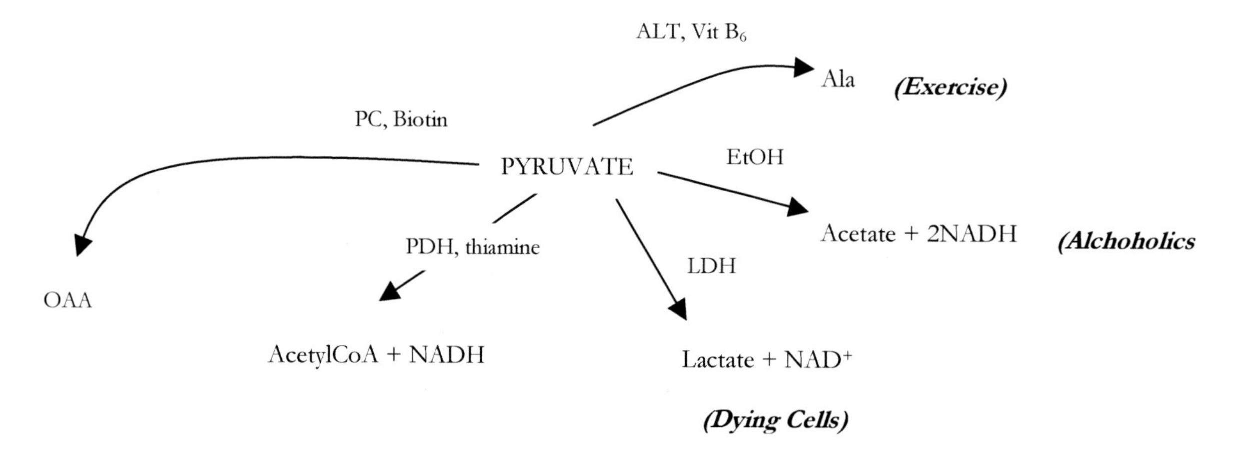

1) Lactate:

- Anaerobic
- Uses LDH
- NADH → NAD^+

2) Alanine:

- Anaerobic, formed due protection from overproduction of lactic acid
- Uses ALT
- Increases during exercise, leaks into blood

3) Acetate:

Ethanol → Acetaldehyde:

- NAD+ → NADH
- Alcohol dehydrogenase

Acetaldehyde → Acetate:

- NAD+ → NADH
- Acetaldehyde dehydrogenase
- Disulfiram and Metronidazole inhibit this enzyme → Acetaldehyde → N/V

Macrosteatosis:

- Large fat droplets
- Atherosclerosis
- High Ketone
- Acetone smell
- Acidosis → GABA connection → heart failure

Microsteatosis:

- Reye's syndrome
- Acetaminophen
- Pregnancy

Alcoholics:

- High NADH; body thinks you're in a high energy state
- Inhibits gluconeogenesis, causing hypoglycemia
- Then use fat, leading to hyperlipidemia/fatty liver
- Tx: Thiamine + Glucose (to prevent lactic acidosis)
- Thiamine deficient alcoholics: have even less AcCoA → more lactic acid

4) OAA:

- Anapleuritic
- Pyruvate Carboxylase
- Cofactor: Biotin

5) Acetyl CoA:

- If O2 is present
- removes CO2
- Uses Pyruvate Dehydrogenase
- NAD+ → NADH

4 Fates of AcetylCoA:

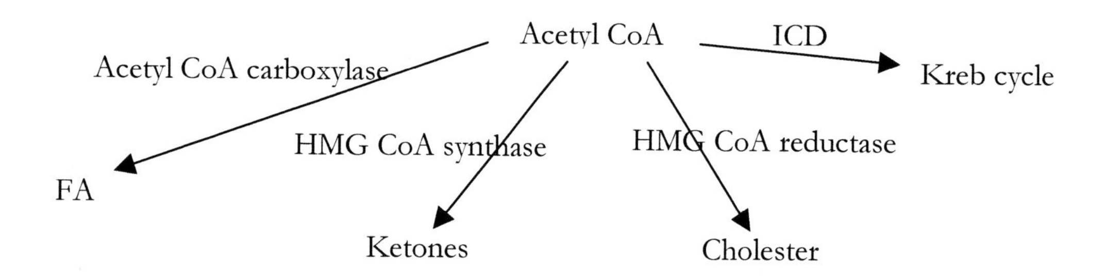

Pyruvate Dehydrogenase:

- Quaternary
- Three proteins
- Five steps
- Five cofactors
- Five vitamins
- Allosteric Activator: Pyruvate
- Allosteric Inhibitor: Acetyl CoA
- Rate limiting enzyme for Krebs cycle
- NAD+ → NADH

NOTES:

NOTES:

Kreb "TCA" Cycle:

"All men who have achieved great things have been great dreamers."

–Orison Swett Marden

Kreb "TCA" Cycle:

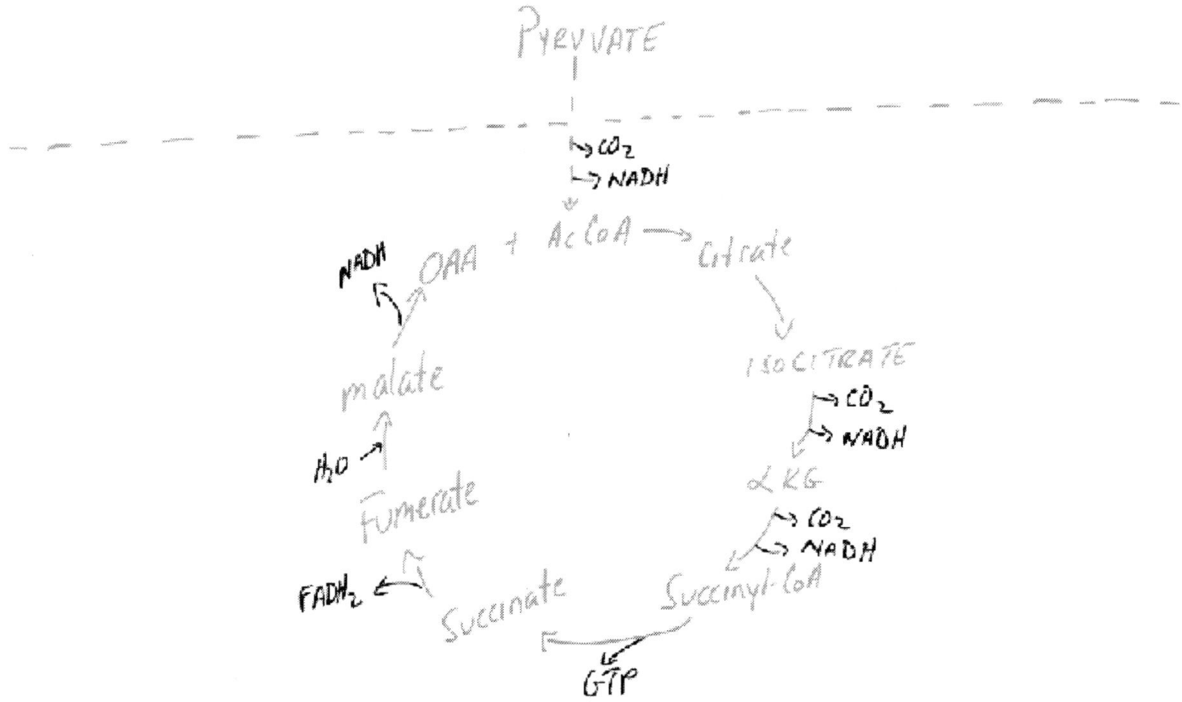

Kreb Cycle Components:

"Cindy Is Kinky, So She Fornicates More Often"

- **C**itrate
- **I**socitrate – inhibits PFK-1
- α-**KG** – transamination
- **S**uccinylCoA – uses odd-chain FA, makes ringed structures (Ex: heme)
- **S**uccinate
- **F**umarate
- **M**alonate
- **O**AA

Important Enzymes:

- **AcetylCoA:** FA synthesis, ketogenesis
- **Isocitrate Dehydrogenase:** rate limiting enzyme
- **Succinate Dehydrogenase:** attached to mitochondria wall, ETC II ($FADH_2$)

TCA Substrates:

- Gly, Ala, Ser "Elastase victims" → Pyruvate
- Lys, Leu, "Bulkies" (Phe,Tyr, Trp) → Acetyl CoA
- **A**sp, **A**sn → O**AA**

- **G**lu, **G**ln → α-K**G**
- Phe, Tyr → Fumarate

<u>OAA + AcCoA → Citrate:</u>

- Citrate synthase

<u>Citrate → Isocitrate:</u>

- Uses aconitase
- Citrate: Allosteric Inhibitor of PFK-1; Allosteritc Activator for FA synthesis
- Citrate shuttle: feeds Acetyl CoA into fatty acid synthesis

<u>Isocitrate → α-Ketoglutarate:</u>

- Uses Isocitrate Dehydrogenase
- Regulatory step
- Releases CO_2
- NAD^+ → NADH
- α-Ketoglutarate is the intermediate for all transaminase reactions
- α Ketoglutarate + Ala → Glu + Pyr
- Glutamate, Aspartate, Alanine build up in concentration during periods of high transamination.
- During amino acid breakdown, Glutamate is still increasing, holding area for amine groups until shunted into Urea cycle.
- 20 transaminases, one for each amino acid; AST/ALT most important
- All transaminases use Vit B_6 as a cofactor → B_6 deficiency causes neuropathy

<u>α-Ketoglutarate → Succinyl-CoA:</u>

- Uses α-Ketoglutarate Dehydrogenase
- Releases CO_2
- NAD^+ → NADH

<u>Succinyl-CoA → Succinate:</u>

- Uses Succinyl-CoA Synthetase
- Split sulphur bond to make GTP
- Succinyl CoA is responsible for:
  1. Heme synthesis
  2. Porphyrin ring substrate
  3. Odd-chained fatty acids feed into Succinyl CoA
  4. Odd-numbered fats break down into Propionic acid
  5. Priopionic acid CoA carboxylated becomes Methylmalonic acid CoA, which is then mutated to succinyl-CoA via Methylmanonyl mutase (Vit B_{12} cofactor).

Succinate → Fumarate:

- Succinate Dehydrogenase is the only Kreb Cycle enzyme firmly anchored to the inner mitochondrial membrane; attached to Complex II
- Forms $FADH_2$

Fumarate → Malate:

- Uses Fumarase
- Adds H_2O

Malate → OAA:

- Uses Malate Dehydrogenase
- NAD^+ → NADH

ATP count:
Glycolysis: 4ATP – 2ATP + 2 NADH = 8 ATP *"8 hrs = 1 work-day"*
Kreb: 2 GTP + 2 $FADH_2$ + 6 NADH = 24 ATP *"24 hrs = 1 day"*
G-3P shuttle: 2 $FADH_2$ => 4 ATP
Malate-Asp shuttle: 2 NADH => 6 ATP

Shuttle Overview: reoxidize NADH → NAD^+ in mitochondria

1) Malate-Asp shuttle: OAA → (AST) → Asp → Malate → Kreb + NADH

- Product ends up in cytosol
- This shuttle is dominant since the G-3P shuttle is wasteful

2) G-3P shuttle: DHAP → Glycerol-3P + $FADH_2$

- Used a great need of energy is needed; i.e. the need supercedes the waste
- Rapid Growth: Ages 0-2, 4-7, puberty
- Rapid Division: pregnancy, cancer, burns, crush injuries

1) Malate-Aspartate Shuttle:

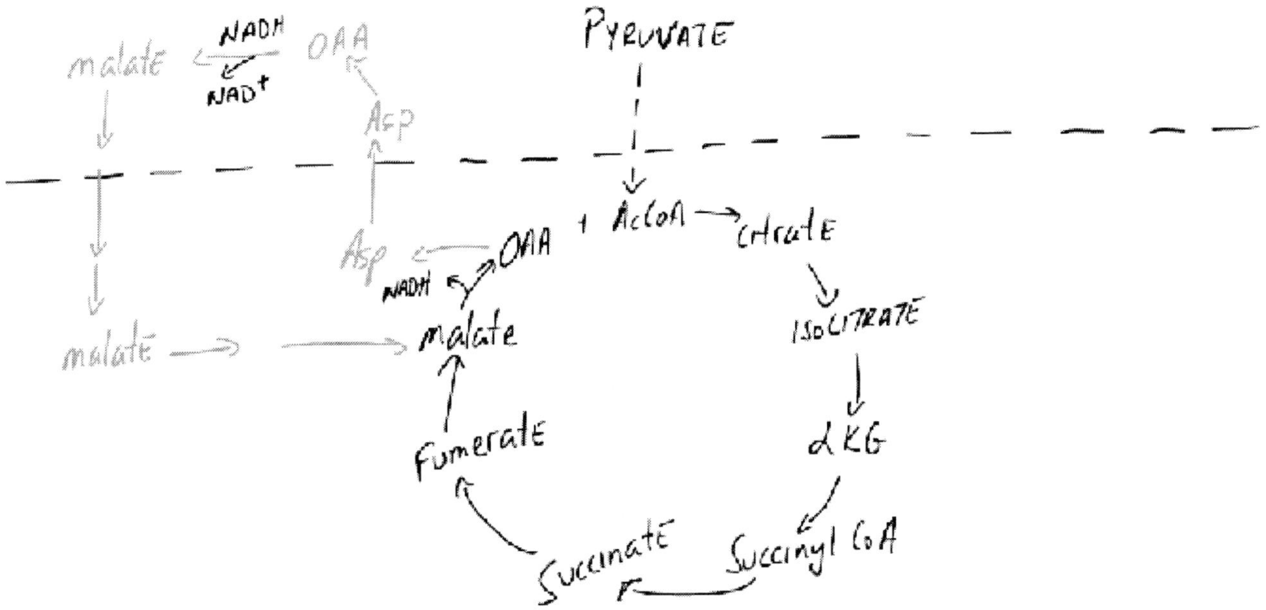

OAA → Aspartate:

- AST – Aspartate Transaminase
- Add NH_3
- Asp travels out to cytoplasm and gets converted back to OAA by AST

OAA → Malate:

- NADH → NAD^+
- Malate travels back into mitochondria; converts to OAA; regains NADH
- No energy wasted

Viral Hepatitis: Virus destroys cell membrane → 1AST, 1ALT leak out
Alcoholic Hepatitis: Alcohol dissolves all membranes → 1ALT, 2AST leak out

2) Glycerol 3-Phosphate Shuttle:

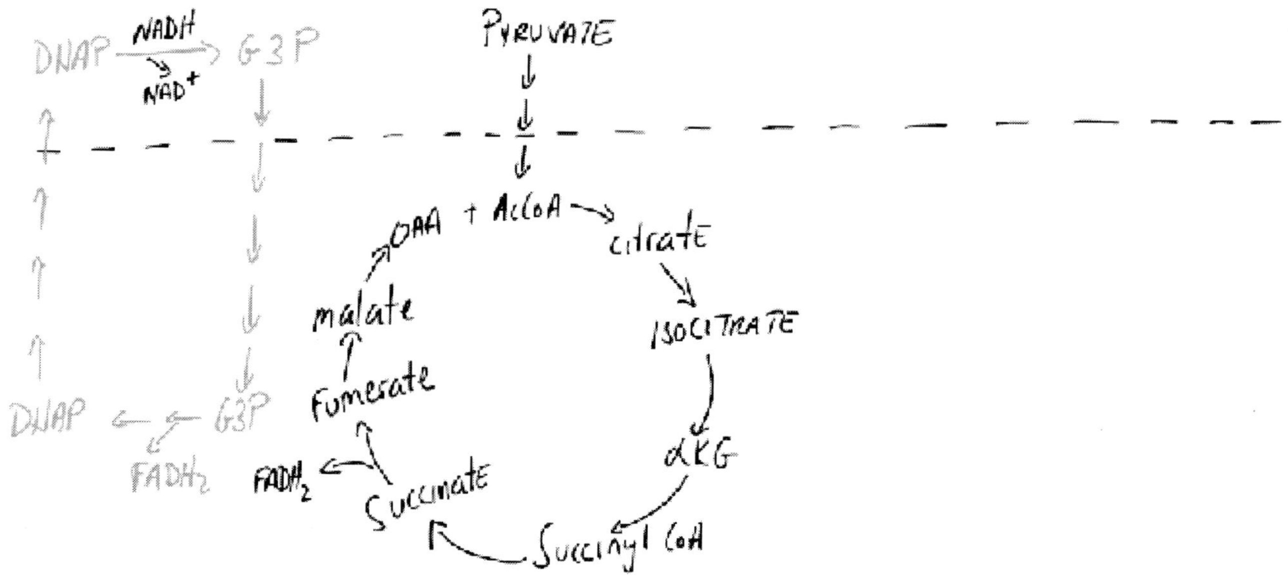

DHAP → G-3-P:

- NADH → NAD^+
- Glycerol-3-phosphate Dehydrogenase

G-3-P → DHAP:

- $FADH_2$ is formed
- 2 ATP is lost due to running the cycle twice.
- The reason for 36-38 ATPs per mole glucose in textbooks depends on which shuttle is used

NOTES:

NOTES:

Proteolysis/ Lipolysis/ Ketogenesis:

"It's not that I'm so smart, it's just that I stay with problems longer."

–Albert Einstein

PROTEOLYSIS:

PYRUVATE

AcCoA + OAA → citrate → ISOCITRATE → αKG → SuccinylCoA → Succinate → Fumerate → malate → OAA

aa + Pyruvate $\xrightarrow{ALT}$ Ala + cc

aa + αKG $\xrightarrow{GGT}$ Glu + cc

aa + OAA $\xrightarrow{AST}$ Asp + cc

LIPOLYSIS:

Fats:

- Synthesis in cytoplasm
- Breakdown in mitochondria

Fatty Acid $\xrightarrow{\text{CAT-I}}$ AcetylCoA + PropionylCoA

Activator = Glucagon

Inhibitor = MalonylCoA

"The Cat lost some Fat"

Citrate Shuttle: FA transport out of the mitochondria

Carnitine Shuttle: FA transport into the mitochondria

1) Citrate Shuttle:

- Feeds Acetyl CoA into fatty acid synthesis
- Citrate moves out to the cytoplasm via the citrate shuttle

2) Carnitine Shuttle:

- Used for long chain fatty acids
- Phosphorylation is the committed step
- Carnitine Deficiency: low energy state, fat build up in cytoplasm

βeta-Oxidation:

The process where fatty acids are broken down to form Acetyl-CoA as an entry into the Kreb cycle.

"When you lose fat => O, HOT!"

1) **O**xidation: 7 NADH → 21ATP
2) **H**ydration
3) **O**xidation: 7 $FADH_2$ → 14 ATP
4) **T**hiolysis: 8 AcCoA → 96 ATP

- Costs 15 ATP, get back 131 ATP
- Ratio is about 9:1 (9 calories per gram of fat)
- Fat repels water, thus able to store a lot of fat without excess water weight

Q: Which layer of the abdomen has the worst wound healing?
A: Adipose because it is always repelling water and has least amount of blood supply

Ketogenosis vs. Ketogenolyis:

AcetylCoA —HMG CoA synthase→ **Acetone + 3-OH butyrate**

Acetone + 3-OH butyrate —Thiophorase (not liver)→ AcetylCoA

Ketones:

1) AcCoA + AcCoA → AcetoAcetylCoA
2) AcetoAcetylCoA + AcCoA → HMG-CoA
- HMG-CoA synthase is the rate limiting enzyme for ketogenesis

3) HMG-CoA → AcCoA + AcetoAcetate
4) AcetoAcetate → Acetone
- Acetone is the only one able to be measured on a dipstick

5) AcetoAcetate → beta-OH Butyrate
- Uses NADH cofactor
- Crosses BBB
- Hydroxy group is less polar than a ketone group

<u>**NOTES:**</u>

Glycogenolysis / Glycogenesis:

"Aerodynamically, the bumble bee shouldn't be able to fly,

but the bumble bee doesn't know it so it goes on flying anyway."

–Mary Kay Ash

Glycogen: Glycogenolysis vs. Glycogenesis

1 gram of Glycogen carries 2-3 grams of water => why your liver is so big

Glycogen = branched glucose polymer, maintains blood glucose levels

- α-1,4 glycosidic bonds
- α-1,6 branches (every 10 residues)

Insulin => glycogen synthesis
Epi => glycogen breakdown

Overview:

G-6P $\xrightarrow{\text{Glycogen synthase}}$ **Glycogen**

Glycogen $\xrightarrow{\text{Glycogen phosphorylase}}$ Glucose (liver) or Lactate (muscle)

Step 1:

Glycogenin-OH + UTP

↓

Glycogenin-O-PO_4^{3-} + UDP

Step 2:

Glycogenin-O-PO_4^{3-} + UDP-Glu

↓

Glycogenin-O-Glu + UTP

Note: every time Glucose is added, UTP is created, driving Glycogen synthesis.

Glycogen Storage Diseases:

Von Gierke: G-6Pase deficiency *"von Gierke's Guts enGorged"*

Cori's: Debranching enzyme deficiency *"Cori is a short name"*

- *Short* branches of glycogen

Anderson's: Branching enzyme deficiency *"Anderson is a long branching name"*

- *Long* chains of glycogen

McArdle's: Muscle Phosphorylase deficiency *"McArdle's Muscle got Mussed up"*

- Muscle cramps w/ exercise
- Rhabdomyolysis

NOTES:

Pentose Phosphate Pathway:

"An expert is a man who has made all the mistakes

which can be made in a very narrow field."

–Niels Bohr

Pentose Phosphate Pathway:

Ribose-5P's job: Nucleotide synthesis
NADPH's job: FA synthesis, RBC membrane repair, kill bacteria

Overview:

G-6P —G-6PD→ Ribose-5P + NADPH —transketolase→ F-6P + G-3P

Glucose → G6P
Glucose ↓ Glycogen
G6P ↓ (→ NADPH)
6 Phosphogluconate
↓ (→ NADPH)
↓
↓
Ribose 5 Phos

G-6PD: Glucose-6 Phosphate Dehydrogenase

- Rate limiting enzyme for Pentose Pathway
- Activator: Glucose-6 phosphate
- Inhibitor: Ribose-5 phosphate

G-6PD Deficiency:

- ↓NADPH → hemolytic anemia, infection
- More common in Mediterranians (protects them from malaria)
- MCC of hemolytic crisis is #1: infection, #2: drugs
- Drugs that oxidize RBC's: Sulfa drugs, Antimalarials, Metronidazole, INH
- Peripheral smear: Heinz bodies

NADPH:

- Fatty acid synthesis
- DNA synthesis
- RBC repair (used by Glutathione)

Sources of NADPH:

- 90% from Pentose Pathway
- 10% from OAA converted to Pyruvate

Ribose-5 Phosphate:

- Nucleotide synthesis
- Form DNA for cellular division
- Deficiency leads to poor wound healing

NOTES:

Amino Acid Synthesis:

"Success is often the result of taking a misstep in the right direction."

–Al Bernstein

Amino Acid Metabolism:

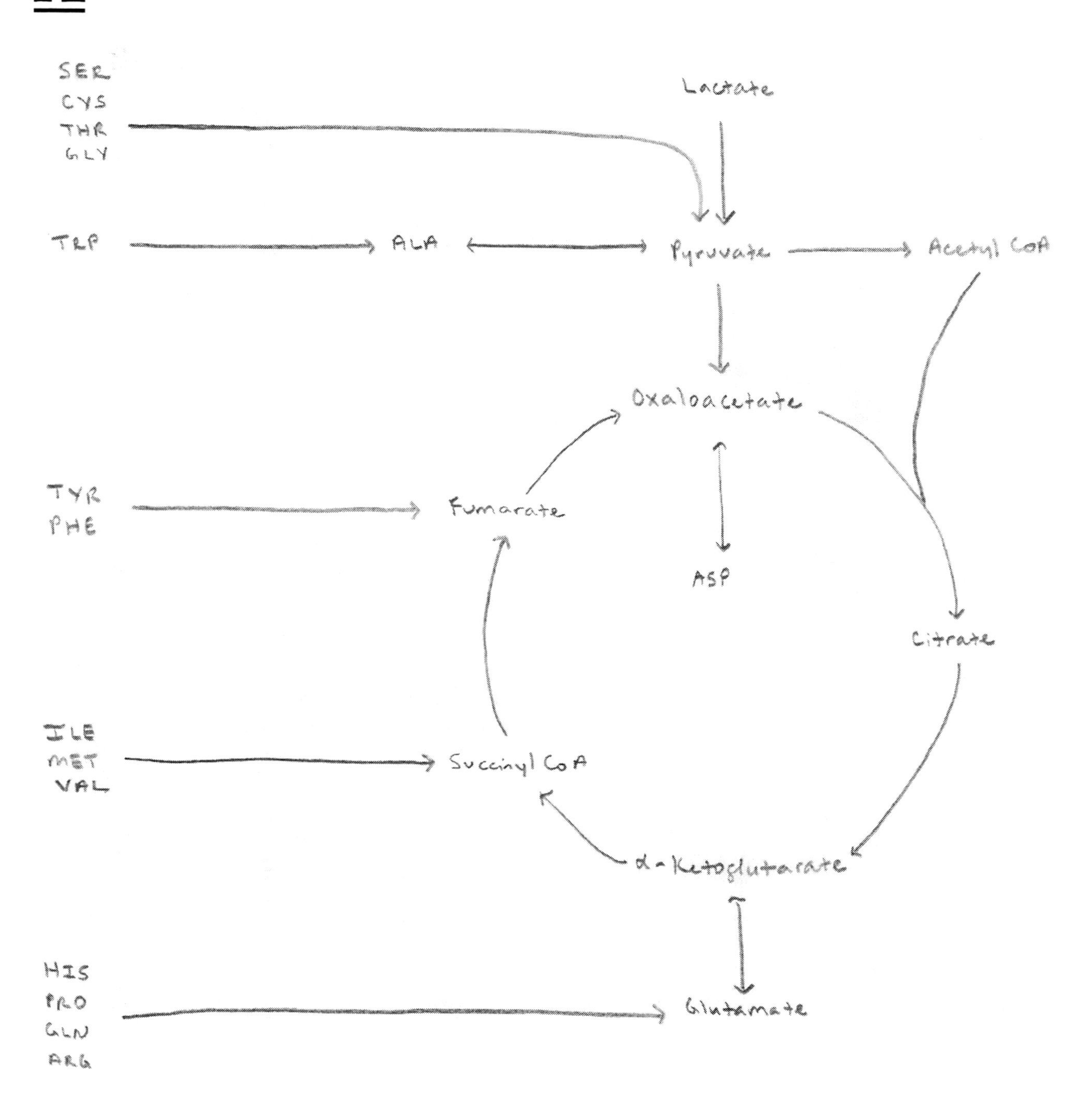

Urea Cycle: removes toxic NH_4^+
RLE: CPS-1
Activator: N-acetylglutamate
Inhibitor: None *(the only pathway that never turns off)*

90% in Liver:

- Excreted in GI tract
- Liver failure leads to death due to GABA connection

10% in Kidney Collecting Duct:

- Leads to increased BUN (dying cells lead to ↑BUN)
- Has glutaminase, which splits off ammonia

"Ordinarily, Careless Crappers Are Also Frivolous About Urination"

- **O**rnithine
- **C**arbamoyl phosphate ← in from mitochondria
- **C**itrulline
- **A**sp ← in
- **A**rginosuccinate
- **F**umarate → out to malate → Kreb cycle
- **A**rg
- **U**rea → out

Glu + NADH → αKG + NH_4^+ + NAD

- Uses Glutamate Dehydrogenase
- Liberates ammonia for Urea Cycle

NH_4^+ + CO_2 + 2ATP → Carbomyl Phosphate

- Carbomyl phosphate synthetase (CPS-1)
- Rate limiting enzyme
- Allosteric activator: N-acetylglutamate

Carbomyl Phosphate → Citrulline

- Ornithine transcarbamylase
-

Citrulline + Asp → Arginosuccinate

- Arginosuccinate synthetase
- Asp (from Mallate-Aspartate shuttle)

Arginosuccinate → Arginine + Fumarate

- Arginosuccinate Lyase

Arginine → Urea + Ornithine

- Urea gets rid of Nitrogen waste and CO_2

Hepatorenal Syndrome: Gln → Glu + NADH → α-KG + NH_4^+

- Liver failure occurs first *"hepato---"*

- High ammonia suppresses glutaminase in kidneys, leading to renal failure
- No way to get rid of ammonia, leading to coma and death
- Kidney can still be transplanted, as long as it gets away from the high ammonia

Urea Cycle Defects:

- Increase in blood NH_4^+ levels → ↑GABA → ↑serum pH
- When trying to detect which enzyme is the defect, pick the earliest enzyme in the list. If Uracil or Orotic Acid mentioned, chose the later enzyme in the list.

NOTES:

NOTES:

Fatty Acid Synthesis:

"Faith consists in believing when it is beyond the power of reason to believe."

–Voltaire

Fatty Acid Synthesis:

Function: fuel, padding, insulation

AcetylCoA —AcCoA carboxylase→ **FA**

Activator: Citrate
Inhibitors: Palmitic acid, MalonylCoA
RLE: Acetyl CoA Carboxylase

Acetyl CoA Carboxylase:

- Rate limiting enzyme for fatty acid synthesis
- Acetyl CoA → Malonyl CoA
- Biotin is cofactor
- Needs ATP
- Activator: Citrate
- Inhibitor: Malonyl CoA and Palmitic Acid

Citrate → Acetyl CoA + OAA:

- Citrate Lyase
- Uses ATP

Fatty Acid Synthase:

- Largest quarternary enzyme in biochemistry
- Seven enzymes
- Ten steps (don't need to know)

Fatty Acid Rules:

How many ATP: C-1
How many rounds: (1/2C)-1 *"2ATP per round"*
How many NAPDH: (1/2C-1) x 2

- C_{16} = max limit "Palmitic Acid"
- Double bonds must be 3C apart (Ex: C_4-C_8)
- No double bonds after C_{10} (Ex: C_{10} okay b/c it is C_{10}=C_{11})
- Saturated (w/ H) → no double bonds
- Unsaturated → it has double bonds

Essential Fatty Acids:

- Linolinic
- Linoleic (used to make arachadonic acid)

- Only source is from diet; body can not make these

Cholesterol:
Function: membrane component, precursor of bile acids, hormones => droplets

AcetylCoA —HMG CoA reductase→ Mevalonate → **Cholesterol**

Inhibitor: Dietary cholesterol

Phospholipids:
Function: components of membranes and lipoproteins

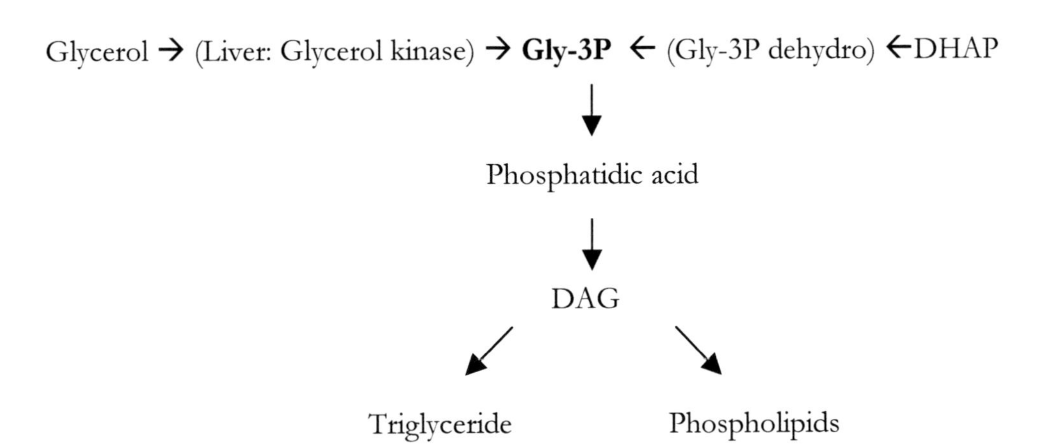

- Lecithin = Phosphatidyl Choline (emulsify fat with bile, also part of surfactant)
- Cephalin = Phosphatidyl Ethanolamine
- Phosphatidyl Isositol
- Phosphatidyl Serine
- Cardiolipin: Anti-cardiolipin Ab => recurrent abortions, clotting then bleeding

Sphingolipids:
Function: component of membranes/neuronal tissue => bilayer vesicles
- Sphingosine → Ceramide → Sphingomyelin + Cerebrosides,Gangliosides
- Sphingosine = PalmitoylCoA + serine
- Cerebroside = Ceramide + UDP-sugar
- Ganglioside = a "gang" of cerebrosides

Lysosomal Storage Diseases:
Tay-Sachs: hexosaminidase A deficiency => blindness, incoordination, dementia
Sandhoff's: hexosaminidase A/B deficiency
Gaucher's: glucocerebrosidase def => wrinkled tissue MP *"wrinkled grouch"*
Neimann-Pick: sphingomyelinase deficiency => zebra bodies (demyelination)
Fabry's (XL): α-galactosidase def => corneal clouding, attacks baby's kidneys
Krabbe's: β-galactocerebrosidase deficiency => globoid bodies

Metachromatic Leukodystrophy: arylsulfatase deficiency => childhood MS
Hunter's (XL): L-iduronosulfate sulfatase deficiency *"X marks the hunter's target"*
Hurler's (AR): α-L-iduronidase deficiency, worse form

More Clues:
Cherry-red macula: Tay-Sachs,Neimann-Pick (hepatosplenomegaly)
Gargoyle-face: Gaucher's, Hurler's *"hurl when you see grouchy gargoyle face"*
X-linked: Fabry's, Hunter's

NOTES:

NOTES:

Nucleotides:

"Love is composed of a single soul inhabiting two bodies."

–Aristotle

NUCLEOTIDES:

Fxn: Carriers (Ex: UDP), Energy (Ex: ATP), 2nd messengers (Ex: cAMP)

1) Purine synthesis and degradation

Low PRPP => salvage pathway (uses Gly, take bases from dying cells)

High PRPP => de novo synthesis (uses Ribose-5P in rapid growth periods)

Gly, Ribose-5P —PRPP, HGPRT→ **AMP/GMP**

AMP/GMP —xanthine oxidase, Mb→ Uric acid

Purine Diseases:

If you see Purines or Gly in urine → pick earlier enzyme

- Lesch-Nyhan: (HGPRT deficiency) => gout, neuropathy, self-mutilation
- Gout: (excess uric acid, *dehydration causes crystals*) => podagra
- SCID: (adenosine deaminase deficiency) => decreases rapidly dividing cells

2) Pyrimidine synthesis and degradation

Gln —Asp transcarbamylase, CPS-II→ Orotic acid —THF→ **UMP**

UMP → CO_2 + Urea

Pyrimidine Diseases:

If you see pyrimidines/orotic acid in urine → pick later enzyme

- White diaper crystals = excess orotic acid

Types of RNA:

rRNA – most abundant, comes from nucleolus

mRNA – most variable, largest *"Big Mama"*

tRNA – smallest (AUG = start codon, UAA, UAG, UGA = stop codons)

SnRNPs

Types of DNA:

A: R hand helix, 10 bp per turn

B: R hand helix, 11 bp per turn – we have this

Z: L hand helix, 12 bp per turn – prokaryotes have this, more compact

Base – closest to neighbor 4 doors down (3.6 bases per turn)

Big Picture:

DNA synthesis → Replication (make 2nd strand) → Transcription (make DNA babies)

Replicate: 5' → 3' (RNA Pol replicates it with "U")

Read: 3' → 5'

DNA synthesis:

G0: do ***nothing*** (cardiac, neurons = permanent cells) (liver, kidney can be bribed to leave)

G1: make all proteins for DNA synthesis

S: do ***synthesis*** (and make centrioles)

G2: make all proteins for mitosis

M: do ***mitosis***

- **I**nterphase: **i**ntermission
- **P**rophase: nuclear membrane dissolves; chromosomes clump, then **p**air up
- **M**etaphase: line up in the **m**iddle
- **A**naphase: pull **a**part, non-disjunction may occur
- **T**elophase: cell **t**ears in two

> **Histones:** wrap around DNA
> H1 → linker protein
> H2a, H2b, H3, H4 → nucleosome

DNA Replication: Prokaryotes

DN**A-A**: uses **ATP** to denature poly**A** tail

SSB: stabilizes 2 **s**ingle **s**trands

Helicase: uses ATP to break bonds

Primase: RNA Pol lays down 8-10 nucleotides

DNA Pol III: 5' → 3' polymerase, has 3' → 5' exonuclease

DNA Pol I: DNA Pol III fxn *plus 5' → 3' exonuclease*

DNA ligase: makes final bond

Topisomerase: removes supercoils

- Topo **I** cuts **one** strand, spins around **once**, removes **one** supercoil
- Topo II can insert a negative supercoil

> ***HIV:*** *only virus to inhibit a proofreader*

DNA Replication: Eukaryotes (multiple replication forks and are bidirectional)

α = prim**a**se

β = DNA Pol I

γ = mitochondrial DNA only *"Gamma = The Geek"*

δ = DNA Pol III leading strand

ε = DNA Pol III lagging strand

> **Heterochromatin** = tightly coiled
> **Euchromatin** = loose (10nm fibers)

DNA Transcription: replicate 1 gene

- Little Ms. Zinc Fingers – the librarian that locates genes when you need to transcribe
- RNA Pol
- Core enzyme (α, β, β' subunits): recognizes –35 sequence (CAAT box)

"The Core can reach far away to -35"

- Holoenzyme (core, σ factor): recognizes –10 sequence (TATA, Pribnow box)

"The Holoenzyme is too fat to reach that far"

- Promoter
- Enhancer
- Initiator
- Repressor – can bounce from one chromosome to another

Methyl Donors:

Bio**tin** – for carboxyla**tion**

THF – for nucleotides

SAM – for all other rxns

Post-Transcription Modification:

1) splice away introns, smoosh exons together
2) add 3' polyA tail → sticks to polyU "Shine Delgarno sequence" on 30S *"Shiny Apple"*
3) add 5' guanosine cap
4) transport to cytoplasm
5) methylate guanosine
6) ready for translation

Euks are monocistronic:

1 mRNA → 1 protein

Proks are **p**olycistronic:

1 mRNA → many proteins

Protein Translation: takes 4 GTP per amino acid

- tRNA 3' end: CCA-aa, uses 2 GTP to activate
- 30S subunit made => initiation factors released (IF_2)
- 50S subunit made => creates A, P sites
- AUG → falls into P site => Met (Euks) or fMet (Proks)
- Elongation factors released (EF_2) – brings tRNA over, uses 1 GTP
- Peptidyl transferase – makes peptide bond
- Translocation – costs another GTP

Mutations:

Silent: same aa

Point: change 1 base

- transition: 1 purine to another
- trans*version:* change families from purine to pyrimidine *"converted"*

Frameshift: insert or delete 1-2 bases => early onset diseases

Missense: mistake => late onset diseases

Nonsense: stop codon (UAA, UGA, UAG) => early onset diseases

"U Are Away, U Go Away, U Are Gone"

Lab Tests:

PCR = most specific test

1) add a ton of primers
2) wash off excess
3) add heat-stable DNA polymerase
4) denature

Lab Tests:

*"SNoW DRoP"**

Southern blot → **D**NA

Northern blot → **R**NA

Label thymidine – DNA

Label cytosine – RNA

Nomenclature:
Pyrimidines: C,U,T *"CUT the Py"*
Purines: A,G
Higher Tm: C-G (more bonds)

U → (CH_3) → T
U → (NH_3) → C

Guanine = Base
Guanosine = Base + Sugar

NOTES:

Immunology:

"The difference between an itch and an allergy is about 100 bucks."

–Anonymous

Immunogen: >6kD Ag that can set off an immune response, looks "different", has variability => tries to decrease immunogenicity by making it more like self:

- Strep will cover itself with basement membrane => immune response to Strep and our BM
- After bacteria is gone, our body will have autoimmune response:
  - Post-Strep GN
  - Rheumatic fever
  - Goodpasture's

Hapten: <6kD Ag that is too small to set off an immune response (Ex: virus)

Antigen: protein (except cardiolipin)
SuperAntigen: crosslinks APC-MHII….TCR β-chain => activates T cells

Carrier Effect: Macrophages make hapten recognizable: (Ex: vaccination)

- Ingest: MP eat hapten
- Phagosome formation
- Digest
- Present
- MHC-II presentation (on β-chain variable region => displace the invariant region)
- V-beta region
- Invariant chain is displaced
- Release IL-1 =>
- Fever (↑BMR to make stuff move faster)
- Non-specific signs of illness
- Recruit T_H to amplify the immune response => secrete other IL's

Thus, MP commit suicide by binding the hapten, then get killed

Acute Phase Reactant: proteins made during an inflammatory response

<u>Fever:</u>
1° degree will ↑HR by 10bpm → immune cells will come faster; IgA secretion
Thus, ↓HR → Heart Block

<u>Heart Block Bugs:</u> *"LSD Loves Company"*

- **L**egionella: pneumonia
- **S**almonella: typhoid fever
- **D**iphtheria: Corynebacterium diphtheriae
- **L**yme disease: Borrelia borgdorfori
- **C**haga's: Trypanosoma cruzi

Limit Infection (except Shigella, who can cause infxn with only 10 bugs):
Detergent: impairs *adhesion* of pathogen

Disinfectant/Antiseptic: *inactivates toxins* by dissolving their anchoring membrane
- Ex: Phenol, Iodine

Sterilization: *kill spores* (121°C/vaporized)
- Autoclaving has an expiration date => release new toxin – bacteria are packin' now…
- Spore = inactive bacteria, can't replicate, can release toxin
- Only two bacteria form spores: **B**acillus and **C**lostridium *"Be Careful of the spores"*

The Two Arms of The Immune Response:
This is a mind-blowing concept – so hold onto your seat. We all know that there are two arms of the immune response, the humoral and cell-mediated. But, if you can decide whether a disease is humoral or cell mediated, you can predict what types of cells will be found there, what type of culture you should order, and whether an antibiotic is going to help or not. So, ask yourself the question: Are B cells and PMNs involved or T cells and macrophages? If you don't know that; is bacteria involved or not? If not, just follow the most common list of cell-mediated killers listed in the most common order of occurrence from viruses to fungi and down…

	Humoral:	**Cell Mediated:** *nutrition affects here first*
Patrols:	Blood → do culture	Tissue → do biopsy
Policemen:	B cells PMNs (TH_2)	T cells Macrophages: • Blood = Monocytes • Brain = Microglia • Lung = T1 pneumocytes • Liver = Kupffer cells • Spleen = RES cells • Lymph = Dendritic cells • Kidney = Mesangial cells • Peyers patches = M cells • Skin = Langerhans • Bone = Osteoclasts • Connective Tissue: 1. Histiocytes 2. Giant cells 3. Epitheloid cells

The Bad Guys:	Bacteria	*"Very Foolish to Meet the Parents Post-Nuptially"* • **V**irus (CMV, EBV = most common) • **F**ungus • **M**ycobacterium • **P**rotozoa – kills you • **P**arasite • **N**eoplasm

NOTES:

NOTES:

Immunodeficiencies:

"A successful man is one who can lay a firm foundation with the bricks others have thrown at him."

–avid Brinkley

How can you have 1 antibody, but not another? T/B cell interaction.

- IL-4
- Class switching
- CD-40: receptor
- Tyr kinase

T CELL DEFICIENCIES: *die of viral infection*

DiGeorge Syndrome: 3rd/4th pharyngeal pouch → no thymus, inf. parathyroids → ↓Ca^{2+}

- Only T-cell deficiency with an electrolyte problem

Chronic Mucocutaneous Candidiasis: T cells can't kill Candida albicans → chronic fatigue

Steroids:

- Kills T cells and eosinophils
- Inhibits macrophage migration
- Stabilizes mast cell membranes
- Stabilizes endothelium
- Inhibits Phospholipase A
- Proteolysis
- Gluconeogenesis
- Upregulates all receptors during stress

Betamethasone – inhaled, induces surfactant
Beclomethasone – induces surfactant
Danazol – tx endometriosis
Dexamethasone – best CNS penetration, tx cerebral edema, tx meningitis (prevents inflamm)
Fludrocortisone – best mineralcorticoid (acts like Aldo)
Fluticasone – nasal spray
Methylprednisolone – IV
Mometasone – nasal spray
Megestrol – tx appetite loss in cancer pts
Prednisone – oral
Triamcinalone – inhaled
Hydrocortisone – topical and injectable
Cyproterone – tx prostate cancer

Cyclosporine: blocks T cell function via calcineurin → can't produce IL

- Gingival hyperplasia, hirsutism, renal failure (PCT)

Tacrolimus: less side effects than cyclosporine

Hairy Cell Leukemia:

- Fried egg, "sun burst" appearance
- t (1:19) → good prognosis
- TRAP + *"Trap the hairy fried eggs"*
- Tx: Cladribine

T-Cell Lymphomas:

- Mycosis Fungoides → rash
- Sezary Syndrome→ found in blood (indented cell membrane = "crenation")

B CELL DEFICIENCIES: *die of bacterial infection*

Bruton's Agammaglobulinemia(XL): kids w/ defective Tyr kinase, arrest at pre-B stage, no Ab

- B cell count is normal, but function is lacking
- Lung/sinus infection

Stimulate B cells:

- Pokeweed mitogen
- Endotoxin

CVID: Bruton's with onset after 1y/o

Leukemias/Lymphomas: usually involve B cells

Plasmacytomas: one osteolytic lesion

Multiple Myeloma: multiple osteolytic lesions, IgG (M-spike), κ light chain (Bence-Jones proteinuria), plasma cell CA, rouleaux (Tx: Melphalan)

Heavy Chain disease: IgA and Multiple Myeloma of GI tract, malabsorption *"Gee, I Am so Heavy!"*

Selective IgA deficiency: transfusion-related **a**naphylaxis (use IgA filter), lung/GI infections

Selective IgG_2 deficiency: recurrent encapsulated infections

Encapsulated Organisms:

Gram +: Strep pneumo

Gram –:

"Some Killers Have Pretty Nice Capsules"

- **S**almonella
- **K**lebsiella
- **H.** influenza B

- **P**seudomonas
- **N**eisseria
- **C**itrobacter

Job-Buckley: common in redhead females w/ fair skin

- Class switch problem => stuck in IgE stage
- Recurrent Staph infection

Common Variable Hypogammaglobulinemia: young adults w/ B cells don't differentiate into plasma cells

T AND B CELL DEFICIENCIES:

Ataxia Telangiectasia: DNA endonuclease defect

- Ataxia (wheelchair-bound), thymus hypoplasia, recurrent sinus infections

SCID: (no adenosine deaminase) => baby dies by 2y/o, "frayed" long bones, no thymus/LN

- Tx: bone marrow transplant

Wiscott-Aldrich (XL): ↓IgM/platelets, eczema, petechiae, ↑lymphoma, lung infections

HIV: RNA retrovirus looks for CD4 receptors, inhibits proofreading, likes acidic medium

- HIV invades T_H and kills them
- HIV is the only virus that does not penetrate cells, it injects its RNA into cells
- CCR4/5 mutation on CD4 cell => HIV can't attach to inject virus into cell

- CD4 receptors:
  - Blood vv => vasculitis/Kaposi's sarcoma ("violaceous" nodules/HSV-8)
  - Brain => CNS lymphoma
  - Genetalia: testicles, cervix, rectum

- HIV reservoir: Lymph node germinal centers (follicular dendritic cells)

- Fastest growing populations:
  - Heterosexual black females (risk factor: bisexual male)
  - Elderly – due to greater detection of past infxn
  - More common: Male → Female (HIV likes mucosa)
  - Lowest risk: Prepubertal female (don't have acidic mucosa yet)

Cancers Common in HIV:

Cervical cancer – very aggressive

Kaposi's sarcoma

CNS Lyphoma

Testicular Lymphoma

HIV Protein:	Function:
GP 41	Portal of entry for RNA
GP 120	Attachment to CD4 receptor
Pol	Integration into our DNA
CCR5	Entry into cell
Reverse transcriptase	Transcription/replication
P17	Assembly
P24	Assembly

HIV Vaccinations:

- dT
- Hep A/B
- Influenza
- Pneumovax
- MMR (B cells work)

HIV Transmission: sex, blood, Mom (pregnancy/breast milk)
AIDS definition: CD4 <200/μL or clinical sx

HIV screening: began in 1985
1) ELISA: detect IgG Ab to p24 Ag (develops 1mo after exposure)
2) Western Blot: must see at least 2 proteins
3) PCR: use if <18 mo (b/c mom's IgG will make ELISA positive) – detects viral RNA

HIV Infections: *most specific acute infection = oral ulcers*
Acanthamoeba: encephalitis
Bartonella Henselae: bacillary angiomatosis *(pathognomonic for AIDS)*
Candida: white eye lesions (Tx: Amphotericin B/Flucytosine)
Cryptococcus: red umbilicated papules (do biopsy), meningitis
- Tx: Amphotericin B/Flucytosine → lifetime Fluconazole

Cryptosporidia: watery diarrhea, partial acid fast
Cytomegalovirus "CMV": big shallow esophageal ulcers, bloody diarrhea, yellow retinitis
- Tx: Ganciclovir or Foscarnet

Herpes Simplex Virus "HSV": small deep esophageal ulcers (Tx: Prednisone)
JC virus: PML=Progressive Multifocal Leukoencephalopathy → brain demyelination (6mo: death)
Microsporidia: diarrhea
Pnemococcus: pneumonia, silver stains, "crushed Ping Pong ball"
Toxoplasmosis: fluffy retina lesions (Tx: Pyramethamine/Folate → lifetime Bactrim)

CD4:	Disease:	Treatment:
<500	TB	INH/Vit B_6
<200	PCP: diagnose w/BAL, follow LDH	Bactrim (IV) or Pentamidine (inhaled) Prednisone – add if hypoxic
<100	Candida	Fluconazole
<100	Toxoplasmosis	Sulfadiezene + Pyramethimene/Folate
<50	MAI	Azithromycin

HIV Post-exposure Prophylaxis: 4 wks *"LIZ"*

Don't need permission to check pt blood if healthcare worker get stuck w/ needle…

- **L**amivudine (3TC)
- **I**ndinavir
- **Z**idovudine (AZT)

HIV Tx:

- 2 reverse transcriptase inhibitors + 1 protease inhibitor
- HAART: if CD4<350 or symptoms

Reverse Transcriptase Inhibitors:

1) Nucleoside analogs: painful neuropathy, pancreatitis

- Zidovudine (AZT): T analog => low Epo, aplastic anemia, myopathy, nosebleeds
- Stavudine (d4T): T analog
- Didanosine (ddI): A analog => pancreatitis ← only purine analog
- Zalcitabine (ddC): C analog => painful neuropathy
- Lamivudine (3TC): C analog

2) Non-nucleoside analogs:

- Nevirapine: rash
- Delavirdine: rash
- Efavirenz: rash, vivid dreams, may be teratogenic

HIV Markers:

Check q3 mo

CD4 count: status of dz

Viral load: progression of dz

Protease Inhibitors: *inhibit assembly, cause back fat pad*

- Indinavir => kidney stones, liver toxicity, thrombocytopenia
- Nelfinavir
- Ritonavir
- Saquinavir => quick resistance

NEUTROPHIL DEFICIENCIES:

Neutrophils: have MPO and NADPH oxidase to kill anything that comes along…

Anaerobes: no SOD

Aerobes: have lots of SOD

$$O_2 \xrightarrow{\text{NADPH oxidase}} \text{Free radical} \xrightarrow{\text{SOD}} H_2O_2 \xrightarrow{\text{MPO}} HOCl$$

$$H_2O_2 \xrightarrow{\text{Catalase}} H_2O + O_2$$

Chronic Granulomatous Disease (XL): NADPH oxidase deficiency

- Recurrent Staph/Aspergillus infections
- Nitroblue Tetrazolium stain negative → yellow

- Tx: INF-γ

Myeloperoxidase Deficiency: Catalase (+) infections (Staph/Pseudo/Neisseria)

Staph/Pseudo Infections: *"ABCD"*
Agranulocytosis: absolute neutropenia
Burn patients
Cystic Fibrosis
Diabetics

MACROPHAGE DEFECTS:
Monocytes: Macrophages in circulation
- Change names in different tissues, uses INF
- Only has NADPH oxidase to kill

Macrophages:
- Kills everything that enters tissues
- Processes Ag → presents to T_H during Ab formation

Extreme Monocytosis (>15%): *"STELS syndrome"*
Syphilis – chancre, rash, warts
TB – hemoptysis, night sweats
EBV – teenager sick for a month
Listeria – baby who is sick

Macrophage Deficiency:
Chediak Higashi *"lazy lysosome syndrome"*: lysosomes are slow to fuse around bacteria
- Oculocutaneous albinism

Poikilocytosis = different shapes
Anisocytosis = different sizes

Sulfa Drugs: *mimic PABA*
- Displace albumin
- Anaphylaxis
- Interstitial nephritis
- Hemolytic anemia
- MetHb
- Kidney stones

Sulfamethoxazole/Trimethoprim "Bactrim": tx UTI
Sulfadiazine/Pyrimethamine: tx burns
Sulfacetamide: eyedrops to prevent newborn Chlamydia
Sulfasalazine: tx UC
Sulfapyrazone: tx UC

NOTES:

Leukocytes:

"If you do not hope, you will not find what is beyond your hopes."

–St. Clement of Alexandra

Meet the Family:

Neutrophil: The Phagocyte (has anti-microbials, most abundant)
Eosinophil: The Parasite Destroyer, Allergy Inducer
Basophil: The Allergy Helper (IgE receptor => histamine release)
Monocyte: The Destroyer => MP (hydrolytic enzymes, *coffee-bean nucleus*)
Lymphocyte: The Warrior => T, B, NK cells
Platelets: The Clotter (no nuclei, smallest cells)

Naming Pattern:
--------blast
Pro-----cyte
--------cyte
Meta---cyte

Blast: baby hematopoietic cell *"Blast them all"*
Band: baby neutrophil (has maximal killing power)

Embryology of Hematopoietic Cells:
Blood Dwellers:
Normoblast → Reticulocyte → RBC
Megakaryoblast → Platelet *"big to little"*
Monoblast → Monocyte → MP

Tissue Dwellers:
Lymphoblast → Lymphocytes →
- *NK cells*
- *T cells* (**T_H=CD4, T_K=CD8**)
- *B cells* => plasma cells

Myeloblast → Band "stab" cell → WBC "leukocyte" → Granulocytes:
- *PMNs* = "neutrophils"
- *Eosinophils*
- *Basophils* => mast cells

WBC: 90% are marginated along blood vessels (most are mature), 10% are in circulation
Under extreme stress the body will demarginate even immature WBC
Steroids/Cortisol/Epi => demargination (w/ low eosinophils, low T cells, high PMNs)
1) Pavementing:
Selectins: select mature WBC out of circulation
Integrins: (via ICAM-1) integrate WBC into endothelium
2) Margination = flatten (Epi and cortisol cause this)
3) Diapedesis = moves like a slinky looking for a break in endothelium
4) Migration = slide into the tissue

Decreased WBCs "Leukopenia":
Virus: Parvo B-19, Hep E, Hep C

Drugs: AZT, Benzene, Chloramphenicol, Vinblastine

Increased WBCs:

+ High PMNs: "Stress demargination"
+ Blasts (<5%): "Leukemoid Rxn = extreme demargination, looks like leukemia" (Ex: burn pt)
+ Blasts (>5%): Leukemia
+ Bands (immature neutrophils, max germ-killing ability): "left shift" => have Infection
+ B cells => Bacterial infection

Agranulocytosis (↓WBC): *stomatitis is earliest sign*

- Carbamazepine
- Ticlopidine
- Clozapine
- PTU

Myelodysplastic Syndromes: ↑stem cells => ↑RBC, WBC, platelets

Polycythemia Rubra Vera: Hct >60%, ↓Epo, *"pruritis after bathing"*

- Gout, splenomegaly, ruddy appearance/cyanosis
- Tx: Phlebotomy

Essential Thrombocythemia: Plt >600k → stroke/DVT/PE/MI, stainable Fe, ↓c-mpl (TPOr)

Myelofibrosis: fibrotic marrow => teardrop cells, extramedullary hematopoiesis, poor prognosis

Aplastic anemia: bone marrow replaced with fat, low retics

- Viruses: Parvo-B19, Hep E, Hep C
- Drugs: AZT, Benzene , Chloramphenicol, Vinblastine

Plasma Neoplasms: *produce lots of Ab*

Waldenstrom Macroglobulinemia: IgM, hyperviscous, M spike

Monoclonal Gammopathy of Undetermined Significance: old people w/ gamma spike

- Peripheral neuropathy

Multiple Myeloma: old people with back pain

- Multiple "punched out" osteolytic lesions
- IgG (M-spike)
- κ light chain (Bence-Jones proteinuria)
- Rouleaux
- Poor prognosis: ↓albumin, ↑vascularity, ↑Ca, ↑LDH, ↑IL-6, ↑Creatinine
- Dx: Serum protein electrophoresis, bone marrow biopsy, skeletal survey
- Tx: Melphalan + Prednisone

Histiocyte (MP) Neoplasms:

Langerhans Cell Histocytosis "Histocytosis X":

- Kids w/ eczema, "punched out" skull lesions, diabetes insipidus, exophthalmos

Leukemias: *Arsenic is toxic to leukemia cells*

Acute: started in bone marrow, squeezes RBC out of marrow

Chronic: started in periphery, not constrained => will expand

Myeloid: ↑RBC, WBC, platelets (↓lymphoid cells) => do bone marrow biopsy

Lymphoid: ↑NK, T, B cells (↓myeloid cells) => do lymph node biopsy

	ALL: *most common*	**AML:** *worst prognosis*	**CML**	**CLL:** *best prognosis*
Age:	0-15	15-30	30-50	>50
Gender:	Male	Male	*Female*	Male
Subtypes:	<u>Morphologic:</u> **L1:** scant cytoplasm **L2:** irregular nuclei **L3** = B-ALL <u>Phenotypic:</u> **B-ALL:** TdT(-) **T-ALL:** mediastinal mass (bad)	**M3 "Promyelocytic Leukemia":** t(15,17) Tx: Retinoic Acid **M5:** bleeding gums **M7:** Down's, anti-platelet Ab	<u>Chronic:</u> ↑WBC <u>Accelerated:</u> ↓RBC, ↓platelets <u>Blastic:</u> ↑blast cells Red plaques	Diffuse lymphadenopathy *Average survival: 3 yrs*
Presentation:	Infxn: Bacterial Bleeding/petechiae Teardrop cells Low energy state Bone pain Thrombocytopenia	Same as ALL	Go everywhere macrophages go	Small mature lymphocytes, "soccer ball" nuclei
Markers:	PAS stain ⊕ TdT ⊕ Calla ⊕ => good chemo response t(4,11): bad prog t(12,21): good prog	Sudan Stain Auer rods (M3) MPO	t(9,22) "Philadel" bcr-abl ↓LAP	(CD 19,20) Smudge cells = *fragile WBCs* Mutated V_h genes: good prog
Tx:	Daunorubicin	Daunorubicin	Imatinib = "Gleevac"	Chlorambucil

Lymphomas: *recurrent urticaria, eosinophilia*

	Hodgkin's: *20-40 y/o*	**Non-Hodgkin's:** *ileum (↑lymphoid tissue), affects immunocompromised*
Staging:	Pathology Staging: I: 1 group of lymph nodes II: 2 groups (same side of diaphragm, 90% cure) III: 2 groups *across diaphragm* IV: metastases (lymph node → organs) A = without symptoms B = sx: weight loss, fever, night sweats	Histology Staging: I: limited to first layer II: second layer III: local invasion IV: metastasis Check for: HIV (ELISA) Colon Cancer (colonoscopy)
Common Types:	EBV a) Lympho Predom: best prognosis b) Lymphocyte Depleted: worst prog c) Mixed lymphocyte/histiocyte d) Nodular Sclerosis: low RS **Reed-Sternberg:** B cells with bad Ig, owl's eyes, CD30, lacunar cells Albumin <4: poor prognosis	**B-cell:** *most* Follicular: t(14,18), bcl-2 Burkitt: t(8,14), c-myc, starry MP, EBV • American kids: abdominal mass • Poor: jaw mass **T-cell:** *rare* Mycosis Fungoides: "bathing suit" rash Sezary syndrome: cerebriform cells = MF mets to blood (Tx: radiation)
Chemo Tx:	MOPP or ABVD	CHOP

NOTES:

Lymphocytes:

"Seek the lofty by reading, hearing and seeing great work at some moment every day."

–Thornton Wilder

Lymphocytes:

Lymph nodes include tonsils and Peyer's patches
Lymph node biopsy => looking for cancer

T cells: Process 1 Ag => mature now
B cells: Take a long time to mature => why most lymphomas are B-cell

	Development site:	Maturation site:	Differentiation site:
B cell:	Bone Marrow	Bursa of Fabricius Equivalent *(unknown etiology in humans, choose lymph node if necessary)*	Lymph Node: *Germinal center "Follicular"*
T cell:	Bone Marrow	Kids: Thymus (Thymosin, TPO) Adults: Lymph Nodes	Lymph Node: *Paracortical*

B-cell maturation:
CD 9/10: Immature B-cell
CD 19/20: Mature B-cell
CD 40: The communication device that T cells use to talk to B cells

IL-4/5: B-cell Differentiation

- MP ingests antigen
- MP forms phagosome
- Lysosome digests antigen
- MP presents antigen
- Antigen binds MHC-II

Stages of B cell Development:

- **Pre-B cell:** heavy chain rearranges → μ-chain in cytoplasm
- **Immature B cell:** Monomeric IgM on surface
- **Mature B cell:** secretes Ab, *has "MD" on surface* due to alternative RNA splicing

IL-3/6: B-cell Proliferation

- Label Thymidine (dividing DNA) to detect *Pokeweed mitogen, Endotoxin*

Cytokines:
IL-1: recruits TH_1 => fever
IL-2: T-cell prolif, *"The Most Potent IL" → recruits everyone*
IL-3: B cell proliferation
IL-4: B-cell differentiation => IgE, class switching
IL-5: B-cell differentiation => class switching for IgA only

MP => IL-1
T_H cells => everything else

T_{H1}: IL2, INF-γ (cell-med)
T_{H2}: IL4,5,10 (humoral)

IL-6: Acute phase reactants
IL-8: PMN recruiter
IL-10: suppresses cell-mediated immunity (T/MP)
IL-12: enhances cell-mediated immunity (T/MP)
IFN-α: inhibits viral replication
IFN-β: inhibits viral replication (used to treat MS)
IFN-γ: activates MP and NK cells, suppresses humoral immunity
TNF-α: weight loss (↑BMR), secreted by MP => cellular death ← DIC/septic shock
TNF-β: weight loss (↑BMR), secreted by T cells => cellular death

<u>T-cell maturation:</u> "The Foreman" – they are the boss, only a few of them in the periphery
Sgt. Thymosin/Lt. TPO direct thymus maturation: *"Some of you will make it, some will not..."*

CD3/4/8: immature T cells

<u>Round 1:</u> Negative selection – get a label or die (T_K-CD28....B7-APC)
CD8 = T_K/T_S (low affinity), live in tissue

- T_S: suppresses infection from spreading
- T_K: kills via:

1. perforin attack
2. send B cell to coat w/Ab → MP eat it "ADCC"

- Responds to MHC-I = self
- All nucleated cells in body express MHC I (except RBCs/Platelets, not recognized as self)

CD4 = T_H (high affinity), live in blood

- Responds to MHC-II = non-self
- TH_1 => cell-mediated
- TH_2 => humoral

<u>Round 2:</u> Positive selection – lymphocytes must walk through the affinity detector
Sheriff T_K cell waits by the lymph node:
Sees speedy RBC without his MHC-1 identification
Catches him when the RBC gets old and slows down (Day 120)

RBC are not recognized as self => involved in all autoimmune diseases
Spleen removes the dead RBC => extravascular hemolysis

<u>NK cells:</u> The "Men In Black" – no one knows where they came from, have lots of guns
CD16/56: NK cells
Lymphocytes just pat your head to see if you have *some* MHC-1 Ag
NK cells require that you have exactly *100 MHC-I antigens* or they will kill you...
<100 MHC-1 => cell was invaded
>100 MHC-1 => cell is cancerous

NK Actions:

1) Apoptosis: cell membrane degeneration
2) Perforation
3) Direct B-cell to coat victim w/ Ab

Immunoprivileged Sites:

no lymphatic flow => no Ag
=> easy to transplant

- Brain
- Cornea
- Thymus
- Testes

Vaccinations:

Vaccine = Ag
Toxoid = inactivated **toxin**

Vaccination Contraindications:

- Previous vaccination: anaphylaxis, encephalopathy
- No MMR if febrile, pregnant, or allergy to Neomycin
- MMR is OK with egg allergy, TB, breastfeeding

Antibody Response Times (Show up/ Peak/ Gone):

	Day:	Week:	Month:	Year:	Year:
IgM:	3	2	2		
1st IgG:	--	2	2	1	
Memory IgG:	3	--	--	5	10

Child Vaccinations: *Pertussis is #1 preventable childhood dz*

Birth: All baby has is IgM (MMR dz is uncommon in US babies)
6 mo: Baby's making IgG now
18 mo: Baby has memory IgG now (more susceptible to MMR now)
6 y/o: Can wait until memory IgG is gone
16 y/o: Not susceptible to HIB/Diphtheria/Pertussis (d is the carrier)

	Birth:	2mo:	4mo:	6mo:	18mo:	6y:	16y:
MMR:					X	x	x
IPV:		x	x	x	X	x	x
HIB:		x	x	x	X	x	
DPT:		x	x	x	X	x	
dT:							x
Varicella:					X		
Hep B:	+ Ig	x			X		

For children <6 y/o with no vaccines, just start schedule now

Adult Vaccinations:

Hepatitis B: 3 shots

- If Ab titer is negative "non-responder" → repeat vaccination
- If stuck w/ infected needle → 2 doses of Hep B Ig

Influenza: >65 y/o, COPD, health care workers, chronic illness
Pneumococcal: >65 y/o, ill, post-splenectomy, myeloma (only give once per lifetime)
Tetanus: every 10y
Varicella: 2 shots (>1 y/o)
Hep A: Drug users, homosexuals, liver dz, clotting factor d/o

Travel Prophylaxis:

- Hep A
- Typhoid
- Yellow Fever
- Meningococcus
- Malaria: Chloroquine

Toxoid: inactivated toxin
Killed vaccines: humoral immunity
Live vaccines: humoral and cell-mediated immunity

Live Vaccines: *Don't give if family has AIDS, cancer, or chronic steroid use*
"Bring Your Own Very Small Virus + MMR"

- **B**CG
- **Y**ellow Fever
- **O**PV (Sabin): oral polio
- **V**aricella
- **S**mallpox
- Rota**V**irus
- **M**easles = Rubeola: shed in stool for 8 wk => tell Mom not to get pregnant now
- **M**umps
- **R**ubella = German 3-day measles

Vaccine Allergies:

Measles: made from eggs
Yellow Fever: made from eggs
Influenza A: made from eggs
Hep **B**: made with **B**aker's yeast

The Immunoglobin:
Isotype: type of heavy chain
Idiotype: site Ag binds to (part of Fab) "hypervariable region"
Allotype: ϰ light chains/ IgG subclass heavy chains; genetic diseases
Xenotype = Heterotype: difference between two species
Epitope: Ag when it is bound to Ab

Fab: binds Ag, variable region (contains heavy/light chains)
Fc: binds complement, constant region (contains heavy chains)

Heavy Chains: M, A, G, E, D
Light Chains: 90% Kappa
10% Lambda

Papain: destroys Ab *"it's a painful death"*
Pepsin/Trypsin: leave Fab alive

Class Switch: T cell talks to B cell and keeps Fab → IL-4 hooks a new heavy chain on

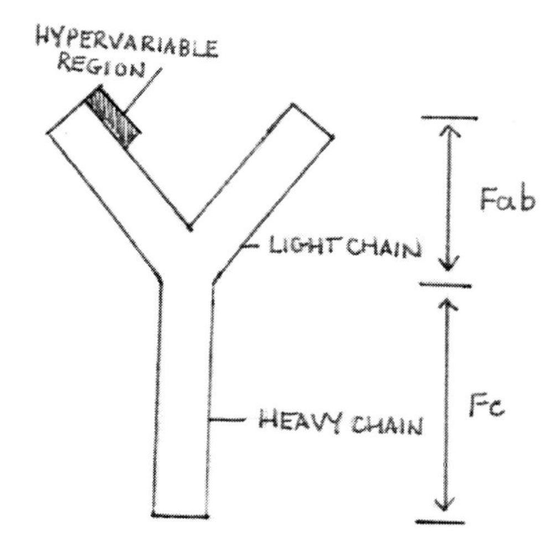

IgM: 1st dude on the scene, most avidity (has 5 arms, most likely to bind)

- Fix complement on C_1 and C_3
- First to be secreted by plasma cells, then switches to other types
- Surface: monomer
- Secretions: pentamer (2 J-chains) *"joiner"*
- Ex: Waldenstrom's hyperviscosity(↑), Wiscott-Aldrich(↓)

IgG: 2nd dude on scene, highest affinity (only 1 arm => has to grab on tightly)

- Fix complement on C_2
- Surface: monomer
- IgG_1 – crosses placenta → erythroblastosis fetalis
- IgG_2 – opsonin → encapsulated infections, most common
- IgG_3
- IgG_4 – not fix complement
- Ex: Multiple myeloma(↑), Selective IgG_2 deficiency(↓), Rheumatoid arthritis

IgA: in mucosal surfaces/secretions (expressed in mother's milk)

- Fix complement on C_1 – uses **al**ternative pathway
- Blood: monomer
- Secretions: dimer
- Ex: Heavy chain dz(↑), Ataxia Telangiectasia(↓), Selective IgA deficiency(↓)

IgE: monomer, binds allergens/parasites → forces mast cells to release:

- Histamine
- SRS-A = Slow Reactive Substance of Anaphylaxis
- ECFA = Eosinophil Chemotactic Factor of Anaphylaxis
- Ex: Job-Buckley syndrome

IgD: surface marker, no known function

NOTES:

NOTES:

Granulocytes:

"The only way of finding the limits of the possible is by going beyond them into the impossible."

–Arthur C. Clarke

D<u>**ifferential:**</u> (the tissue dwellers)

"Never Let Mom Eat Beans"

- **N**eutrophils: 60% (< 2500 => worry about Staph/Pseudo – or fungus if fever persists)
- **L**ymphocytes: 30%
- **M**onocytes: 8%
- **E**osinophils: 2%
- **B**asophils/**B**ands: <1%

Gram +

- Thick peptidoglycan wall – attracts PMNs
- Exotoxin (except Listeria) *"turn the + sideways to make an x"*
    - Heat stable: can travel through bloodstream, cause sepsis
    - Heat unstable: can't travel through the bloodstream
- Have **t**eichoic acid (not toxic) => they can't make you very sick

Gram -

- Thin PG wall
- Outer membrane – creates a periplasmic space
- May have capsule
- Endotoxins: *"LOral contraceptives the crystal violet out"* => stains red
  Lipid A (same for all bugs) – toxic part => sepsis (part of LPS)
  O Ag (different for every family) Ex: O157:H7 Jack-in-the-box EHEC strain
  Core Ag (different for every family member) *"Core = who you are individually"*

Note: If outer membrane destroyed → release endotoxin → pt gets worse at first w/ antibiotics
Note: Only Neisseria can release toxin while it's still growing
Ex: Give dexamethasone before antibiotics to prevent complications of bacterial meningitis

Gram stain:

- Crystal violet – binds peptidoglycan => stains Gram + purple *"Positive = Purple"*
- Iodine – seals it in
- Wash excess away w/ EtOH
- Red Saffranin => stains Gram – red

Acid Fast = Ziehl-Neelson stain: stains mycolic acid => pathogen is pink (otherwise = blue)

- Acid fast: Mycobacteria
- Partially acid fast: *"Not Concise"*
- Gram +: **N**ocardia (attacks diabetic lungs/sinus)
- Protozoan: **C**ryptosporidia (stored in intestinal brush border) => watery diarrhea in AIDS

The Allergic Response:

Mast Cells: *Giemsa stain*

1st Date w/ Ag: Body doesn't react → it makes IgE → binds mast cells

2nd Date w/ Ag: IgE (Fab portion) binds Ag, crosslinks → release:

- Histamine: (Tx: Histaminase)
  - Rubor: red
  - Calor: heat
  - Dolor: pain
  - Tumor: swell up
- SRS-A: Slow Rxn Substance for Anaphylaxis = Leukotriene $C_4D_4E_4$ (Tx: Aryl sulfatase)
  - *SRS-A is the most potent bronchoconstrictor…*
- ECF-A: Eosinophil Chemotactic Factor of Anaphylaxis (Tx: Heparin)

Note: Eosinophils release histaminase, aryl sulfatase, and heparin

Extreme Eosinophilia:

"NAACP"

Neoplasm (lymphoma)

Allergy

Addison's disease (no cortisol → relative eosinophilia)

Collagen vascular disease

1st Gen Anti-histamines: H_1 receptor blockers, anti-cholinergic (hot dry skin)

- Diphenhydramine "Benadryl"
- Dimenhydrinate "Dramamine" – tx motion sickness
- Meclizine – best to tx vertigo
- Hydroxyzine – tx stress-induced hives
- Phenylpropanolamine – taken off market b/c of arrhythmias
- Phentermine – taken off market b/c of pulmonary and cardiac fibrosis

2nd Gen Decongestant: α_1-agonists

- Pseudoephedrine "Sudafed" – tx stress incontinence, used to make methamphetamine
- Ephedrine – OTC

3rd Gen Anti-histamines: H_1 receptor blockers => 1 per day, less sedating

- **T**erfenadine – taken off market b/c of **T**orsade w/ macrolides
- Loratidine – good for outdoor allergies
- Desloratidine
- Citirizine – good for indoor/outdoor allergies
- Fexfenodine – good for indoor/outdoor allergies
- Astemazole
- Citirazine

NOTES:

Hypersensitivities:

"Do not wait to strike till the iron is hot; but make it hot by striking."

–William B. Sprague

Hypersensitivities: *(+) direct Coomb's*

"ACID"	**Type 1:**	**Type 2:**	**Type 3:**	**Type 4:**
Nickname:	**A**naphylaxis/Atopy	**C**ytotoxic/ADCC	**I**mmune Complex	**D**elayed/Cell-mediated
Disease:	Urticaria "hives" Angioedema Anaphylaxis Food allergies	Autoimmunes Hemolytic anemia Goodpasture's Grave's Rh/ABO	SLE (exception) RA (exception) Serum sickness Cryoglobulinemia	Contact dermatitis TB skin test Sarcoidosis Chronic transplant rejection
Cell:	Mast cell	T_K/MAC	------	Langerhans cell
Mediator:	IgE	IgG/IgM	Ag-Ab	Hapten
Pathophys:	1) Ag → IgE to bind mast cell 2) Ag crosslinks IgE → mast cell release	Fix complement deliberately: Ag binds Ab → recruit killers	Fix complement accidentally: Ag-Ab ppts → RBCs take to spleen	Hapten binds Langerhans-MHCII to lymph: T_{H1} to skin: 1) INFγ → MP 2) CD8 → T_K

The Complement Cascade: *Need complement to destroy encapsulated organisms*

I) Classic Pathway: uses Ag-Ab

Ag-Ab: activates C1 → C1q

C1q cuts (via C1 Esterase):

- C4 => C4a + C4b (C4 is the first to disappear => can test for this)
- C2 => C2a + C2b

Forms "C3 convertase": C4b2a

Forms "C5 convertase": C4b2a3b

Use C3 and C5 convertase:

- C3 => C3a "anaphylatoxin" + C3b "opsonin – coats encapsulated organisms"
- C5 => C5a "anaphylatoxin" + C5b "part of MAC complex"

MAC attack: C5b-C9

- C8 – aligns capsule for destruction (holds 'em down)
- C9 – perforates the capsule (kills 'em)

II) Alternative pathway: uses IgA

C3 → C3b → C3 and C5 convertase

Diseases:

C5-9 Deficiency: encapsulated infections (Neisseria = biggest)

- Recurrent meningitis, trunk petechiae
- Tx: Vancomycin + Ceftriaxone (Dexamethasone if hypotensive)

Hereditary Angioedema (AD): C_1 Esterase Inhibitor deficiency, low C4

- Recurrent facial swelling
- Primary: in kids, born with deficiency
- Secondary: in adults, usually due to ACE-I
- Prophylaxis: Androgens
- Tx: Metoprolol/Epinephrine (no ACE-I)
- Anaphylaxis Tx:

1) Epinephrine (1:1,000) – 0.5mL q15min x 3
2) Benadryl – 50mg IV
3) Steroids – will take 8-12hr to work
4) Dopamine

<u>ACE Inhibitors</u>: ↑K, have sulfur, ↓heart failure mortality, ↓DM proteinuria, tx CHF
Inhibit the C_1 Esterase inhibitor => facial swelling (angioedema)
↑**Bradykinin** => dilate veins "venodilate" (↓preload), cough
↓**ATII** => dilate arteries "vasodilate" (↓afterload)

- Captopril
- Enalopril
- Lisinopril
- Rinilopril

Angioedema Sx: throat swelling or vocal cord edema (Tx: intubate, steroids, racemic Epi)

<u>ATII receptor blockers:</u> *no cough*

- Losartan – pee out uric acid (tx HTN with gout)
- Valsartan
- Irbesartan

<u>Sulfur Drugs:</u>

Urticaria, SJ

G-6PD hemolysis

- ACE-I
- Celocoxib
- Loop diuretics
- Thiazides
- Sulfonylureas
- Dapsone

<u>High Yield Antibody Review:</u>
Anti-Rh Ab: Warm hemolysis
Anti-platelet Ab: ITP
Anti-IF Ab: Pernicious anemia
Anti-parietal cell Ab: Type A Gastritis

Anti-gliadin Ab: Celiac sprue
Anti-reticulin Ab: Celiac sprue
Anti-tissue transglutaminase Ab: Celiac sprue
Anti-endomysial Ab: Dermatitis herpetiformis
Anti-mitochondrial Ab: 1° Biliary Cirrhosis
p-ANCA Ab: Poylarteritis Nodosa, 1° Sclerosing Cholangitis
c-ANCA Ab "anti-proteinase Ab": Wegener's
Anti-C3 convertase Ab: MPGN type II
Nephritic factor: MPGN type II

ASO Ab: Post-Strep Glomerulonephritis
Anti-GBM Ab: Goodpasture's

Anti-TSH receptor Ab: Grave's
Anti-ganglioside Ab: Guillain-Barre
Anti-microsomal Ab: Hashimoto's
Anti-thyroglobulin Ab: Hashimoto's
Anti-peroxidase Ab "TPO": Hashimoto's
Anti-centromere Ab: CREST
Anti-RNP: Mixed connective tissue disease
Anti-topo I "Anti-Scl70" Ab: Scleroderma
Anti-histone Ab: Drug SLE *"HIPPPE"*
ANA: SLE sensitive
Anti-Smith Ab: SLE specific
Anti-ds DNA Ab: SLE w/ renal dz
Anti-neuronal Ab: SLE cerebritis
Anti-SSA "Rho" Ab: Neonatal SLE (heart block)
Anti-cardiolipin Ab: repeat miscarriages, false (+) VRDL (Tx: ASA or Heparin if pregnant)

Anti-Jo-1 Ab: Dermatomyositis
Anti-melanocyte Ab: Vitiligo
Anti-BMZ Ab: Dermatitis herpeteformis
Anti-desmosome Ab: Pemphigus vulgaris
Anti-hemidesmosome Ab: Bullous pemphigoid

Anti-myelin Ab: Multiple sclerosis
Anti-Ach receptor Ab: Myasthenia Gravis
Anti-Ca channel Ab: Lambert-Eaton syndrome
Anti-SSB "La" Ab: Sjogren's
Rheumatoid Factor (IgM against IgG Fc): Rheumatoid arthritis
Anti-SM Ab: Autoimmune hepatitis type I (females)
Anti-LKM Ab: Autoimmune hepatitis type II (kids)
Heterophile Ab: EBV Mononucleosis
Anti-GAD Ab: DM type I
Donath-Landsteiner Ab: paroxysmal cold autoimmune hemolysis

Transplants: *Blood is the most common transplant*
Syngeneic: Twin→ Twin
Autograft: Self → Self
Allograft: Human → Human
Xenograft: 1 species → another species (Ex: pig → human)

Blood Transfusion Reactions:

1) Febrile: ↑temp

- Due to donor RBC
- Tx: Acetaminophen

2) Anaphylaxis: itch/wheeze

- Due to host plasma proteins
- Tx: Epi/Steroids/Anti-histamines

3) Incompatibility: hypotension/back pain

- Due to human error
- Tx: IVF, stop transfusion

HLA Typing: Good match: >60%

- Sibling: best chance
- Parents: not good donors b/c they won't match >50%

Serology:

I) Blood Components:

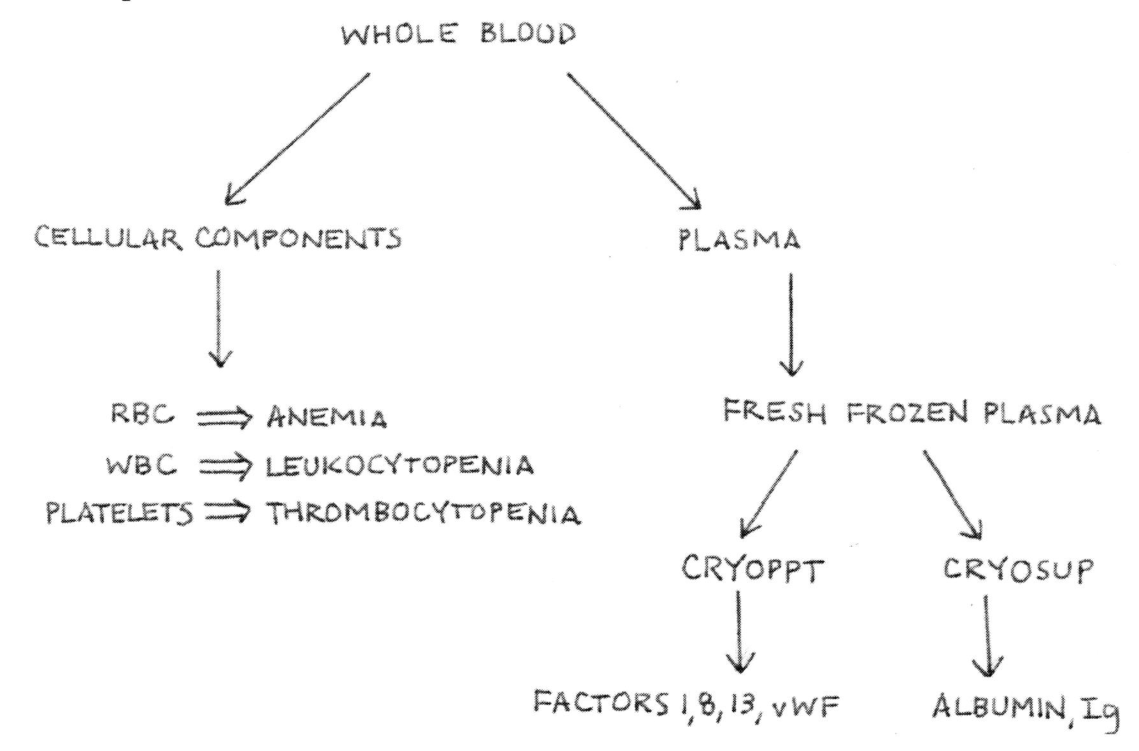

II) RBC Agglutination studies: RBC$^{(-)}$ normally repel each

Coombs test => Ab involved

- Direct: on surface => hemolytic anemias *"Direct people show their feelings on the surface"*
- Indirect: in serum
- RBC Phenotype => Pt Ag

- Ab screening => Pt Ab
- Crossmatch => Pt pre-formed Ab against Donor Ag (past transfusion)
- Mixed lymphocyte reaction: Pt reaction against donor's lymphocytes

Type and cross – you know you can use that blood, save it for specific pt
Type and match – type it and wait
Forward typing – uses **Ab** to detect Ag ***"Fabulous"***
Backward typing – uses Ag to detect Ab

Transfusion Infxns:

- HIV
- Hep B,C,D
- CMV
- EBV
- Syphilis
- Malaria
- Babesiosis

Who is not the father?

1) ABO => who is not the daddy
2) Rh
3) Minor antigens
4) Immunoglobulin allotypes

1) ABO Blood Typing:

- These Ag are sugars
- You inherit the enzyme that attaches the sugar on your RBCs

The Ag Perspective:
Blood type A => have A Ag
Blood type O = no Ag "universal donor"
Blood type AB = both Ag "universal recipient"

	Type A:	Type B:	Type O:	Type AB:
Glycoprotein:	Galactos**amine** Transferase	Galactose Transferase	Fucose	Both transferases
Ag:	A	B	None	A, B
Ab:	anti-B	anti-A	anti-A, anti-B	None
Genotype:	AA or AO	BB or BO	OO	AB

2) Rh System:
Rh(+): has D Ag => "D"
Rh(-): not have D Ag => "d"

Hemolytic Disease of Newborn:
Worry about Rh(–) Mom with Rh(+) Dad
Placenta tears, baby's blood → Mom → Ab → attacks fetus

- Test: Indirect Coomb's (if Mom has high titer Rh Ab, then it's too late)
- RHOGAM = anti-D IgG
- 1st Dose: 28 wk gestation
- 2nd Dose: 72 hrs post-delivery (if baby is Rh+)
- Other Dose: Amniocentesis, Spontaneous abortion, Ectopics, Trauma

If Mom=OO, Dad=AB =>Mom *will* have Ab to baby's blood

- HDN Tx: intra-uterine blood transfusion → Phenobarbital (breaks down fetal bilirubin)

Kleihauer-Betke test: # fetal RBCs in Mom => Rho-Gam dosage (or 300 μg if Mom bleeds <30mL)

Transplant Rejection:

Hyperacute: within hours *(preformed Ab)*

- Tx: Remove organ

Acute: within weeks *(T-cells, MP)*

- Tx: Prednisone, Cyclosporine, anti-lymphocyte Ab, FK-506

Chronic: within months-years (Fibroblasts)

- Tx: Remove organ

Graft vs. Host: bone marrow transplants *(T_K, MP): "The Lone Lymphocyte gets revenge..."*

Transplant Problems:

- Viral infection: CMV, EBV, HPV => genital warts
- Cyclosporine => gout
- Graft arteriosclerosis

NOTES:

Normal Flora:

"We are what we repeatedly do. Excellence, therefore, is not an act but a habit."

–Aristotle

How do I know if it's Gram + or - ?

- Exposed to O_2 => Gram +
  Ex: Staph aureus on skin
- Exposed to O_2, yet hiding => Gram + anaerobe
  Ex: P. acnes in hair follicle
- Not exposed to O_2 => Gram negative (most are anaerobic)
  Ex: E. coli in small intestine

Clues for Anaerobes:

- Gas
- Bad odor

Facultative Aerobe: prefers O_2, can survive without
Facultative Anaerobe: prefers no O_2, can survive with

How did that bug get a new infectivity?

- Transduction – in the wild => virus gives plasmid to bacteria (bacteriophage vector)
- Transformation – in hospital => 1 bacteria dies, another bacteria picks up its plasmid
  - Natural Transformers:
  - H. influenza
  - N. gonorrhea
  - H. pylori
  - S. pneumoniae
- Conjugation – bacteria w/ pili have sex
- Transposon = "jumping gene"

Adequate Sputum Sample:

<10 Epithelial cells/ low power field
>25 Leukocytes/ low power field

High Yield Review:

Skin Bugs:	**Clues:**
Staph aureus	Cellulitis
Strep pyogenes	*"LINES"*
Staph epidermidis	Central lines or catheter infections
Propionibacterium acnes	Acne

Eye Bugs:	**Clues:**
Staph aureus	Stye
Hemophilus aegyptus	Swollen eyeball w/ pus
Francisella tularensis	Red eye, swollen glands, rabbits, deer
Pseudomonas	Contact lens keratitis
Chlamydia	Retina cotton exudate
Bacillus cereus	Drug user proptosis, eye ring abscess

Ear Bugs:	Clues:
Pseudomonas aeruginosa	Otitis externa
Strep pneumoniae	Otitis media

Mouth Bugs:	Clues:
Actinomyces	Sulfur granules
Fusobacterium	Painful ulcers, gum pus
Strep mutans	Dental cavities
Strep viridans	Subacute bacterial endocarditis
Strep salivarius	Cold agglutinin test
Eikenella	Human bite (Tx: Augmentin)

Throat Bugs:	Clues:
Strep pyogenes	Rheumatic fever
Group C strep	Pharyngitis
N. gonorrhea	Pharyngitis w/ oral sex
Corynebacterium diphtheriae	Chinese letters, gray pseudomembrane, suffocation

Lung Bugs:	Clues:
Strep pneumonia	Gram + diploccoci
H. influenza B	Thumb sign epiglottitis
N. catarrhalis	Mucus
Chlamydia psittaci	Parakeets, parrots
Chlamydia pneumonia	Staccato coughing, elementary and reticulate bodies
Mycoplasma pneumonia	Bullous myringitis, ground-glass CXR
Legionella pneumonia	AC ducts, silver stains lung, CYEA
Pneumocystis jirovecii	AIDS, premies, silver stains
Clostridium botulism	Canned food, honey, inhibits ACh release
Clostridium tetani	Rusty nail wounds, inhibits Gly release
Bordetella pertussis	Whooping cough
Bacillus anthracis	Black eschar, woolsorter's lung disease

Stomach Bugs:	Clues:
H. Pylori	Peptic ulcers

Colon Bugs:	Clues:
Salmonella	Raw chicken&eggs, turtles, rose typhoid spots
Campylobacter jejuni	Raw chicken&eggs, very bloody diarrhea
Clostridium perfringens	Holiday ham&turkey, DM gas gangrene
Bacillus cereus	Rried rice
Listeria monocytogenes	Raw cabbage, hot dogs, spoiled milk, migrant workers
Staph aureus	Milk
Vibrio parahaemolyticus	Raw fish
Hepatitis A	Shellfish
Vibrio vulnificus	Oysters
Shigella	Day care outbreaks => seizures
Vibrio cholera	Rice water diarrhea
Clostridium difficile	Explosive diarrhea
Yersinia enterolitica	Presents like appendicitis + Reiter's
Strep bovis	Colon cancer
Clostridium septicum	Colon cancer
Bacteroides fragilis	Post-op pelvic abscess

Urinary tract Bugs:	Clues:
E. coli	Raw hamburger
Proteus mirabilis	Staghorn calculus
Klebsiella pneumonia	Alcoholics, currant jelly sputum
Staph saprophyticus	Young girls/College girls UTIs
Enterococcus	Nitrite negative

Blood Bugs:	Clues:
N. meningitidis	DIC, Waterhouse-Friedrichson (adrenal hemorrhage)
N. gonorrhea	Fitz-Hugh-Curtis (liver hemorrhage)
Brucella	Farmers, veterenarians, spiking fever 5x/day, butchers
Pasteurella multocida	Cat/dog saliva (Tx: Amoxicillin-Clavulanic Acid
R. rikettsii	Ticks => Rocky mountain spotted fever
R. akari	Mites => Ricketsial pox, fleshy papules
R. typhi	Fleas, endemic typhus
R. prowazekii	Lice, epidemic typhus
R. tsutsugamushi	Mites => scrub typhus
Coxiella burnetii	Dusty barn => Q fever, lung disease

Lymph Bugs:	**Clues:**
Yersinia pestis	Rats, fleas => Bubonic plague
Bartonella henselae	Cat scratch => single painful lymph node in armpit (Tx: Bactrim or Erthythromycin)

NOTES:

Gram Positives:

"Our greatest glory is not in never falling, but in rising every time we fall."

–Confucius

Classification Simplified:

Gram (+) Cocci

Catalase (+):

- Coag (+): Staph Aureus
- Coag (-):
  - N (+): Staph Epidermidus
  - N (-): Staph Saprophyticus

Catalase (-):

- α green:
  - Diplococci, Quellung (+), Optochin (+), Bile Salt (+): Strep Pneumo
  - (-): Strep Viridans
- β clear:
  - B (+): Strep Pyogenes "Group A"
  - B (-): Strep Agalactiae "Group B"
- γ red (no hemolysis):
  - Hemolytic on Blood Agar: Strep Bovis
  - Nitrite (-): Echinococcus "Group D"

Gram (+) Rods:

- Clostridium
- Corynebacterium
- Listeria
- Bacillus

I) Gram + Cocci:

Enterococcus (nitrite negative) *"nothing => enter-o"*

- Faecalis
- Faecium
- Infectious endocarditis after GI/GU problems; UTIs
- Tx: Vancomycin

Diarrhea 0-8 hrs :

(preformed toxin)

- Staph aureus
- Bacillus cereus
- Clostridium

Staph Aureus (Gram + cocci in clusters): most common (except *"LINES"*)

- Coagulase +, Catalase +
- Cellulitis: flat, red, blanching (infection of epidermis)
- Omphalitis: umbilicus cellulitis
- Panniculitis: abdominal ring of cellulitis
- Mastitis: breast infection
- Balanitis: penis head infection
- Carbuncle: small area of infection "boil"
- Furuncle: carbuncle w/ hair follicle in it
- Folliculitis: furuncle infection (pus)
- Fasciitis
- Paronychia: infection of skin around nail margin
- Osteomyelitis: except sickle cell and salmonella
- Toxins:
  - Enterotoxin => symptoms <8 hrs
  - Elastase => acute bacterial endocarditis, pneumonia (2wks after flu)
  - Collagenase: eats collagen
  - Lecithinase: eats connective tissue
  - Lipase: eats fat => panniculitis
  - β-lactamase: eats penicillin
  - Coagulase: eats through clots

Bacterial Pigments:

Pseudomonas: gold & green

Staph aureus: gold

Staph epidermidis: white

Staph saprophyticus: none

Bacterial Endocarditis:

Acute: Staph aureus (attacks tricuspid valves)

Subacute: Strep viridans (attacks mitral valves)

Tx: Nafcillin + Gentamycin for 4 wks

Staph Epidermidis (Gram + clusters): under epidermis

- Coagulase +, Catalase –, Novobiocin senstitive
- Vegetations on tricuspid
- Central lines, V-P shunts
- Tx: Vancomycin
- Tx: Linezolid – for VRE (usually found in armpits or groin)

Staph Saprophyticus:

- Coagulase –, Novobiocin resistant
- UTI in females (5-10, 18-24 y/o)
- "Honeymooner's cystitis"

Diplococci:

Gram –: N. gonorrhea

Gram +: Strep pneumo

Strep Agalactiae "Group B Strep"

- Child meningitis

Strep Pharyngitis

- Anterior cervical nodes
- Tx: Penicillin G (1.2 million units IM)

Strep Pneumoniae = pneumococcus (*only Gram + diplococci*)

- #1 cause of sinusitis, otitis, bronchitis, pneumonia *(rusty colored sputum)*
- Has IgA protease, capsule
- Otitis media (red bulging tympanic membrane)
- Pneumococcal vaccine – *covers 23 strains*
- >2 y/o sickle cell pts
- >65 y/o
- End organ failure
- Asplenic

Gram + Capsule:

- Strep Pneumo

Gram – Capsules:

"Some Killers Have Pretty Nice Capsules"

Salmonella

Klebsiella

H. influenza B

Pseudomonas

Neisseria

Strep Pyogenes = "Group A Strep" (Gram + chains)
"LINES"

- **L**ymphangitis – *red streak* (infection follows lymph channels)
- **I**mpetigo – *honey crusted lesions* (but…*bullous* impetigo is Staph => pealing skin)
    - Tx: Cephalexin + topical Mupirocin
- **N**ecrotizing fasciitis = *"flesh-eating bacteria"*
    - Toxins: Streptolysin-O, -S
    - Virulence factor: M protein (M12 => PSGN)
    - Capsule has hyaluronic acid
    - Tx: Debridement + Clindamycin
- **E**rysipelas – red, shiny, swollen, does not blanch, infection of subcutaneous fat
- **S**carlet fever – sandpaper rash, strawberry tongue, involves palm/soles

Strep Viridans = "Group D"

- Can progress to subacute bacterial endocarditis → attacks damaged valves

II) Gram + Rods:

Actinomyces:

- Face fistulas (eats face off)
- Sulfur granules
- Tx: Penicillin x 12mo

Bacillus Anthracis: "box-car like" spore

- Toxin: Edema Factor, Protector factor, Lethal factor, has D-Glu (humans have L-amino acids)
    1) Cutaneous: black eschar, doesn't spread, malignant pustule
    2) Pulmonary: *"woolsorter's disease"* (nomads), lung necrosis => drown in own blood
- No person-person transmission
- Tx: Ciprofloxacin

Bacillus Cereus: fried rice, pre-formed toxin *"B serious about fried rice"*

Clostridium Botulinum: adult canned food (toxin), kids <6mo honey or molasses (spore)

- Toxin inhibits pre-synaptic release of ACh => die of respiratory failure, no gag reflex
- Test: toxin in stool
- Black tar heroin IV drug abusers
- Flaccid descending paralysis (normal sensation)
- Tx: 1)Antitoxin, 2)Penicillin w/ intubation

Clostridium Difficile => pseudomembranous colitis (explosive diarrhea)

- Associated w/ antibiotic use (Clindamycin)
- Dx: stool cytotoxin
- Tx: oral Vancomycin or Metronidazole

Rheumatic Fever Diagnosis:

need at least 2 "SPECC"

Tx: Penicillin

Subcutaneous nodules

Polyarthritis

Erythema Marginatum (red margins)

Chorea (Sydenham's)

Clostridium Septicum (Gram + spore) => Colon CA

- Black pigment

Clostridium Perfringens: anaerobe, found in soil/feces

- Gastroenteritis – holiday ham/turkey ← enterotoxin
- Gas gangrene in legs of diabetics (poor blood supply) ← α toxin
- Dry gangrene – necrosis (Tx: immediate amputation)
- Wet gangrene – gas emboli to RV => outlet obstruction (Tx: lay pt on left side, pound on right)

Clostridium Tetani – "tennis racquet" shape, dirty wounds (rusty metal)

- Toxin: inhibits Gly release in spinal cord => *"lockjaw"*, respiratory failure
- Risus sardonicus => look like a clown
- Strychnine has the same mechanism

1) Antitoxin Ig
2) Toxoid (if haven't had in last 5 yrs).
3) Glucocorticoid
4) Diazepam (decrease muscle spasms)
5) Penicillin

Urease + :

"Urease PPUNCH"

Proteus

Pseudomonas

Ureoplasma

Nocardia

Cryptococcus

Corynebacterium Diphtheriae (Gram + rod) *"Gray Chinese corn gets stuck in throat"*

- Looks like "Chinese letters"
- Toxin ADP-ribosylates EF_2 (like pseudo) → stop protein synthesis
- Vascular gray membrane (don't scrape!), wraps around trachea => suffocate
- Heart block => prolonged PR interval, recurrent laryngeal nerve palsy
- Pathogenic strain: has a temperate bacteriophage (Elek test)
- Tx: Anti-toxin (tetanus), then antibiotics

Simple Gram –

- H. influenza
- E. coli

Listeria Monocytogenes: only Gram + w/ endotoxin, comma shaped, tumbling

- Neonates => meningitis w/ granulomas, abortions
- Adults => gastroenteritis, heart block

- Raw cabbage (migrant workers)
- Hot dogs
- Soft cheese
- Spoiled milk *(unspoiled milk = Staph aureus)*
- Tx: Macrolides, Ampicillin

Nocardia:

- Attacks immunocompromised pts
- Tx: Bactrim or Minocycline x 6mo

Propionibacterium Acnes (Gram + anaerobe): in hair follicles

- White comedone => black if popped (oxidation)
- Likes propionic acid in the sebaceous gland
- Progesterone stimulates propionic acid production
- Hates O_2 => Tx:

1. Antibacterial washes: Benzoyl peroxide
2. Abrasive scrubs
3. Minocycline
4. Retinoic Acid (Vit. A) "Accutane" – rapid turnover of skin
   SE: Photosensitivity, Hyperlipidemia (fat soluble), NTD

Heart Block Infections:

Sx: low pulse, high temp

"Don't TeLL Chaga"

- **D**iphtheria
- **T**yphoid fever (Salmonella)
- **L**egionella
- **L**yme disease
- **C**hagas

IgA Protease Bugs:

(can survive in mouth)

- ⇨ sinusitis
- ⇨ otitis media
- ⇨ pneumonia
- ⇨ bronchitis

1. Strep pneumo.
2. H. influenza
3. Neisseria catt.

NOTES:

NOTES:

Gram Negatives:

"Great spirits have always encountered violent opposition from mediocre minds."

–Albert Einstein

Classification Simplified:

Gram (-) Rods

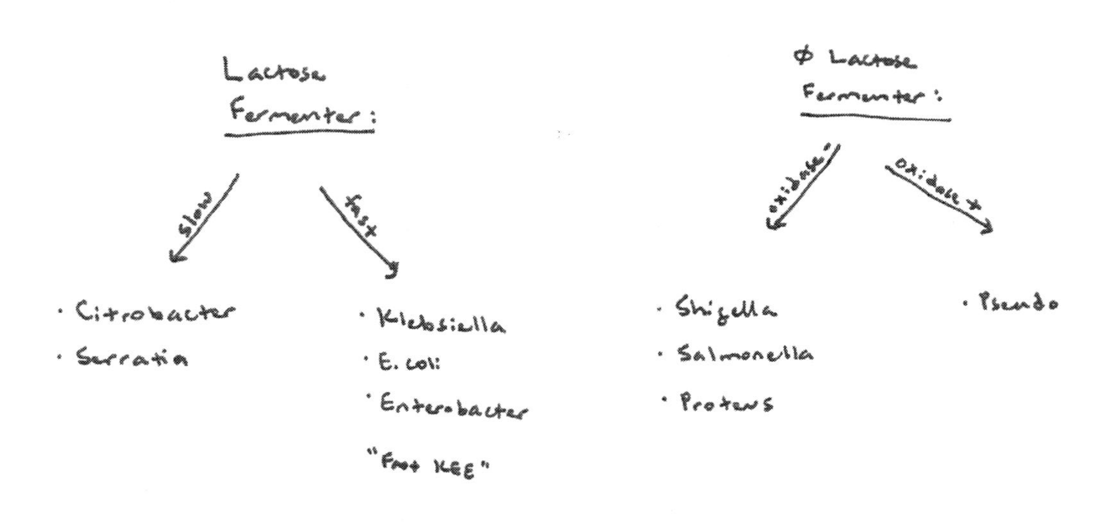

Cocci:

Mal (+)	N. meningiditis
Mal (-)	N. gonorrhea

Coccoid Rods:

- H. influenza
- Brucella
- Bordatella
- Pasteurella

I) Gram – Cocci:

Eikenella Corrodens:

- Human bites
- Tx: Amp-Sulbactam

Neisseria Catarrhalis: *formerly "Moraxella catarrhalis"*

- Loves mucus → attacks respiratory tract; normal respiratory flora
- #3 cause of sinusitis, otitis, bronchitis, pneumonia
-

Ixodes Tick Diseases:

- Lyme disease – migrating target lesion
- Babesiosis – hemolytic anemia, malaria sx
- Ehrlichiosis – puncture wound near eye, dog lick

Neisseria Gonorrhea: Gram – diplococcus

- Thayer-Martin media, IgA protease
- Pili have phage variation => can't kill 'em
- Urethritis, tenovitis, synovitis (wrist/ankle)
- Neonatal blindness, teenager septic arthritis
- Fitz-Hugh-Curtis: pus drops through the Fallopian tube onto the liver

Neisseria Meningitidis: (only Gram – diplococci), *biggest capsule,* IgA protease, ferments maltose

- Has pili => respiratory entry
- Only bacteria to release its toxin while multiplying in log phase → DIC
- Waterhouse-Friderichsen: pus drops through the Fallopian tube onto adrenal glands => adrenal hemorrhage, DIC, purpura, hypotension, shock

II) Gram – Rods:

Bacteroides Fragilis: post-op pelvic abscess

- Tx: Metronidazole/Clindamycin/Cefoxitin

Campylobacter Jejuni: comma "seagull" shaped, *raw chicken and eggs*

- Very bloody diarrhea
- Inactivated by gastric juices, can ppt Guillain-Barre
- Tx: Ciprofloxacin

E. Coli: appendicitis, cholecystitis, SBP, UTI, ascending cholangitis
Makes:

- Biotin
- Vit B_4 = pantothenic acid
- Vit B_9 = folate
- Vit. K => bleeding if suppressed by antibiotics
- Helps absorb Vit B_{12}
- Pink on MacConkey's agar
- Toxin => ADP ribosylation of Gs (like cholera)
- Tx: Ciprofloxacin

*E**I**EC* = **in**flammatory => loose stools
*E**T**EC* = **t**raveler's => rice water diarrhea
*E**H**EC* = **h**emorrhagic, *raw **h**amburger*, verotoxin => renal failure

- Endemic HUS – a few people
- Epidemic HUS = O157:H7 strain – lots of people

*E**P**EC* = pathogenic => newborn diarrhea

Haemophilus Aegyptus: (Gram – pleomorphic rods)

- True "pink eye" = red, swollen conjunctiva w/ pus, looks like eyeball is falling out

Fusobacterium Necrophorum:

- Lemierre's syndrome: jugular vein thrombophlebitis

Haemophilus Influenza *(only Gram – pleomorphic rod)*

<u>80% Non-typable:</u>

- #2 cause of sinusitis, otitis, bronchitis, pneumonia
- Satellite phenomenon: grows near Staph aureus

G-protein Disrupters:

- Pertussis: inhibits Gi
- Cholera: stimulates Gs
- E. coli: stimulates Gs

UTI/ Prostatitis:

1. E. coli
2. Proteus
3. Klebsiella
4. Pseudomonas

Cats:

- Cat scratch – Bartonella henselae *"the cat scratched Bart"*
- Cat saliva – Pasturella Multocida
- Cat **pee** – Toxo**p**lasmosis
- Cat **caca** – Toxocara cati

The Rebels :

Gram + endotoxin: Listeria

Gram – exotoxin: H. influenza
Bordetella pertussis

- Needs Factors V, X to grow
- Tx: Cefuroxime

20% Type B: has polysaccharide capsule: 5 carbon ribose

- Meningitis
- Epiglottitis
- Sepsis

Helicobacter Pylori: more common w/ poor sanitation

- Can survive acid pH b/c of urease: $NH_3 + HCl \rightarrow NH_4Cl$

Klebsiella Pneumonia (capsule) => currant jelly sputum, UTI

- Also causes pneumonia *(especially in alcoholics, DM, HIV)*
- Loves lung fissures, upper lobe cavitations

Proteus Mirabilis: (urease +) => NH_4^+ urine, UTI, swarming growth

- Struvite stones
- Tx: Norfloxacin

Pseudomonas Aeruginosa => "swimmer's ear" = otitis externa

- Pushing tragus => pain, pus
- Conjunctivitis, hot tub folliculitis
- Cholesteatoma – eats bone
- Ecthyma gangrenosum (necrotic blister): ↓EF-2 (like diphtheria)
- Gold & green pigments, anaerobic, grape-like smell
- Nail punctured sneakers, loves rubber
- Whirlpool folliculitis => butt crack and soles of feet infected
- Hospital plastic (48 hrs) => attacks CF, DM, burn pt, neutropenic pts
- Malignant otitis externa "red, swollen, tender, lifted ear" → kills rapidly
- Ecthyma granulosum – black necrotic tense bullae (don't pop!); same enzymes as Staph
- ADP ribosylation of EF-2

Salmonella: *raw chicken and eggs, turtles,* has capsule => H_2S

- Loves to hide in the gallbladder if treated (Shock Tx: Dexamethasone x 48hr)
- Attacks sickle cell pts => osteomyelitis
- Salmonella Dysenteriae – most common world wide
- Salmonella Enteritidis – most common in US
- Salmonella Typhi => typhoid fever (rose spots, heart block, intestine "fire")

Serratia Marcescens: "marachino cherry" pigment

Shigella (Gram – rod): day care outbreaks, infects M cells, destroys 60S ribosome

- Shiga toxin => seizures, loose stools a few days later

Cryoglobulinemia:
Palpable purpura
"I AM HE"
Influenza
Adenovirus
Mycoplasma
Hep B, C
EBV

Monocytosis:
Sx: Granulomas
"STELS"
Syphilis
TB
EBV
Listeria

- Tx: Ciprofloxacin

1) Shigella Dysenteriae – most common in the world
2) Shigella Sonnei – most common in US

Vibrio Cholera: poor sanitation => rice water diarrhea (lose isotonic plasma)

- Stimulates G_s via ADP-ribosylation => high cAMP *"turns the on on"*
- Tx: WHO oral rehydration, Tetracycline, Doxycycline

Vibrio Parahaemolyticus: raw fish

Vibrio Vulnificus:

- Raw oysters, fisherman
- Cellulitis in swimmer's cuts; cuts while walking on the beach

Yersinia Enterocolitica: presents like appendicitis

- Loves to attach to ileum => mesenteric adenitis
- Reiter's syndrome: "post-infectious arthritis"
- Tx: Ciprofloxacin

Yersinia Pestis: rats and fleas => Bubonic plague (New Mexico)

- Bubbo = infected lymph node, pus
- Keep pt in strict isolation (Tx: Streptomycin)

Comma shaped bugs:

*"HaLV-C => comma"**

H. pylori

Listeria

Vibrio

Spore Formers (D-Glu):

Occurs during lag phase

- Bacillus anthracis
- Clostridium botulinum

III) Coccoid Rods:

Bordetella Pertussis (Gram – w/ exotoxin)

- Inhibits Gi (ADP-ribosylates G_i) => high cAMP
- Lymphocytosis, rarely causes re-infection
- *Whooping cough* => suffocation

1) Catarrhal stage – mucus in lungs
2) Paroxysmal stage – "staccato" coughing
3) Convalescent stage – symptoms improve

- Test: Immunofluorescence (nasopharyngeal wash)

Tx: Erythromycin (must treat family), DTP vaccine

Big Mama anaerobes:

- Bacteroides fragilis
- Strep. bovis
- C. septicum

Tx:

- Clindamycin
- Cefoxitin
- Metronidazole

Brucella: vets/farmers (from animal placentas), undulated fever (fever spikes 5x/day)

- Brucella **A**bortus = cows *"killed the cow and **a**te it"*
- Brucella **S**uis = pig *"**s**ooey"*
- Brucella **M**elitensis = goats *"**M**illy the goat"*

Francisella Tularensis (Gram – coccobacillus): *rabbit ticks, deer ticks* => tularemia

- Sx: ulcers at tick bite site (skin, eyes) w/ draining lymph node
- Tx: Streptomycin

NOTES:

Atypicals:

"Knowing is not enough; we must apply. Willing is not enough; we must do."

–Johann Wolfgang von Goethe

Atypicals: *No cell wall → cell-med immunity*

Chlamydia Pneumonia: parasite w/ DNA&RNAs:

- Elementary body – infectious phase *"evil elementary body"*
- **R**eticulate body – **r**esting phase, can't divide
- Sx: stacatto coughing

Chlamydia Psittaci: parakeets and parrots

- Pneumonia, splenomegaly, meningoencephalitis
- Tx: Doxycycline

Chlamydia Trachomatis: can't make ATP

- Neonatal blindness (inturned eyelashes => cornea ulcer)

Walking Pneumonia:

Sx: dry cough

Tx: Fluoroquinolones

"Come My Love for a walk"

0-2 mo: **C**hlamydia pn.

10-30 y: **M**ycoplasma pn.

>40 y: **L**egionella pn.

AIDS/Premies: Pneumo Jirovecii

Legionella Pneumonia *"Old silver AC duct"*

- Sx: disorientation, pneumonia, N/V/D, ↑LFTs
- Loves heating and air-conditioning ducts (standing water) = *"highrise building syndrome"*
- Silver stains in lung, grows on charcoal yeast agar extract *(CYAE)*
    - Pontiac fever = mild fever alone
    - Legionnaire's disease = full blown pneumonia
- Tx: Erythromycin

Mycoplasma Pneumoniae (obligate aerobe): mycolic acid, no cell wall

- *Bullous myringitis* = bullae on eardrum
- Smallest extracellular bacteria, can't make cholesterol
- Attaches via P_1 protein
- Cryoglobulinemia, erythema multiforme
- CXR: interstitial, ground-glass, reticulo-nodular pattern
- Tx: Erythromycin

Silver Stainers:

- PCP – lung
- Legionella – lung
- Bartonella – lymph node

NOTES:

NOTES:

Mycobacteria/ Spirochetes/ Rickettsia:

"Happy are those who dream dreams and are ready to pay the price to make them come true."

–Leon J. Suenes

Meet the Mycobacteria:

These are non-bacteria w/ mycolic acid and a peptidoglycan wall
Phantom lesion – hyaline cartilage calcification

Mycobacterium ulcerans – ulcers

Mycobacterium marinum – fish tanks

Mycobacterium scrofulaceum – supraclavicular lymph nodes, +PPD (Tx: excise LN)

Mycobacterium leprae = Leprosy (Hanson's disease) => hypopigmentation + no sensation

- Tx: Rifampin + Dapsone – inhibits PABA

Mycobacterium avium intracellulare – AIDS pt w/ gastroenteritis → SOB

- Tx: Clarithromycin, Azithromycin

Mycobacterium tuberculosis: night sweats, hemoptysis, weight loss

1° TB: asymptomatic => perihilar Ghon focus (Ghon complex if it has spread to lymph)

- MP lysosomal fusion impaired b/c of "cord factor" => MP calls for IL-1
- IL-1 recruits cell-mediated TH_1
- Forms "granuloma" = MP w/ TH_1 cells around it => INF to transform MP
- IL-2 recruits more MP
- IL-12 promotes cell-mediated recruitment
- INF interferes with protein synthesis → cells die, then calcify
- MP and T cells => TNF-α

2° TB: apex cavitation

- T cells are decreased (poor nutrition, steroids, chemo)
- TB breaks from granuloma → airway (more O_2) → cough up blood to upper lobes

Miliary TB:

- Cough into GI tract → whole body (fat soluble)
- Lymph nodes, Peyer's patches (ileum) try to trap TB → obst. in ileum → absorbed →
- Brain: loves post fossa (CN3, then CN9-12) => hydrocephaly
- Spinal cord: "Pott's disease" => compression fractures
- Kidney: sterile pyuria (WBC in urine w/ negative cultures => didn't test for TB)
- Psoas muscle: "cold abscess", no pus

- Adrenal gland: dies
- Heart: constrictive pericarditis

Granulomas:

Sx: erythema nodosum

- TB
- Sarcoidosis
- Syphilis
- Histiocytosis X
- The Ellas

PPD test: *negative AFB x 3 → non-infectious*

- Test #1 (dilute) – if has sx, need to prove TB
- Test #5 – for screening
- Test #250 (conc) – for AIDS pts

Positive PPD: (induration, not erythema) – check 24-48 hrs (T cells here)

- 15mm (normal population)
- 10 mm (overcrowding: prisons/NY city, health care workers, IV drug users)
- 5 mm (immunosuppressed/steroids, kids<4 y/o, TB household members)

TB Drugs:

"RIPE" → all cause liver failure

Rifampin: inhibits RNA Pol => orange secretions, revs up p450, myositis

INH: inhibits mycolic acid synthesis, use for prophylaxis if >35 or high risk

SE: Fat soluble => myositis, depletes Vit B_6, SLE, inhibits p450, seizures

Pyrazinamide: ↑uric acid

Ethambutol: inhibits arabinogalactan (cell wall) => impaired color vision

Latent TB Tx: *(+)PPD/(-)CXR or BCG history*

- INH/Vit B_6 x 9mo

Active TB Tx:

- "RIPE" x 2mo → check CXR → "RI" x 4mo

Other TB Tx:

Miliary TB: "RIPE" x 12mo

HIV (+): Replace Rifampin with Rifabutin

Pregnant: Replace Pyrazinamide with Ofloxacin (check LFTs monthly)

Leprosy: armadillos (Tx: Dapsone)

1) Tuberculoid:

- TH_1, Langerhan's/epithelioid
- No acid fast stain
- Local infection

2) Lepromatous:

- TH_2, foamy MP
- Acid fast stain
- Erythema nodosum

- Hematogenous spread
- Sensory neuropathy of ears/nose/distal extremities

Meet the Spirochetes:

Treponema Pallidum *"rule of 6's"*

- 1° syphilis: painless chancre (1-6 wks)
- 2° syphilis: palm/sole rash, ischemic stroke, transverse myelitis (6 wks)
- 3° syphilis: painful neuropathy/cardiopathy (6 yrs)
  - Attacks dorsal columns → Tabes dorsalis *"lancinating/shooting pain"*
  - Attacks Edinger-Westphal nucleus → Argyll-Robertson pupil: reacts poorly to light, but well to accomodation
  - Attacks aorta → Obliterative endarteritis *"tree bark appearance"*
- Neonatal syphilis: loves bones
  - Flat forehead
  - Snuffles – nasal bone gone "saddle nose"
  - Hutchinson's teeth – sharp
  - Saber tooth shins – anterior leg bowing
  - Rhagade's – mouth fissure

Syphilis testing:

- Dark field microscopy – most specific/sensitive
- Blood tests: FTA-ABS or TPI – specific, IgM (IgG can stay positive forever)
  RPR/VRDL – *sensitive* => use for screening (can stay positive for 1 yr)

Syphilis Tx:

- 1° syphilis: Benzathine Penicillin G IM 1.2 million units x 1
- 2° syphilis: Benzathine Penicillin G IM 2.4 million units x 1 *"double it"*
- 3° syphilis: Benzathine Penicillin G IM 2.4 million units x 3wk *"give it three times"*
- Neonatal: 50,000 units/kg/day

Jarisch-Herxheimer: ↑fever after treating syphilis with penicillin due to released spirochetes

Treponema Pallidum Variant → Bijel: non-venereal condyloma lata in kids
Treponema Pertenue → Yaws: raspberry ulcers *"you Yawn when you Pretend"*
Treponema Carateum → Pinta: red scaly patches change into white spots *"Carry the Pintas"*

Borrelia Borgdorfori: ixodes tick => **Lyme disease**

- 1° stage: rash = erythema chronicum migrans *(only disease with this rash)*
- 2° stage: neuro/carditis: heart block, Bell's palsy, meningitis → do LP (Tx: Ceftriaxone)
- 3° stage: arthritis

- Test: Synovial Fluid ELISA → Ab to Borrelia (IgM and IgG)
- Tx: Doxycycline x 30 days (or Amoxicillin for kids)

Borrelia Recurrentis: cyclic fever (1/wk for 5 wks)

- Brill-Zinsser disease = pathogen hides in lymph nodes, comes out 1/wk slightly mutated

Leptospira Interrogans: rat/wild boar urine => sewage workers
"After they interrogated the rats, they escaped through the sewer"

- Shepherd's crook shape
- Fort Brag's fever
- Weil's disease = infectious nephritis and hepatitis
- Dx: (+) Macroscopic slide agglutination test
- Tx: Penicillin

M**<u>eet The Rickettsia:</u>** likes to invade blood vessels

Tx: Chloramphenicol
Rickettsia Rickettsii – tick => Rocky mountain spotted fever (palm/sole rash)
Rickettsia Akari – mites => Rickettsial pox (fleshy papules and vesicles)
Rickettsia Typhi – fleas, starts in armpit => Endemic typhus *"ty-flea"*
Rickettsia Prowazekii – lice, starts on body => Epidemic typhus
Rickettsia Tsutsugamushi – mites => scrub typhus
Coxiella Burnetii – dusty barn => Q fever, lung disease

NOTES:

Fungi/ Parasites/ Protozoa:

"We are all inventors, each sailing out on a voyage of discovery, guided each by a private chart, of which there is no duplicate. The world is all gates, all opportunities."

–Ralph Waldo Emerson

Meet the Fungi:

- Like warm, moist, sugar => Tx: cool and dry
- Ergosterol membrane
- Methenamine silver stains

I) Superficial fungi:

Piedra "little black balls on hair shaft" – fungus eats keratin off hair shaft (Tx: cut hair off)

II) Deep fungi:

Candidiasis – white cheesy itchy discharge, oral thrush, +KOH pseudohyphae

- Tx: 150mg Fluconazole

III) Cutaneous Fungi:

- Topical Clotrimazole – for most
- Griseofulvin+Selenium – for capitis and versicolor (in stratum corneum)

Tinea capitis – flaky crusty lesions of scalp, hair loss

- Caused by **T**richophyton (Tx: **T**erbinafine)
- Kerion = tinea capitis + infection *"boggy"*

Tinea barbae – razor bump infection on chin *"think barber"*
Tinea corporis – ringworm on body/face (from cats)
Tinea intertrigo – skin touches skin (armpits, under breasts) => skin peels off
Tinea versicolor – on back, "spaghetti meatball" distribution, worse in heat (Tx: Ketokonazole)
Tinea unguium – **un**der nail => discolored nails
Tinea nigra – flaking palms => dark lines
Tinea manis – pealing between finger webs "maceration"
Tinea pedis – pealing between toes "athlete's foot" (Tx: Tolnaftate cream)
Tinea cruris – on groin "jock itch"

Loeffler Syndrome:

Sx: pulm eosinophilia

"NASSA"

Necator americanus
Ancylostoma duodenale
Schistosomiasis
Strongyloides
Ascaris lumbricoides

IV) Systemic fungi: inhale spores (Tx: Itraconazole)

Histoplasma – bat poop (Mississippi river), lives in MP, oral ulcers *"history in Mississippi"*
Blastomyces – pigeon poop (NY), **b**road-**b**ased hyphen, rotting wood in beaver dams *"NY blast"*
Coccidioides – soil (Arizona), desert bump fever, budding yeast
Paracoccidioides – looks like a ship's wheel (S. America)
Rhizopus & Mucor – mucormycosis (eats eyes, nose, sinus)

- Dx: Orbital CT
- Tx: Surgery + Amphotericin B

Sporothrix schenckii – rosebush prick (Tx: topical KI)
Cryptococcus neoformans – AIDS pt

- Urease positive, stains w/ India ink, encapsulated yeast

PIE Syndrome:

- Aspergillus
- Loeffler's
- Churg-Strauss

- Lung Tx: Fluconazole
- Meningitis Tx: Amphotericin B/Flucytosine → lifelong Fluconazole

Aspergillus – moldy hay/basement, invades blood vv., fungal ball, mimics asthma, hempotysis

- Allergic rxn => Pulmonary Infiltrate w/ Eosinophilia *"PIE syndrome"*
- Tx: Steroids (or Voriconazole if necrotizing)

Anti-Fungals:

1) Polyenes – bind ergosterol => pores in fungal wall
Topical: Nystatin
Systemic: Amphotericin B – attacks fungi ergosterol and human cholesterol

- SE: hyperK(systemic toxicity) and hypoK(renal failure), SIADH

2) Azoles – inhibit ergosterol synthesis
Topical:

- Miconazole
- Clotrimazole
- Econazole

Systemic:

- Ketoconazole – inhibits p450, inhibits 5-α reductase => gynecomastia, excess menstruation
- Itraconazole
- Fluconazole – crosses BBB *"flew to the brain"*

3) Microtubule inhibitor:

- Griseofulvin – fat soluble

4) Anti-metabolites:

- Terbinafine – tx cutaneous fungi
- Flucytosine – stops DNA replication (give w/Amphotericin B)

Meet the Parasites:

Sx: Gastroenteritis (duodenum) => microcytic anemia
Lab: Eosinophils, T cells, MP, low volume state

Liver Flukes: ↑ALT/AST (Tx: Praziquantel)
Schistosoma: walking barefoot in a swamp, **snail** is carrier

- Mansoni => liver cancer, have lateral spine, pipe-stem fibrosis
- Haematobium => bladder cancer (SCC), Egypt

Toxocara Cati: cat poop
Toxocara Cani: dog poop
Echinococcus: raw lamb/dog poop

- Hydatid cyst w/ eggshell calcifications
- Liver abscess
- (+) Casoni test

Clonorchis Sinensis = Opisthorchis: likes biliary tract (↑alk phos) => cholangiosarcoma

Hookworms: hook into intestine wall => diarrhea *"can't keep a NEAT AS"*
Tx: *now, then repeat 1wk later to kill hatched worms*
Me**bend**azole – most of them (paralyzes microtubules) *"hookworms bend"*
Pyrantel **P**amoate – tx **P**inworm
Thiabendazole – tx Strongyloides

Necator Americanus – most common worm in US, infective stage = metacercariae
Enterobius Vermicularis – pinworm => pruritis ani (ass itching)
Scotch tape test: only female can make it from cecum to anus to lay eggs
Ancylostoma Duodenale => duodenal obstruction
Trichuris Trichiura – whipworm => rectal prolapse, tenesmus
Ascaris lumbricoides – roundworm, swallow and cough, ingest egg from human feces
Strongyloides stercoralis – threadworm => dermatitis, pulmonary eosinophilia

Tapeworms: Tx: Niclosamide – inhibits oxidative phosphorylation
Diphyllobothrium Latum – raw fish, eats Vit. B_{12}
Taenia Saginata – raw beef *"saged"*
Taenia Solium – raw pork (cysticercosis = larva swims in aqueous humor)
Trichinella Spiralis – raw pork, muscle pain, periorbital edema "*spirals under muscles*"

Roundworms:
Onchocerca – blackfly, river blindness, moving knuckle nodules

Meet the Protozoans:

Skin:
Leishmania Donovani – sandflies, attacks skin and nostrils, rash "Gulf War syndrome"
Leishmania Rhodiensis – sandflies, attacks organs "kala-azar"

Brain:
Toxoplasmosis – cat urine => multiple ring-enhancing brain lesions
- Tx = Pyrimethamine (inhibits DHF reductase) + Sulfadiazine (mimics PABA)

Naegleria Fowleri – swamp trauma
- Fulminant meningoencephalitis – eats through cribiform plate into brain => die in 48 hrs

Trypanosoma Brucei – tsetse fly => African sleeping sickness
(↑GABA)

Eye:
Acanthamoeba – in contact lenses, eats through cornea, HIV

Malaria protection:
Sickle cell: AAs
G-6PD: Mediterraneans

encephalitis
Ehrlichia – dog **lick** => ixodes tick => puncture wound near eye (Tx: Doxycyline)

Heart:
Trypanosoma Cruzi (reduviid bug) => **C**haga's: eats ganglia => heart block, achalasia, Hirschsprung's

Lung:
Pneumocystis Jirovecii – AIDS pts, silver stains lung
- Tx: Bactrim, Pentamidine

The Multiples:
Liver abscesses: Entamoeba histolytica
Cerebral abscesses: Citrobacter
Lung aneurysms: Osler-Weber-Rendu

GI:
Giardia lamblia – hiker's, mountain streams, Army boot camp => duodenum
- String test – swallow into duodenum, pull it up and see bugs
- Adheres via a "ventral sucking disc", crescent-shaped protozoa, watery diarrhea
- Tx: IV Metronidazole (or Paromomycin if pregnant) – confirm stool Ag before tx

Entamoeba histolytica – eats RBC => multiple liver abscesses, *"Erlenmeyer flask lesions"*
- "falling leaf" motility on wet mount
- "anchovy paste" liver abscesses, dysenteric diarrhea
- Tx: Metronidazole (8wks),

Microsporidia – most common diarrhea in AIDS pts
Cryptosporidia – *watery* diarrhea in AIDS pts, partially acid fast

Fever:
4-5/ day = Brucella
Every 2 days = Plasmodium sp.
Every 3 days = Plasmodium malaria
Every wk = Borrelia recurrentis

GU:
Trichomonas vaginalis: green frothy discharge (Tx: 2g Metronidazole)

Blood:
Babesiosis – ixodes tick, similar to malaria (East coast), co-infection with Lyme dz
- live in RBCs, tetrads on smear
- Tx: Quinine + Clindamycin

Plasmodium Malaria – invade old RBC, fever every 3days "quartian"
Plasmodium Falciparum – fatal, black water fever, anopheles mosquito
- attacks RBCs => massive hemolysis => urine black w/ bilirubin

Plasmodium Vivax/Ovale – chronic malaria in liver
- attacks reticulocyes (virgin RBCs)

Malaria Tx: oxidize RBC membrane
- Quinine – tinnitus, resp depression
- Mefloquine – good liver penetration
- Primaquine – prevents relapse
- Chloraquine – kills RBCs

Prophylaxis: 1 pill/wk (-2 to +4 weeks) => need 11 tablets for a 5-wk trip

Lymphatics:
- **Wucheria Bancrofti** – elephantiasis (no tx)

NOTES:

Viruses:

"Let him who would enjoy a good future waste none of his present."

–Roger Babson

Meet the Viruses:

- 1-3 wks to replicate
- Most common route of viral transmission = contact => washing hands ↓transmission
- Tropism => viruses have receptors for certain cells, enter via endocytosis
- Genetic drift – spontaneous mutation (Ex: Influenza B)
- Genetic shift – gene reassortment (Ex: Influenza A, Rotavirus)

	RNA virus:	**DNA virus:**
Envelope:	Enveloped (except "CPR" **c**alcivirus, **r**eovirus, **p**icornovirus)	Naked (except herpes, pox, hepadenovirus)
Strands:	Single stranded Except: • Reovirus (ds) • Rotavirus (segmented ds)	Double stranded Except: • Parvovirus (ss) • Hepadenovirus (segmented)
Replication:	Cytoplasm (except retrovirus)	Nucleus (except poxvirus)
Polymerase:	RNA polymerase (RNA dep: retrovirus) DNA polymerase (rvs transcriptase)	RNA polymerase (DNA dep: poxvirus)
Assemble:	cell membrane	nuclear membrane
Notes:	(-) strand → (+) mRNA before tln Assembled on cell membrane (+) strand: illness takes <1 wk (Ebola) (-) strand: illness takes 1-3 wks	"intranuclear" inclusions DNA viruses cause cancer b/c they replicate in the nucleus

Viral Entry Into The Cell:

1) **Invasion:** virus enters body (viremic, asymptomatic)
- **Inhibitor:** γ-Globulins

2) **Adhesion** to their receptors (viremic, asymptomatic)

3) **Penetration** via receptor-med. endocytosis – except HIV

4) **Uncoat:** take capsule off to expose RNA or DNA (Eclipse)
- **Inhibitors:** *tx Influenza A*
- Amantidine – Anticholinergic/DA release
- Rimantidine – quick resistance
- Selegeline – $MAOI_B$

5) Replication (Eclipse): inserts his DNA/RNA into your genome

- **Inhibitors:**
- Guanine Analogs:
  - Acyclovir – take 3-5 times per day
  - Pencyclovir – take 3x/day
  - Demcyclovir – take 3x/day
  - Valcyclovir – take 1x/day, better compliance (not a cure)
  - Gancyclovir – tx CMV retinitis
  - Famcyclovir – tx shingles
- Adenosine Analog: Vidarabine
- Thymidine Analogs: Idoxuridine, Trifluridine

6) Assembly: package the virus (viremic, symptomatic)

7) Lysogeny: virus explodes out of the cell (viremic, symptomatic)

- RNA → destroys cell membrane
- DNA → destroys nuclear membrane
- Wrap themselves with cell membrane => autoimmune diseases

Viral Associated Cancers:

EBV: Nasopharyngeal CA, Burkitt's lymphoma
HPV 16,18,31,45: Cervical CA
Hep B: Liver CA
HIV: Kaposi's sarcoma, CNS/testicular lymphomas
HTLV-1: **T** cell leukemia

Encephalitis: Headache, ataxia

- **Arbovirus:** equine mosquitos => headache, ataxia *(frontal lobe) "more e's => more lethal"*
- **HSV-1:** hemorrhagic encephalitis, seizures, EEG slow wave complexes *(temporal lobe)*

Meningitis: headache, photophobia, stiff neck (infection of pia and arachnoid)

- **B**rudzinski's sign: **b**end neck => knee flexion
- **K**ernig's sign: flex **k**nee => pain, resistance

Viral Meningitis:

Kids: Arbovirus
Adults: HSV-2 (Tx: Acyclovir)

Bacterial Meningitis:

Tx: Vancomycin + Ceftriaxone
Prophylaxis: Rifampin or Ciprofloxacin
Most: Strep pneumo
0-2 mo: Group **B** Strep/**E.** coli/**L**isteria → can cause deafness *"Baby BEL"*

10-21 y/o: N. meningitides

Meningitis CSF:

- PMNs → bacteria
- T cells/MP → non-bacterial
- Normal glucose → viral (usually Enteroviruses)
- Low glucose → fungus, TB
- Protein → TB
- (+) Quelling: Pneumococcus
- Geimsa stain: Trypanosomes
- India ink: Cryptococcus
- Gram stain: Bacteria
- Wet mount: N. Fowlerii
- PCR: HSV
- AFB: TB, Nocardia

Common Cold Viruses:

Coronavirus (spring) *"drink a Corona beer on spring break"*
Rhinovirus (summer) => runny nose
Adenovirus (fall) => red eyes, necrotizing bronchiolitis, *"swimming pool conjunctivitis"*
Influenza virus (winter) => Staph aureus follows it => pneumonia (an orthomyxovirus)

- **H protein:** penetrate cells ← vaccine
- **N protein:** explodes out of your cells
- Tx: Amantadine, Rimantadine
- Tx: Oseltamivir "Tamiflu", Zanamivir – neuraminidase inhibitors

Parainfluenza: most common
Herpesvirus: painful ulcers of gums, keratitis (cornea will be destroyed by steroid tx!)

Other Viruses: *alphabetical order*
CMV: vision loss w/ "floaters" (neonates, AIDS pts)

- Tx: Ganciclovir or Foscarnet

Coxsackie B:

- ST Depression => myocarditis
- ST Elevation => pericarditis

Coxsackie A:

- Hand-Foot-Mouth dz

Coxsackie B:

- Endocarditis
- DM type1

EBV Mononucleosis: teenager w/ sore throat, fatigue, splenomegaly

- + Monospot test: may be negative for first week
- + Heterophile Ab
- Posterior cervical lymphadenopathy
- No Ampicillin (=> skin rash if have EBV due to circulating PCN Ab)

- Avoid contact sports (can rupture spleen)
- Tx: NSAIDs

HSV-1:
- Gingival stomatitis (gum ulcers)
- Herpetic keratitis => dendritic spine *cornea "shattered window" on fluorescein stain*
- + Tzanck prep
- Tx: no steroids

Rabies: hydrophobia, laryngospasm
- Bat exposure (aerosolized bat poop)
  Tx: Rabies vaccine "toxoid" x 5 doses + Ig
- Dog/cat/wild animal bites (raccoon/skunk/fox)
  Tx:
  1) Tetanus toxoid x 3 doses (if >5 yrs)
  2) Tetanus Ig (if dirty wound)
  3) Amoxicillin/Clavulanate
- Rodent bites (mice/rats/squirrels/rabbits) → no rabies
- Virus → unmyelinated nerves → CNS (hippocampus) → peripheral nerves, Negri bodies

Human Herpes Viruses:
1: HSV1 (oral) => fever blisters, corneal blindness, herpetic whitlow (dentist finger pustules)
2: HSV2 (genital) => genital ulcers, neonatal herpes (via birth canal)
3: Varicella *"chickenpox"* → Zoster *"shingles"* => encephalitis in AIDS pts
4: Epstein-Barr virus => mononucleosis, Burkitt's lymphoma, oral hairy leukoplakia (AIDS pt)
5: CMV => fetal blindness, pneumonia (transplant pts)
6: Sixth Disease => roseola in kids (fever → rash)
7: Pityriasis Rosea: herald patch → then follows skin lines *"C-mass tree appearance"*
8: Kaposi's Sarcoma (AIDS pts) – purple, pink, brown nodules all over body

Herpes Virus Tx:
- Ganciclovir (G) – tx CMV
- Idoxuridine (T)
- Vidarabine (A)
- Foscarnet (phosphonate)
- Acyclovir (G) – kills renal tubules
- Valacyclovir – don't use w/ Cimetidine (↑CNS toxicity)

Meningitis:
0-2mo: *"baby BEL"*
- Group **B** Strep
- **E**. coli
- **L**isteria

>2mo: Strep pneumo
10-21y/o: N. meningitidis

Hepatitis:
Core Ag – gone before pt has symptoms (2 mo)
Core Ab – past infection (stays positive for life, "natural immunity")
Surface Ag – current infection (or recent immunization if surface Ag only)

Surface Ab – vaccination
E Ag – transmissibility/infectivity
E Ab – low transmissibility
Window period – core Ab only (no Ag) "equivalence zone"

Hepatitis: virus, EtOH, Acetaminophen, Aflatoxin

- Surface Ag >6 mo
- Elevated enzymes >6 mo
- Chronic persistent: nothing wrong with liver
- Chronic active: will have fibrosis → can lead to cirrhosis, cancer
- Icteric (jaundice) phase of hepatitis => bile plugs in canaliculi
- Tx: Interferon or Amantidine/Rimantidine

	Hep A/E	**Hep B** (DNA virus) rvs transcriptase	**Hep C**	**Hep D** "defective"
Incubation:	2-6 wks	2-6 months	20-30 years	
Transmission:	Fecal-oral *"Vowels from the bowels"*	1. IV 2. Blood 3. Sex 4. Vertical (mom-baby)	1. Blood 2. Transfusion 3. IV 4. Sex (not vertical)	Post-Hep B ↑AST, ALT
Chronic active:	No	10% risk	70% risk	
Cancer:	No	Highest risk	Less than B	
Tx:		INF + Lamivudine	INF + Ribavirin	
Miscellaneous:	▪ Hep E attacks pregnant women/ Asians ▪ Shellfish => Hep A	▪ Dane particle = DNA ▪ Baby must get vaccine and IgG at birth		

Autoimmune Hepatitis:
Type I: anti-SM Ab (young women)
Type II: anti-LKM Ab (kids)

Hep A:
IgG => past
IgM => current

High-Yield Review:

Non-Bacteria/Disease:	Clue:
Mycobacterium Marinum	Fish tanks
Mycobacterium Scrofulaceum	Supraclavicular TB in kids
Mycoacterium Leprae	Leprosy: hypopigmentation + no sensation
Mycobacterium Avium Intracellulare	AIDS pt w/ gastroenteritis and SOB
Mycobacterium Tuberculosis	Night sweats, hemptoysis, weight loss
Miliary TB	Hydrocephaly, compression fx, pericarditis
Tuberculoid Leprosy	TH1, no acid fast stain
Lepromatous Leprosy	TH2, acid fast stain, erythema nodosum
Treponema Pallidum	Syphilis
1° Syphilis	Painless chancre (1-6wk)
2° Syphilis	Palm/Sole rash (6wk)
3° Syphilis	Tabes dorsalis, obliterative endarteritis, Argyll-Robertson pupil
Treponema Pallidum Variant	Bijel: non-venereal condyloma lata (kids)
Treponema Pertenue	Yaws: raspberry ulcers
Treponema Carateum	Pinta: red scaly patches → white spots
Borrelia Borgdorfori	Lyme disease (ixodes tick)
1° Lyme disease	Erythema chronicum migrans
2° Lyme disease	Heart block, Bell's palsy, meningitis
3° Lyme disease	Arthritis
Borrelia Recurrentis	Cyclic fever, Brill-Zinsser disease
Leptospira Interrogans	Rat urine, sewage workers, Fort Brag's fever, Weil's disease (nephritis/hepatitis)
Piedra	Little black fungal balls on hair shaft
Candidiasis	White cheesy discharge, oral thrush
Tinea Capitis	Flaky crusty scalp
Kerion	Tinea capitis + infection
Tinea Corporis	Ringworm (from cats)
Tinea Intertrigo	Skin peels off under breasts or armpits
Tinea Versicolor	V-shape on back, spaghetti-meatball distrib
Tinea Unguium	Discolored nails "under nail"
Tinea Nigra	Flaking palms, dark lines
Tinea Manis	Pealing between finger webs
Tinea Pedis	Pealing between toes "Athlete's foot"
Tinea Cruris	On groin "Jock itch"
Systemic Candidiasis	T-cell defect
Histoplasma	Bat droppings in Midwest, no capsule
Blastomyces	Pigeon poop in NY, broad-based hyphen
Coccidiodes	Las Vegas, desert bump fever
Paracoccidiodes	S. America, ship's wheel

Rhizopus & Mucor	Mucormycosis (eats nose, eyes, sinus)
Sporothrix Schenckii	Rosebush prick
Cryptococcus Neoformans	AIDS meningitis, +India ink
Aspergillus	Fungal ball, moldy basement/hay, fungal keratitis (Tx: Natamycin)
Schistosoma	Walking barefoot in swamp, snails
Schistosoma Mansoni	Liver cancer
Schistosoma Haematobium	Squamous cell bladder cancer
Toxocara Cati	Cat poop
Toxocara Cani	Dog poop
Echinococcus	Raw lamb, hydatid cyst, eggshell calcification
Clonorchis Sinensis	Opisthorchis: Cholangiosarcoma in biliary tract
Necator Americanus	Most common hookworm in US
Enterobus Vermicularis	Pinworm, pruritis ani
Ancylostoma Duodenale	Duodenal obstruction
Trichuris Trichiura	Whipworm, rectal prolapse, tenesmus
Ascaris Lumbricoides	Roundworm, human feces
Strongyloides Stercoralis	Threadworm, dermatitis, pulm eosinophilia
Diphyllobothrium Latum	Raw fish, eats Vit B12
Taenia Saginata	Raw beef
Taenia Solium	Raw pork, cysticercosis
Trichinella Spiralis	Raw pork, spirals under muscles
Leishmania Donovani	Sandflies, skin/nostrils,Gulf War Syndrome
Leishmania Rhodiensis	Sandflies, attacks organs
Toxoplasmosis	Cat urine, ring-enhancing brain lesions
Naefleria Fowleri	Swamp trauma, meningoenchephalitis
Trypanosoma Brucei	Tsetse fly, African sleeping sickness
Acanthamoeba	Contact lenses, eats through cornea
Ehrlichia	Dog lick, ixodes tick, puncture wound near eye
Trypanosoma Cruzi	Reduviid bug, Chaga's, heart block, Achalasia, Hirschsprung's
Pneumocystis Jeruvechii	AIDS pneumonia, silver stains
Giardia Lamblia	Hikers, mountain streams, string test
Entamoeba Histolytica	Anchovy paste liver abscesses, Erlenmeyer flask lesions
Microsporidia	Most common AIDS diarrhea
Cryptosporidia	Watery diarrhea in AIDS, partial acid fast
Trichomonas vaginalis	Green frothy vaginal discharge
Babesiosis	Ixodes tick, malaria-like, lives in RBCs
Plasmodium Malaria	Invade old RBCs, fever q3days
Plasmodium Vivax/Ovale	Invade virgin RBCs, chronic liver malaria
Plasmodium Falciparum	Anopheles mosquito, black water fever, die
Wicheria Bancrofti	Elephantiasis

Arbovirus	Encephalitis, ataxia (kids)
Adenovirus	Swimming pool conjunctivitis
Influenza virus	Cold followed by Staph aureus infection
Parainfluenza	Most common
Herpes	Painful gum ulcers, keratitis
Cytomegalovirus	Vision loss with "floaters"
Coxsackie B	Pericarditis, myocarditis
Epstein-Barr virus	Sore throat, fatigue, splenomegaly, +Monospot test, +Heterophile Ab
Rabies	Hydrophobia, laryngospasm, Negri bodies
HHV-1: HSV-1	Fever blister, herpetic whitlow (dentist finger pustules), corneal blindness
HHV-2: HSV-2	Genital ulcers, neonatal herpes during birth
HHV-3: Varicella	Chickenpox, AIDS encephalitis
HHV-4: Epstein-Barr	Mono, Burkitt's lymphoma, Oral hairy leukopenia (AIDS pts)
HHV-5: Cytomegalovirus	Fetal blindness, attacks transplant pts
HHV-6: Sixth Disease	Roseola in kids (fever → rash)
HHV-7: Pityriasis Rosea	Herald patch, follows skin lines "Christmas tree appearance"
HHV-8: Kaposi's Sarcoma	AIDS pts nodules: purple/pink/brown
Hepatitis A	Shellfish, fecal-oral
Hepatitis E	Attacks pregnant women, fecal-oral
Hepatitis B	DNA virus, get from needles, cancer risk
Hepatitis C	Get from blood, chronic active
Hepatitis D	Follows Hep B, high AST and ALT

NOTES:

Antibiotics:

"Doctors are men who prescribe medicines of which they know little, to cure diseases of which they know less in human beings of whom they know nothing."

–Voltaire

Pharmacokinetics: what the body does to drugs

1) Absorption

Alimentary:

- Oral – uses small intestine (large surface area)
- Buccal – goes into venous circulation
- Sublingual – goes into SVC (bypass liver)
- Rectal – 50% bypass liver

Parenteral:

- IV – 100% goes into circulation
- IM
- Subcutaneous
- Intrathecal – into subarachnoid space

Inhalation – pulmonary drugs
Topical – localized disease
Transdermal – sustained release

2) Distribution

Diffusion: high → low conc, may be facilitated
Active transport: against concentration gradient, ATP → ADP

3) Metabolism – lipids pass through cell membrane, polar eliminates quicker
Phase I rxn (liver ER): redox and hydrolysis => polar groups
Phase II rxn (liver cytosol): conjugation => add glutathione, acetic acid, sulfate
Biotransformation factors: Genetics, age, gender, liver disease, P450
Zero-order Kinetics – *metabolism independent of concentration*

- Ex: Phenytoin, Chemo drugs
- Ex: EtOH (100mg/dL/hr): 1 glass wine, 1 shot whiskey, 2 cans of beer

1st-order Kinetics – *constant drug percentage metabolism over time*

- Ex: 10% of drug (conc=100mg/dL) eliminated every 2 hours:
- T=0 hrs: [D]=100mg/dL
- T=2 hrs: [D]=90 mg/dL
- T=4 hrs: [D]=81 mg/dL

4) Excretion

Excretion = removal of drug from body via urine, feces, respiration, skin, breast milk
Secretion = transport drug to another compartment

Pharmacodynamics: what drugs do to the body
Receptor interactions

- Transmembrane (Ex: Insulin)
- Ligand-gated ion channels (Ex: BZ, ACh)
- Transcription factors (Ex: Steroid hormone receptors)
- Second-messengers: (Ex: cAMP, cGMP, IP_3)

Mechanism of action

- Agonist – activates receptors
- Antagonist – inhibits receptors

Drug Dosage: **the link b/w kinetics and dynamics**

- $t_{1/2} = (.693)(V_d) \div$ clearance
- V_d: total drug ÷ plasma conc (large V_d => most of drug is sequestered)
- Loading dose: (desired plasma conc)(V_d)
- Maintenance dose: (desired plasma conc)(clearance)
- Steady-state plasma conc (C_{ss}): availability rate = elimination rate, *takes 4.5 half-lives*
- Clearance: volume of plasma cleared of drug
- Excretion rate: (clearance)(plasma conc) – rate of elimination
- TI: toxic dose ÷ therapeutic dose (high TI => safe drug) = LD_{50}, ED_{50}
- Peak level: 4 hrs after dose (too high => decrease dose)
- Trough level: 2 hrs before dose (too high => give less often)

Antibacterials:

Bacteriostatic: Protein synthesis inhibitors (except aminoglycosides)
Bacteriocidal: all the rest

p450-dependant drugs: *levels rise if you inhibit p450*

"Women's DEPT"

Warfarin
Digoxin
E$_2$
Phenytoin
Theophylline

Inhibit p450:

"Frequently I Do SMACK Grapefruit juice"

Fluoroquinolones
INH
Diltiazem
Sulfa drugs
Macrolides
Amiodorone
Cimetidine

Hospital Abscesses:

Day 1-3: Staph aureus – lots of O_2
Day 4-7: Strep viridans – no enzymes
Day >7: Anaerobes – PMNs

Staph Drugs: (at least one)

- Amoxicillin + Clavulanate
- Ampicillin + Sulbactam
- Methicillin
- Naficillin
- Cephalosporins
- Vancomycin
- Macrolides

Ketoconazole
Grapefruit juice

<u>p450 Inducers:</u>
"Get ABC/PQRS"
Griseofulvin
Alcohol
Barbiturates
Carbamazepine
Phenytoin
Quinidine
Rifampin
St. John's wort

<u>Pseudo Drugs:</u> (at least two)
Start anti-fungal if fever >48hr

- Ticarcillin + Clavulinic acid
- Piperacillin + Tazobactam
- Carbenacillin
- Ceftazidime or Cefepime
- Vancomycin
- Fluoroquinolones
- Aminoglycosides

<u>DNA Synthesis Inhibitors:</u>
1) Fluoroquinolones: inhibit TopoII (DNA gyrase), decreased w/ antacids
Tx: Gram +/- , Atypicals (not anaerobes)
SE: Inhibit rapidly dividing cells => anemia, UTI, dry skin, fetal growth retardation
CI: Pregnancy, Kids

- Levofloxacin
- Ciprofloxacin – best Pseudomonas coverage
- Norfloxacin – tx UTI only
- Trovafloxacin – taken off market due to hepatic necrosis
- Ofloxacin – tx Gonorrhea (1-dose), increasing resistance
- Gatafloxacin – tx Gonorrhea (1-dose), increasing resistance

<u>Myositis:</u>
"RIPS"

- **R**ifampin
- **I**NH
- **P**rednisone
- **S**tatins

2) RNA Polymerase Inhibitors: *"The 5 R's of Rifampin"*
SE: **R**evs up P450, **R**ed-orange secretions, **R**esistance if used alone
SE: Fat soluble => myositis (**r**ips up muscle), hepatitis
Tx: TB, N. Meningitidis and HI-B prophylaxis (give to close contacts)

- **R**ifampin (inhibits β subunit of RNA Pol)

<u>Disulfiram Rxns:</u>

- Metronidazole
- Cephalosporins
- Procarbazine

3) Sulfa Drugs: inhibit folate synthesis
Tx: Gram+, simple Gram –
SE: G-6PD hemolytic anemia, porphyria, MetHb

- Sulfamethoxazole – inhibits DHP synthetase, used for kids
- (Trimethoprim – inhibits DHF reductase, kernicterus, renal failure, no sulfur)
- **Sulfadiazine – tx burn pts (cream)**
- **Sulfacetamide – tx Chlamydia eye infections, prevent neonatal blindness in 3rd world**
- **Sulfasalazine – tx UC**
- **Sulfapyrazone – tx UC**

Cell membrane disrupters:
Polymixins: cationic detergents
Tx: Gram –

Metronidazole => O_2 free radicals: *"GGET on the Metro"* => disulfiram rxn (N/V/D w/ EtOH)
- Inhibits acetaldehyde dehydrogenase => hemolytic anemia due to oxidation
  Tx: **G**iardia, **G**ardnerella, **E**ntamoeba, **T**richomonas, all anaerobes

Protein Synthesis Inhibitors: *"buy AT 30, CEL at 50"*
30S Inhibitors:
1) Aminoglycosides – inhibit IF_2; bacteriocidal
Tx: All Gram –
SE: ototoxicity, nephrotoxicity, neurotoxicity (inhibits presynaptic Ca influx) => can't use w/ NM
"GNATSS"
- **G**entamicin – neuropathy
- **N**eomycin – topical (including gut surface) => rash, kills NH_4-producing bacteria
- **A**mi**k**acin – hepatic excretion => OK for **k**idneys, long $t_{1/2}$
- **T**obramycin – tx CF pts
- **S**treptomycin – tx TB pts, tx tularemia
- **S**pectinomycin – fomer gonorrhea tx *"now just a spectator"*

Dysgeusia:
- Metronidazole
- Clarithromycin
- Li

2) Tetracyclines – block **t**RNA
Tx: all Gram + , simple Gram –, Atypicals (not Staph aureus), Rickettsia
SE: photosensitivity (wear SPF15), errosive esophagitis, binds Ca^{2+}=> don't use w/ Tums or milk
SE: permanent grey teeth, revs up P450, Fanconi's (if old drug), negates Oral contraceptives
CI: Pregnancy, kids
- Minocycline – tx acne *"mean-o teenagers have acne"*
- Doxycycline – hepatic excretion
- Demecocycline – blocks ADHr => nephrogenic DI (like Li)
- Oxytetracycline

Triple Antibiotic:
"Brand New Potient"
Bacitracin
Neomycin
Polymyxin D

50S Inhibitors:
1) Chloramphenicol "CAM" – blocks peptidyl transferase
- Tx: all Gram +, Rickettsia
- SE: aplastic anemia, inhibits ETC complex IV, grey-baby syndrome

2) Macrolides – block translocation
- Tx: all Gram +, simple Gram –, Atypicals
- SE: inhibit P450, Torsade w/ 3rd generation anti-histamines
  - Clarithromycin – dysguisia (metallic taste)
  - Erythromycin – GI upset, tx Legionella pneumonia
  - Azithromycin – longest $t_{1/2}$, tx MAI (AIDS pts), good for pregnant women

3) Lincosamides – block translocation

- Tx: all Gram +, simple Gram – , all Anaerobes
- SE: Pseudomembranous Colitis
- Clindamycin
- Lincomycin – not used now

Chlamydia 1-dose Tx:

- Azithromycin (1g)
- Or 2g to tx GC

Cell wall synthesis inhibitors:

1) Penicillins:

Tx: Gram + (not Staph), simple anaerobes
SE: anaphylaxis, non-specific rash, drug fever
SE: hemolytic anemia, interstitial nephritis, bone marrow suppression

Simple:

- Penicillin G: shot
- Penicillin V: oral *"remember the terminal sulcus V line in the back of the tongue?"*
- Benzathine: long-acting
- Procaine: short-acting

Extended Spectrum: also cover E. coli, H. influ + Staph aureus (β-lactamase inhibitors)

- Amoxicillin + Clavulinic acid
- Ampicillin + Sulbactam: => J-H rxn (EBV, CMV, Syphilis), #2 interstit. nephritis, tx pregnant UTIs

Anti-Staphylococcal:

- Methicillin: IV => #1 PCN causing interstitial nephritis
- **Na**fcillin: IV – high **Na** load (do not use w/ Conn's, arrhythmias, HTN, seizures)
- Oxacillin: oral
- Cloxacillin: oral
- Dicloxacillin: oral

Anti-Pseudomonal:

- Ticarcillin + Clavulinic acid (also covers *S. Aureus*)
- Piperacillin + Tazobactam (also covers *S. Aureus*)
- Carbe**na**cillin: high **Na** load
- Mezlocillin

Staph/Pseudo Attack:

- DM
- CF
- Burn pts
- Neutropenic pts

2) Cephalosporins:

Tx: all Gram +, simple Gram -, simple anaerobes
SE: same as penicillins (15% crossover with penicillins for anaphylaxis)
More Gram – coverage as you progress through generations

Vancomycin tx:

- MRSA
- Staph. epidermidis
- Enterococcus

1st Gen:

Cefazolin – parenteral (IV)
Cephalothin – interstitial nephritis
Cephalexin

2nd Gen: *"... foxy family wearing fur"*

Cefaclor – use in kids => erythema multiforme, urticaria, bone pain, serum sickness
Cefotetan – inhibits Vit K => bleeding

Ce**fox**itin – excellent complete anaerobic coverage
Ce**fam**andole – inhibits Vit K => bleeding
Ce**fur**oxime – tx epiglottitis

<u>3rd Gen</u>:

Ceftriaxone – cross BBB, hepatic excretion (not <2m/o), tx meningitis
Cefixeme – oral
Cefotaxime – renal excretion, tx meningitis <2m/o
Cef**tazi**dime – best Pseudo coverage *"pseudo tazo tea"*
Cefoperazone – dec. Vit K
Moxalactam – dec. Vit K

<u>4th Gen:</u>

Cefepime

<u>Gonorrhea 1-dose Tx:</u>

"Try to fix the fox with floxs"

- Cef**tri**axone
- Ce**fix**eme
- Ce**fox**itin
- Ciprofloxacin
- Ofloxacin
- Gatifloxacin

<u>3) Carbapenems:</u> everything (except atypicals)

- Imipenum + Cilastatin (inhibits renal enzymes to decrease seizure incidence)
- Meropenum

<u>4) Monobactams:</u> Gram – rods

- Aztreonam – good for penicillin allergies

5) Vancomycin – irreversibly blocks formation of cell wall peptide bridges
Tx: all Gram +
SE: ototoxicity, nephrotoxicity, red man syndrome

Quick Antibiotic Review:

Drug Class:	**Specific Drugs:**	**Bacterial Coverage:**
DNA Synthesis Inhibitors:		
1) Fluoroquinolones	--floxacin	Gram +/-, Atypicals
2) RNA Pol Inhibitors	Rifampin	TB, Meningitis/HIB prophylaxis
3) Sulfa drugs	Sulfa--	Gram +/Simple Gram –
Protein Synthesis Inhibitors:		
1) Aminoglycosides	Gentamicin, Neomycin, Amikacin, Tobramycin	All Gram –
2) Tetracyclines	Minocycline, Doxycycline, Demecocycline	All Gram +, Simple Gram –, Atypicals, Rickettsia
3) Chloramphenicol	Chloramphenicol	All Gram +, Rickettsia
4) Macrolides	Clarithromycin, Erythromycin, Azithromycin	All Gram +, Simple Gram –, Atypicals
5) Lincosamides	Clindamycin, Lincomycin	All Gram +, Simple Gram –, Anaerobes

Cell Wall Synthesis Inhibitors:		
1) Simple Penicillins	Penicillin G, Penicillin V	Gram +, Simple Anaerobes
2) Extended Spectrum Penicillins	Amoxicillin + Clavulinic acid, Ampicillin + Sulbactam,	+E coli, H influ + Staph aureus
3) Anti-Staphylococcal	Methicillin, Nafcillin, Oxacillin, Cloxacillin	Gram +, Simple Anaerobes
4) Anti-Pseudomonal	Ticarcillin + Clavulinic acid, Piperacillin + Tazobactam	Gram +, Simple Anaerobes, Staph
5) Cephalosporins: more Gram – w/ generation	Cef--	Gram +, simple Gram -, simple anaerobes
6) Carbapenems	Imipenem + Cilastatin, Meropenem	everything (except atypicals)
7) Monobactams	Aztreonam	Gram – rods
8) Vancomycin	Vancomycin	Gram +, Pseudo
Others:		
Polymixins (cationic detergent)	Polymixins	Gram –
Metronidazole	Metronidazole	Giardia, Gardnerella, Entamoeba, Trichomonas, all anaerobes
VRE drugs	Linezolid	Vancomycin-resistant drugs

Toxicology:

Drug Overdose:	**Antidote:**
Acetaminophen	1) Charcoal (if <2hr) 2) N-acetylcysteine (2-10hr)
Benzodiazepines	Flumazenil
β-blockers	Glucagon
Calcium Channel Blockers	Calcium Chloride + Glucagon
CO *(cherry red lips)*	100% O_2
Cu, Gold, Cd, Mercury, Arsenic	Penicillamine
Cyanide *(almond odor)*	Amyl Nitrite, Na thiosulfate
Diphenhydramine	Physostigmine
Ethylene Glycol *(maltese cross)*	EtOH, Fomepizole
Fe	Deferoxamine
Heparin	Protamine sulfate
HF	Ca carbonate
Levothyroxine (T_4)	PTU

Lithium	Sodium polystyrene sulfonate
MetHb	Methylene blue
Nitrites	Methylene blue
Organophosphates	Atropine + Pralidoxime *"2-PAM"*
Opioids	Naloxone IV
Pb	Succimer or BAL
Salicylates *(epigastric pain)*	Charcoal or $NaHCO_3$
TCAs	Bicarbonate
t-PA, Streptokinase	Aminocaproic acid
Theophylline	Esmolol
Warfarin	Vit K (takes 6hr)+FFP (lasts 6hr)

Herbal Side Effects:

Herb:	**Used To Treat:**	**Side Effect:**
Evening Primose Oil	Breast Pain	Bloating, GI upset
Garlic	Lower Cholesterol	Platelet Dysfunction, Bleeding
Ginseng	Improve Memory	Stephen-Johnson, Psychosis
Kava	Insomnia, Anxiety	Hepatotoxicity
Gingko Biloba	Improve Memory	Platelet Dysfunction, Bleeding
Licorice	Expectorant	HTN
Saw Palmetto	BPH	HTN
St. John's Wort	Depression	P450 Inducer

NOTES:

Biostatistics:

"Do not put your faith in what statistics say until you have

carefully considered what they do not say."

–William Watt

Central Tendency of Distributions:

Mean: average ← sensitive to outliers
Median: middle value ← use w/ skewed population
Mode: most frequent value

Endemic: localized epidemic
Pandemic: world-wide epidemic

Variability:
1 STD: 68%
2 STD: 95%
3 STD: 99%

Chi^2: compare percentages
T-test: compare 2 things
ANOVA: compare 3+ things

Standard Error of Mean = $S/\sqrt{N}$ => precision of mean
S = standard deviation, N = sample size
Skewing: shift to right (mean > median > mode)

1 → 2 Question Survey:
Decrease: Sn, NPV
Increase: Sp, PPV

Contingency Tables:

	D+		D-	
T+	Sn	a	b	FP
T-	FN	c	d	Sp

St = a/ a+c = TP/ all diseased (people that have dz) ***"STD: SensiTivity → have Disease"***

- Sensitivity=85% → 85% of diseased people have positive test

Sp = d/ b+d = TN/ all non-diseased (people that don't have dz) *"SpIn"*

- Specificity=85% → 85% of non-diseased have negative test

PPV = TP/ all positives (increases w/ prevalence) Ex: (+) ELISA → Do you have HIV?
NPV = TN/ all negatives (probability of not having a dz if have negative test)

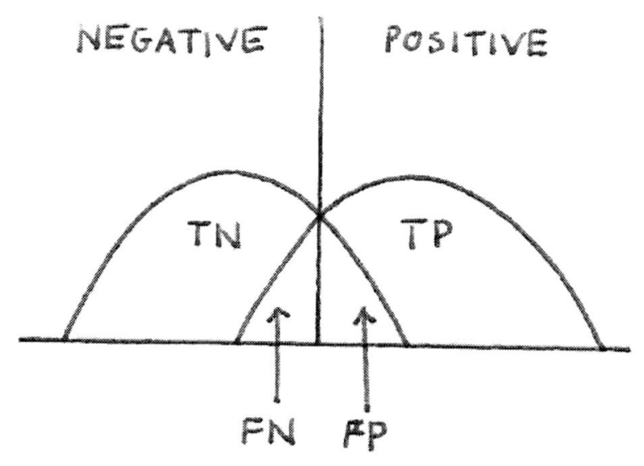

Incidence = new cases/ total population
Prevalence = all diseased/ total population (Ex: improved quality of care)

Odds Ratio = (ad + bc) cross product (diseased are x times more likely to see risk factor)

- OR <1 => protective factor
- OR >1 => risk factor

Correlation Coefficient:

- **0:** nothing
- **+:** correlation
- **-:** negative correlation

CI = 95% => 95% sure it lies within the interval (cannot include 1.0)
RR = exposed/ unexposed (risk of getting dz w/ known exposure)
NNT = number needed to treat to change 1 life
p value <0.5 = random chance that you will be wrong 1 time out of 20

Hypothesis Testing:
Null hypothesis => nothing's happening
Power: 1-β = probability of detecting a true intervention ← improve by ↑size of study
Effect size = how different two groups are
α: type I error (FN), P value error *"too optimistic"*
β: type II error (FP), Power error *"too pessimistic"* ← small samples

Prevention:
1°: ↓Incidence
2°: ↓Prevalence
3°: Slows disease progression

Bias:
Accuracy: validity *"truth"*
Precision: reliability *"keep making the same mistake"*
Admission rate: hospitals have certain populations
Confounding bias: forgotten variable => use matching, restriction, randomization
Hawthorne effect: the "watched" change their behavior

Lead time: time between diagnosis and treatment
Observer bias: the observers have knowledge about control and study samples
Recall bias: inaccurate recall of past events
Respondant: subjective diagnosis
Sample distortion: sample does not represent population
Selection: who is in or out
Unacceptability: subjects lie

<u>I) Clinical Trials:</u> experimental
Phase I: Toxicity *"hurt pt?"*
Phase II: Efficacy *"help pt?"*
Phase III: Comparison *"any better?"*
Phase IV: Post-marketing surveillance *"can they screw it up?"*

<u>II) Observational Studies:</u>
1) Cohort study: Prospective *"a Cohort uses Roman Rulers to march Forward"*
- Observe a "cohort" (group of people with similar characteristics) to see how many develop a specific disease after exposure to a risk factor.
- Determines incidence: new cases of disease
- Uses Relative Risk
- Occurs in community
- Ex: finds new cases of common disease

2) Case control: Restrospective *"What are the rare Odds that the Case was Retro?"*
- Select subjects with a disease, compare to controls, and study the differences
- Uses Odds Ratio
- Occurs in hospital
- Has more selection bias
- Ex: finds risk of developing rare disease

3) Cross sectional: snaphot in time
- Determines prevalence: total cases of diseased
- Ex: polls/surveys

<u>III) Descriptive Studies:</u>
1) Case report: describe an unusual pt
2) Case series report: describe several unusual pts
3) Consensus panel: panel of experts provides a recommendation
4) Clinical wisdom: "I think…."

IV) Meta-Analysis: tries to combine data from many trials

NOTES:

NOTES:

Step 2 CS:

"Medicine is learned by the bedside and not in the classroom.

Let not your conceptions of disease come from words heard

in the lecture room or read from the book. See, and then

reason and compare and control. But see first."

–William Osler

Adult Vital Signs:

- Resp rate = 14-20 bpm
- Temp = 98.6°F (37°C)
- Pulse = 60-100 bpm
- BP = 120/80 (+20/+10 increases stage, need 3 readings to diagnose)

Scales:

- Heart murmur: 1-6 (Grade IV has thrill)
- Muscle strength: 0-5 (normal = 5)
- Reflexes: 0 to 4+ (normal = 2+)
- Pulses: 0 to 4+ (normal = 2+)
- PMI: normal = "dime size" (2cm)
- JVD: normal = 3-8 cm
- Edema: 1+ to 4+ (2-8mm)

Radiating Pain:

to Jaw: MI
to Left Shoulder: MI or Spleen rupture
to Scapula: Gall bladder problems
to Trapezius: Pericarditis
to Groin: Ureter stones, Osteoarthritis of hip

Signs:

Blumberg's sign: rebound tenderness => peritonitis
Brudzinski's sign: bend neck, see knee flexion => meningitis
Chadwick's sign: purple vagina => pregnancy
Chandelier sign: cervical motion tenderness => PID
Chovstek's sign: tap facial nerve => muscle spasm
Courvoisier's sign: palpable gall bladder => cancer (pancreatic)
Cozen's sign: dorsiflex wrist causes pain => lateral epicondylitis *"tennis elbow"*
Cullen's sign: bleeding around umbilicus => hemorrhagic pancreatitis
Gottran's sign: scaly patches on MCP, PIP joints => dermatomyositis
Hamman's sign: crunching sound on auscultation => subcutaneous emphysema
Homan's sign: pain in calf w/ dorsiflexion => DVT
Kehr's sign: left shoulder pain => spleen rupture
Kernig's sign: flexing knee causes resistance or pain => meningitis
Kussmaul's sign: big neck veins w/ inspiration => pericardial tamponade
Levine sign: hold clenched fist over sternum => angina
Murphy's sign: pressed gallbladder causes pt to stop breathing => cholecystitis
Obturator sign: inward rotation of hip causes pain => appendicitis

Phalen's test: flexing wrists for 60s causes paresthesia => carpal tunnel
Psoas sign: extending hip causes pain => appendicitis *"move leg toward ass"*
Rovsing's sign: palpating LLQ causes RLQ pain => appendicitis
Tinel's test: percussing median nerve causes 1st-3rd finger pain => carpal tunnel
Trousseau's sign: BP check causes carpal spasm => hypocalcemia
Turner's sign: bleeding into flank => hemorrhagic pancreatitis

History:
CC: *Age, gender, presenting complaint*

HPI: *"OPQRST"*
- **O**nset? (When start? Had this before?)
- **P**rovocation **(**Anything make it better or worse?)
- **Q**uality (Scale:1-10; stabbing vs. dull pain)
- **R**adiation
- **S**ymptoms (Fever, SOB, HA, etc.)
- **T**iming (How long? How often? When start?)

PMH:
- Illness
- Injuries
- Surgeries
- Medications (OTC meds, vitamins, herbs)
- Allergies (What happens?)

SH:
- Job, Stress
- Who lives at home with you?
- Tobacco, EtOH, Drugs, Diet

FH:
- Age of relatives
- Illnesses of relatives

Sexual Hx:
- # Sexual partners
- Gender of partners
- Condom use
- STD hx
- Women: LMP/ GxPxAx

Health Maintenance:

- Diet/ Exercise
- Injury prevention (seat belts, bicycle helmets)
- CA screen (Mammogram/ Pap or Prostate/ Guaiac/ Sigmoidoscopy)
- EtOH, Tobacco, Drugs
- CAGE (2/4 => problem)
  - Have you ever felt the need to **C**ut down?
  - Do you feel **A**nnoyed by people asking?
  - Have you ever felt **G**uilty about it?
  - Have you ever had an **E**ye opener drink?

Pediatric History:

Birth Hx:

- How long in hospital/ Complications
- Prenatal care/ Drug use
- Vaginal vs. C/S delivery
- Amniocentesis
- Full term

Feeding Hx:

- Breastfeeding vs. Formula
- Appetite
- Vitamins

Medical Hx:

- Pediatrician/ Last checkup
- Immunizations
- Illness/ Meds

General:

ROS:

- ΔWeight, fever, chills, night sweats

Skin:

ROS:

- Itch, rash
- Growths

Physical Exam:

- Inspect skin, hair, nails
- Palpate lymph nodes

HEENT:

ROS:

Head: HA, hair loss

Eyes: Δvison, blurring, diplopia, scotoma (see spots), photophobia

Ears: Δhearing, tinnitus, vertigo, discharge

Nose: epistaxis, discharge, obstruction, sinus pain

Throat: abnormal taste, bleeding gums, oral ulcers, hoarseness, neck stiffness, jaw pain

Physical Exam:

Head: Inspect, palpate

- Sclera: color
- Pupils: symmetric, reactive, extra-ocular movements
- Visual acuity: Snellen eye chart

Eyes: PERRLA

Hypertensive Retinopathy: *neovascularization => DM*

- Grade I – arteriole narrowing, copper wiring
- Grade II – A-V nicking
- Grade III – hemorrhage
- Grage IV – papilledema

Ears:

Rinne test: tuning fork on mastoid, then near ear

Weber test: tuning fork on top of head, then near ear

Normal: AC > BC in both ears

Abnormal:

- Right bone conduction problem => BC > AC on right *(lateralize to right side)*
- Left neural problem => AC > BC on both sides *(lateralize to right side)*

Nose: Inspect with otoscope, palpate sinuses

Throat:

Mouth: Inspect tonsils, use tongue blade

Palpate neck: Thyroid, lymph nodes, ROM

Neural Hearing Loss:

- Infection
- Trauma
- Tumor

Conductive Hearing Loss:

- Cerumen, foreign body
- Otitis externa
- Tympanic membrane perforation

Ear Diseases:

TMJ: ear pain worse w/ chewing

Otosclerosis: genetic hearing loss in 30's

Meniere's: excess endolymph =>

Heart:

ROS:

- Chest pain, palpitations
- Orthopnea, PND
- Edema

Physical Exam:

- Inspect (clubbing, cyanosis, edema)
- Palpate PMI
- Percuss
- Auscultate *"APT-M"* (murmurs, rubs, gallops)

- Auscultate carotid arteries w/ breath held
- Palpate arteries:
    - Carotid
    - Brachial
    - Radial
    - Renal
    - Iliac
    - Femoral
    - Popliteal
    - Posterior tibial
    - Dorsalis pedis

Notes:
Correct BP cuff: Length = 80%, Width = 40% of arm circumference
Systole = Phase 1 Korotkoff
Diastole = Phase 5 Korotkoff
Stethoscope findings:

- Lean forward: ↑aortic murmurs
- Left lateral decubitus position: ↑mitral murmurs
- Inspire: R sided problems louder
- Expire: L sided problems louder
- Valsalva: ↑HCM, ↓AS

Heart Sound:	Notes:	Patho:
S_1	M/T closed	Normal
S_2	A/P closed, $S_2 > S_1$	Normal
Physiologically split S_2	Inspiration only	Normal (pulmonic area)
Paradoxically split S_2	Expiration only	L BBB
Fixed split S_2		ASD
Widely split S_2		R BBB, PS
S_3	*"SLOSH(S_1) -ing(S_2) in(S_3)"*	CHF, "ventricular gallop"
S_4	*"a(S_4) STIFF(S_1) wall(S_2)"*	AS, VH, HTN
Systolic Murmurs:		
Ejection murmur	hear S_1, S_2	AS, HCM
Holosystolic	sounds blowing	MR, VSD
Diastolic Murmurs:		
Blowing	S_3 time, high pitch blowing	AR
Rumble	S_4 time, rumble with OS	MS
Continuous Murmurs:		
Friction rub (hold breath)	Inspiration and Expiration	Pericarditis
Friction rub (breathing)	Inspiration and Expiration	Pleuritis
Continuous	Machine-like	PDA, OWR, VHL

Pulse: QRS complex

- Irregularly irregular => A Fib
- Regularly irregular => PVC

Increased JVP => RV failure (add 5cm to sternum)

- Large **a** wave => **a**trial conctraction (TS, RVH, Pulm HTN)
- Large c wave => RV contraction (after carotid pulse)
- Large v wave => atrial filling (TR)
- Kussmaul's sign (↑JVP w/ inspiration) => constrictive pericarditis

PMI: (5th MCIS)
No PMI: MS
Sustained PMI:

- AS (parvus et tardus)
- HTN

Increased PMI:

- DCM (pulsus alternans)
- HCM (trifid PMI w/ bifid carotid pulse)
- AR (Waterhammer bounding pulse)
- MR

Lungs:
ROS:

- Cough, wheeze
- Dyspnea, pleurisy (pain w/ deep breathing), hemoptysis

Physical Exam:

- Inspect
- Thoracic Expansion
- Tactile Fremitus *"say 99"*
    - ↑: Consolidation
    - ↓: Pleural effusion
- Percuss:
    - Hyper-resonance => air (pneumothorax, emphysema, asthma)
    - Flat, dull => fluid, solid (pleural effusion, consolidation)
- Auscultate:

Lung sounds:	Notes:	Pathology:
Normal:		
Bronchial	Exp > Insp (loud)	Normal
Vesicular	Insp > Exp (soft)	Normal
Bronchovesicular	Insp = Exp	Normal
Tracheal	Insp = Exp (loud, high pitch)	Normal
Continuous:	*musical, longer*	
Wheeze	whistle	Asthma
Rhonchi	snore	Bronchitis, pulm. edema
Stridor		Vocal cord obstruction
Egophony	e → a change	Consolidation
Adventitious:	*intermittent*	
Fine crackles	Late insp *"slow Velcro ripping"*	Pulmonary fibrosis
Coarse crackles	Early insp *"horse races"*	Pulmonary edema
Pleural friction rub	Insp and Exp	Pleuritis

Abdomen:

ROS:

- Pain, jaundice
- Δ Appetite, N/V, dysphagia, heartburn
- Δ Bowel Habits: constipation, diarrhea, color
- Rectal pain

Physical Exam:

- Inspect
- Auscultate
- Percuss: Abdomen, Liver
- Palpate light → deep: Liver, Spleen, Kidneys
- Rebound tenderness
- Murphy's sign

Urinary Tract:

ROS:

- Urgency vs. decreased, pain
- Night frequency, incomplete voiding, decreased stream
- Blood, pus, stones
- Flank pain

Physical Exam:

- Tap kidneys

Sexual :

Female ROS:

- Pelvic pain, discharge
- Contraception, sexual dysfunction, dyspareunia
- Age at menarche, menstrual cycle length, LMP or menopause

Male ROS:

- Penis pain, discharge, sores
- Testes pain
- Sexual dysfunction

Endocrine:

ROS:

- ↑urination, thirst, hunger
- Temperature intolerance

Hematologic:

ROS:

- Easy bruising, bleeding, paleness
- Claudication (pain w/ walking)

Musculoskeletal:

ROS:

- Backache, muscle ache, cramps
- Joint pain, swelling, stiffness

Physical Exam:

- Fingers: Make fist, spread fingers
- Wrists: Flex/Extend, Phalen/Tinel tests
- Shoulder: ROM (touch opposite shoulder)
- Knee: Anterior/Posterior drawer signs
- Spine: Gait, reflexes, straight leg raise, tenderness

Neurologic:

ROS:

- Numbness, tingling
- Seizures, paralysis, weakness
- Tremor, gait disturbances
- Syncope

Physical Exam:

Cranial Nerves:

CN 1 (olfactory): Smell

CN 2 (optic):

- How many fingers do you see?
- Wiggling fingers from side – hemianopsia
- Optic discs – papilledema

CN 3 (oculomotor):

- Penlight – check pupils on both sides
- Accomodation – move your finger to their nose (far → near)
- Down and out eyes
- Ptosis – weak levator palpebrae

CN 4 (trochlear):

- Follow "H" with eyes
- Nystagmus: hold finger at ends of "H" and watch for wiggling eyes
- Diplopia when looking down

CN 5 (trigeminal: V_1=opthamic; V_2=maxillary; V_3=mandibular):

- Motor = mastication mm.
- Sensory =
  - Mesencephalic – proprioception (jaw jerk)
  - Main sensory – light touch (cornea reflex)
  - Spinal – pain, temp

CN 6 (abducens): Abduct eyes past midline

CN 7 (facial):

- Puff cheeks
- Raise eyebrows, close eyes tight
- Smile, frown

CN 8 (acoustic): hear rubbing fingers behind ear

- Vestibular nucleus = equilibrium
- Cochlear nucleus = hearing

CN 9 (glossopharyngeal):

- Dry mouth
- Decreased gag reflex, dysphagia

CN 10 (vagus):

- "Ahhh" – soft palate rise
- Hoarse voice
- Uvula – deviates to opposite side

CN 11 (accessory):

- Shrug shoulders
- Turn head against resistance

CN 12 (hypoglossal):

Nerve Reflexes:

L4: knee jerk, foot dorsiflex

L5: big toe dorsiflex

S1: ankle jerk, foot eversion

S2-4: anal wink

- Stick tongue out and wiggle – deviates to weak side

Reflexes:

- Biceps/ Triceps/ Brachioradialis
- Patellar/ Achilles
- Babinksi – toe pulls up after scratching lateral sole of foot
- Clonus – jerk foot up

Motor (Tone/Strength):

- Arms: push in, pull out
- Wrists: pull up, push down
- Hands: spread fingers
- Legs: kick out, pull in
- Ankles: push on the gas

Sensory:

- General: sharp vs. dull
- Light touch – cotton
- Discrimination – trace number in palm, 2 coins, 2 paper clips
- Vibration – tuning fork
- Positional – move fingers up and down

Coordination:

- Alternating movements – flip hand on thigh (cerebellum)
- Point-to-point: (cerebellum)
    - finger to nose to finger
    - drag heel down shin
- Romberg: stand with eyes closed, palms up, arms out (DCML tract)
- Pronator Drift: arms drift w/ eyes closed (hemiplegia)
- Asterixis: hand flap (hepatic encephalopathy)
- Gait: walk on heels, toes, squat, rise

Psychiatric:

- A/O x 3
- Speech: fluid, goal oriented
- Mood: euthymic
- Judgement, Insight
- Suicidal, Homicidal ideation

Mini Mental Status Exam: *".com"*

- **C**oncentration (serial 7's or world backwards)
- **O**rientation to person, time, place
- **M**emory (repeat 3 items later)

Common Presentations:

Abdominal Pain:

RUQ:

- Cholecystitis: Murphy's sign
- Liver

LUQ:

- Spleen
- Stomach
- Diaphragm irritation (w/ deep breathing)

RLQ:

- Appendicitis: pain at McBurney's point (Signs: Psoas, Rovsing, Obturator)
- Intestines
- Kidney
- Female organs

LLQ:

- Diverticulitis
- Intestines
- Kidney
- Female organs

Epigastric:

- Pancreas: pain radiates to back, (Signs: Turner's, Cullen's)
  - Gallstones => acute
  - EtOH => chronic
  - Depression => cancer
- Esophagitis
- AAA
- Ulcer

Diffuse:

- IBD
- Gastroenteritis
- Peritonitis: Guarding, Blumberg's sign, hypoactive bowel
- Mesenteric infarction: pain out of proportion to findings → do CT

Anasarca:

1) Heart:

- CHF: dyspnea, orthopnea, JVD
- Pericarditis

2) Liver:

- Cirrhosis

3) Kidney:

- Nephrotic syndrome: dark urine, proteinuria

- Glomerulonephritis
- Renal failure

4) Hematologic:

- Anemia

Back Pain:

1) Bones:

- Disk herniation
- Stenosis/ Ankylosing spondylitis (better w/ activity)
- Osteomyelitis

2) Joints:

- Arthritis

3) Nerves:

- Sciatic nerve irritation
-

Criteria for Hospitalization:

- Pain > 20 min or at rest
- Age > 65
- EKG: Q wave, TWI, ↓ST

Chest Pain:

1) Heart:

- MI: L side, radiates to jaw or shoulder
- AAA: knife-like tearing pain
- Pericarditis: sharp pain radiating to trapezius

2) Lungs:

- PE: tachypnea, tachycardia
- Pneumothorax: one side
- Pneumonia: cough

3) GI:

- GERD
- Cholecystitis: worse with food, N/V

4) Musculoskeletal:

- Muscle strain
- Chostochondritis: tender to palpation
- Herpes zoster: pain before vesicles, dermatome distribution

5) Psychiatric:

- Anxiety/ Panic

Killer Chest Pain:

- MI
- AAA
- PE
- Spont PTX

Cough:

1) Heart:

- CHF

2) Lungs:

- Allergies
- Asthma: cold and exercise make it worse
- Sinusitis: face pain

Criteria for Hospitalization:

"ADMIT NOW"

- **A**ge > 65
- **D**ecreased immunity
- **M**ental status changes
- **I**ncreased A-a gradient
- **T**wo or more lobes involved
- **N**o home
- **O**rgan failure
- **W**BC < 4,000 or >30,000

- Bronchitis: morning cough
- Bronchiectasis: blood, mucopurulent sputum
- Pulmonary fibrosis: DOE, rales, cor pulmonale
- Pneumonia
- TB: hemoptysis, night sweats
- Cancer: hemoptysis, weight loss

3) GI:

- GERD: night cough, "heartburn"

4) Drugs: ACE-I

5) Psychiatric: stress, occurs during day

Diarrhea:

1) Infection:

- Viruses
- Bacteria:
  - E. coli
  - Shigella
  - Salmonella
  - Yersinia enterocolitica
  - Campylobacter
  - C. difficile
- Parasites:
  - Giardia
  - Amoeba
  - Cryptosporidium

2) Inflammation:

- IBD
- Ischemic bowel

3) Drugs:

- Laxatives, antacids w/ magnesium, antibiotics, colchicine

4) Toxins:

- Heavy metals, seafood, mushrooms

5) Neuropathy:

- DM

***"BRAT"* Diet:**

- **B**ananas
- **R**ice
- **A**pplesauce
- **T**oast

Dyspnea:

1) Heart:

- MI: occurs in DM without chest pain
- CHF: orthopnea, PND, wheezing,
- Pericardial tamponade: pulsus paradoxus

2) Lungs:

- Asthma

- Pneumothorax: acute dyspnea
- PE: ↑V/Q scan, D-dimers
- Bronchospasm (COPD, asthma): wheeze
- Aspiration: difficulty swallowing, loss of consciousness
- Pneumonia: fever, chills, productive cough
- ARDS: low pO_2, low CO_2
- Obstruction: localized wheezing

3) Metabolic:
- Sepsis: respiratory alkalosis
- Metabolic acidosis

4) Hematologic:
- Anemia: bleeding

5) Psychiatric:
- Anxiety

Fatigue DDx:
- Depression
- Anemia
- Hypothyroid
- Mononucleosis

Hematochezia:

1) Upper GI Bleed: "*Mallory's Vices Gave An Ulcer*"
- **M**allory-Weiss tear
- **V**ariceal bleeding
- **G**astritis
- **A**V malformation
- **U**lcer (peptic)

2) Lower GI Bleed: "*Can U Cure Aunt Di's Hemorrhoids?*"
- Ulcerative **C**olitis
- **U**pper GI bleed
- **C**ancer
- **A**ngiodysplasia
- **Di**verticulosis
- **H**emorrhoids

Hematuria:

1) Hematologic:
- Sickle cell anemia

2) Kidney:
- Kidney stones
- Pyelonephritis
- Goodpasture's
- Wegener's

3) Prostate:
- Infection
- Trauma

4) Drugs:
- Rifampin, anthracyclines, anti-coagulants

5) Food:
- Blackberries, rhubarb, beets

Hemoptysis:

1) Heart:
- Mitral stenosis: dyspnea, orthopnea

2) Lungs:
- Bronchitis
- Bronchiectasis
- Pneumonia
- PE
- TB

Joint Pain:
- Osteoarthritis: PIP/DIP joints, pain worse with activity
- Rheumatoid Arthritis: MCP/PIP joints, pain worse in the morning
- Medial epicondylitis = *"golfer's/ pitcher's elbow"*: flex wrist
- Lateral epicondylitis = *"tennis elbow"*: Cozen's sign (dorsiflex wrist)
- Tendonitis: Thompson test – squeeze back of calf, no foot plantarflex
- Polio: Trendelenburg gait (superior gluteal nerve → gluteus mm.)
- Cubital tunnel syndrome: percuss ulnar nerve => 4th, 5th finger pain
- Carpal tunnel syndrome:
    - Tests: Phalen's, Tinel's => 1st–3rd finger pain
    - Tx: wrist splint, steroid injection, cut flexor retinaculum
- Septic arthritis
- Gout/Pseudogout

Local Swelling:
- Ascites: fluid wave, jaundice, spider angioma
- Lymph blockage
- DVT, thrombophlebitis: usually in leg
- Hypothyroid: pretibial, periorbital myxedema

Palpitations:

1) Heart:
- MI
- CHF

2) Heme:
- Anemia

3) Thyroid:

- Hyperthyroidism

4) Psychiatric:

- Anxiety

5) Drugs:

- caffeine, bronchodilators, digitalis, anti-depressants, thyroid meds

Syncope:

1) Cardiac:

- Obstruction: Stenosis, HCM, PE, atrial myxoma
- MI
- Orthostatic hypotension

2) Neurologic:

- Stroke

3) Drugs:

- Vasodilators, α-blockers, diuretics, nitrates, sedatives, cocaine, EtOH

4) Situational:

- Vasovagal: micturition, defecation, coughing, carotid sinus sensitivity
- Subclavian steal

Vertigo:

1) Ear:

- Labyrinthitis
- Meniere's disease: vertigo, tinnitus, hearing loss
- Acoustic neuroma

2) Brain:

- Migrane H/A
- Brainstem injury

NOTES:

Step 3 CCS:

"The only equipment lack in the modern hospital? Somebody to meet you at the entrance with a handshake!"

–Martin H. Fischer

O<u>**verall Strategy:**</u> *write down the following so you can find it quickly…*

Differential Diagnosis:
Allergies:
Counseling:

<u>Then, follow this order for each case:</u>

1) ABC's: stable or not?
2) PE: focused unless pt has a general complaint
3) Labs/Rad
4) Triage/Meds
5) Counsel/Follow-up appt

<u>Computer Hints:</u>

- Press 'Tab' key to get to 'OK' button
- Type first 3-5 letters, then press 'OK' to pull up a list to choose from
- Multiple orders: press 'Enter' after each, then 'OK' to execute them all
- For results: click 'Next available result', don't advance the clock yourself

<u>General Orders:</u>

Everyone: CBC/CMP/TSH
Females: β-HCG
Infection: CXR/UA/Cultures/Tylenol
Cardiac: EKG/CXR/Echo/Cardiac enzymes
Pulmonary: ABG/EKG/CXR
GI: Guaiac
Renal: UA
Heme: Guaiac/PT/PTT/Type and match
Ob-Gyn: β-HCG/Pap/Prolactin/ABO/Rh

<u>Antibiotics:</u>

Cellulitis: Clindamycin + Ciprofloxacin
Prostatitis: Ciprofloxacin + Ampicillin + Gentamycin
Urosepsis: Ciprofloxacin + Ceftriaxone
Diverticulitis: Metronidazole + Ciprofloxacin
Appendicitis: Metronidazole + Cefazolin
Perforated Ulcer: Metronidazole + Ampicillin + Gentamycin
Cholecystitis: Cefuroxime
Sickle Cell: Azithromycin + Cefuroxime
Septic Arthritis: Azithromycin + Ceftriaxone
Community-Aquired Pneumonia: Azithromycin + Ceftriaxone
Bacterial Meningitis: Vancomycin + Ceftriaxone

Bacterial Endocarditis: Vancomycin + Gentamycin
Pelvic Inflammatory Disease: Doxycyclin + Cefoxitin
Febrile Netropenia: Ceftazidime

Emergency Room:

1) Pulse ox/Oxygen
2) IV access/NSS (0.9%NaCl)
3) Vitals/Continuous (BP/Cardiac)

Ward Admission: *"NAD HIC"*

1) **N**PO/Foley/Urine outputs
2) **A**ctivity
3) **D**iscomfort Meds: Mepiridine/Phenergan
4) **H**ome meds
5) **I**nterval PE/Hx
6) **C**onsults

ICU Admission:

1) Prophylaxis: Pantoprazole/Pneumatic compression stockings
2) Interval PE/Hx
3) Send to ward → then home

Discharge:

1) D/C IV access, D/C all tubes
2) D/C IVF
3) Change all meds to PO
4) Diet: disease-specific
5) Follow-up appt: 2wk unless specified

Counseling (at the 5min warning):
"X-RATED SEX"
X: safe seX
Rx: medication compliance/side effects
Alcohol: limit intake
Tobacco: cessation
Exercise: program
Directives: advance
Seat belt: when driving
Eat: diet
X: no illegal drugs

Preventative Medicine: *this changes every few years; stick with most current reference…*
21 y/o: Cholesterol/Pap smear
40 y/o: Glucose/Mammogram/Prostate exam

50 y/o: Colonoscopy
65 y/o: TSH/Female Pelvic exam

NEUROLOGY:

ALZHEIMER'S:

Labs: TSH, Vit B_{12}, Folate, UA
Tests: Head CT
Rx:

- Donepezil (tx Alzheimer's)
- Olanzapine (tx delusions)
- Buspirone (tx anxiety)
- Fluoxetine (tx depression)
- Triazolam (tx insomnia)

Location: Office
Counsel: No driving, Advance directives
Follow-up: 2mo

BACTERIAL MENINGITIS:

Labs: Blood C&S, UA/C&S, PT/PTT,
CSF: Cells/Glucose/Protein/C&S + fungal/Gram/Meningococcal Ag
Tests: EKG, LP
Rx: Acetaminophen (rectal), Dexamethasone → Vancomycin/Ceftriaxone
Contact Prophylaxis: Rifampin or Ciprofloxacin
Consult: ID
Location: Ward (isolation)

SUBARACHNOID HEMORRHAGE *"worst HA of my life"*

Labs: ESR, PT/PTT
Tests: EKG, Head CT (no contrast), Angiography (cerebral arteries)
Rx:

- Ketorolac (tx pain)
- Nimodipine (prevent vasospasm)
- Phenytoin (prevent seizures)
- Labetalol (if HTN)

Consult: Neurosurgery
Location: ICU (Neurochecks q1hr)

TRANSIENT ISCHEMIC ATTACK:

Labs: PT/PTT, Lipid profile
Tests: EKG/Echo, CXR, Head CT (no contrast), Carotid Doppler
Rx: ASA, Pneumatic compression stockings
Consult: Vascular Surgery (for carotid endarterectomy if 70-99% stenosis)

Location: Ward

PSYCHIATRY:

ALCOHOL WITHDRAWAL:
Labs: ABG, EtOH/Urine tox, Blood C&S/UA, Mg/Phos/Glucometer, PT/PTT
Tests: EKG, CXR, Head CT
Rx: NG tube suction, Thiamine → Dextrose 50%, Folate, Lorazepam
Location: ICU

DEPRESSION:
Labs: TSH, Vit B_{12}, Folate
Tests: EKG
Rx: Fluoxetine, Suicide contract
Consult: Psychiatry
Location: Office
Follow-up: 6wk

NARCOTIC OVERDOSE:
Labs: ABG, Cardiac enzymes, EtOH/Urine tox/Acetaminophen, Blood C&S/UA, Mg/Phos/Glucometer, PT/PTT
Tests: EKG, CXR, Head CT
Rx: NG tube suction, Naloxone, Thiamine → Dextrose 50%, Charcoal, (Gastric lavage if needed)
Location: ICU

PANIC ATTACK: *palpitations, chest pain, dyspnea*
Labs: TSH, β-HCG, UA/Urine tox, Glucometer
Tests: EKG, CXR
Rx: Alprazolam + Fluoxetine
Location: Office

CARDIOLOGY:

AORTIC DISSECTION: *sudden severe abdominal pain radiating to back*
#1: IVF
Tests: US (abdomen) if stable
Rx: Vascular Surgery Consult (stat, for AAA repair)
Labs: Type x cross, PT/PTT
Pre-op: Analgesia, Consent, NPO?Foley, UA/CXR (portable), CBC/CMP, EKG
Location: ER → OR

ATRIAL FIBRILLATION:
Labs: Cardiac enzymes, TSH/Free T_4, UA, PT/PTT
Tests: EKG, CXR (PA/Lateral), Echo

Rx: Cardizem, Metoprolol (if HR >120), Heparin/Plt/Guaiac
Location: Ward

CHF EXACERBATION:
Labs: Cardiac enzymes
Tests: EKG, CXR, Echo, Daily weights
Rx: **A**SA, **β**-blocker, **C**holesterol drug, **D**igoxin, **E**nalapril, **F**urosemide/KCl *"ABCDEF"*
Location: Ward

COMPLETE HEART BLOCK: *can occur secondary to trauma*
Labs: Cardiac enzymes, PT/PTT
Tests: EKG, portable CXR, Echo
Rx: Atropine (stat), Pacemaker (temp transcutaneous)
Consult: Cardiology
Location: ER → Cardiac Cath lab

HYPERTENSION:
Labs: TSH, UA, Lipid profile
Tests: EKG
Rx: ↑Fiber/↓Na Diet, Exercise, Meds as appropriate
Location: Office

HYPERTENSIVE EMERGENCY: *diastole >120*
#1: Head CT (↑ICP or Stroke)
Labs: UA, PT/PTT
Tests: EKG, CXR
Rx: Arterial line, Nitroprusside → Metoprolol
Follow-up (1wk): Lipid profile
Location: ICU

INFECTIVE ENDOCARDITIS: *splinter hemorrhages, painful finger nodes*
Labs: Blood C&S, UA/C&S/Urine tox, PT/PTT
Tests: EKG, CXR, Echo
Rx: IVF/Central line, Vancomycin + Gentamycin, Acetaminophen
Location: Ward
Follow-up (1wk): Hepatitis panel, HIV

MYOCARDIAL INFARCTION: *ST elevation*
Labs: Cardiac enzymes q8hr x 3, Lipid profile, LFT
Tests: EKG, CXR
Rx: O_2, ASA, Metoprolol, Nitroglycerin, Heparin/Plt/Guaiac, Morphine
Discharge meds: ASA + Metoprolol + Atorvastatin
Follow-up: Cardiac Cath (if not improving), Cardiology Consult, Stress test
Location: Ward

PERICARDIAL TAMPONADE: *hypotension, muffled heart, ↑JVD*
Rx: Pericardiocentesis (stat), CT Surgery consult, Swan-Ganz catheter
Labs: ABG, Pericardial fluid: cell count
Tests: EKG, CXR, Echo
Location: ICU

PERICARDITIS: *worse w/ inspiration and leaning forward*
Labs: Cardiac enzymes, Blood C&S, ESR
Tests: EKG, CXR, Echo, Pulse ox, ABC
Rx: O_2, ASA, Pericardiocentesis (if febrile)
Location: Ward

UNSTABLE ANGINA: *chest pain at rest, ST depression*
Labs: Cardiac enzymes, Lipid profile, PT/PTT
Tests: EKG, CXR, Echo
Rx: ASA, NGN, Metoprolol, O_2, Heparin/Plt/Guaiac, Morphine
Location: ICU
Follow-up (4wk): Stress test, Angiography (coronary)

PULMONARY:

ASTHMA EXACERBATION:
Labs: ABG, FEV_1/PEFR (peak expiratory flow rate)
Tests: EKG, CXR
Rx: Albuterol, Methylprednisolone → Ipratropium
Location: Ward
Discharge: Beclomethasone (inhaled)

BRONCHIOLITIS:
Labs: ABG, RSV Antigen
Tests: CXR
Rx: Albuterol, Humidified air, Suction airway, Chest PT, Ribavirin (if resp failure)
Location: Ward (isolation)
Follow-up: 24hr

COPD EXACERBATION:
Labs: ABG, PEFR/FEV_1
Tests: EKG, CXR
Rx: O_2, Albuterol, Ipratropium, Prednisone, Levofloxacin
Location: Ward
Follow-up: Influenza/Pneumococcal vaccinations, Home O_2 (if $pO_2<55$)

CROUP: *barking cough*
Labs: CBC
Tests: Neck x-ray
Rx: Cool mist → Dexamethasone (IM) → Epinephrine (inhaled)
Location: ER (observe for 4hr)

CYSTIC FIBROSIS:
Labs: Sweat Cl>60, Fecal fat, Blood C&S, Sputum: Gram/C&S
Tests: CXR, Sinus x-ray
Rx: D5NS, Albuterol, N-acetylcysteine, Tobramycin + Piperacillin, Chest PT
Discharge: Pancreatic enzymes, Influenza/Pneumococcal vaccinations
Location: Ward

LUNG CANCER:
Labs: ABG/PFT, Blood C&S, Sputum: Gram/C&S/Cytology
Tests: EKG, CXR (PA/Lateral), spiral CT (chest)
Mets: Ca/AlkPhos, CT (A/P), Brain MRI, Bone Scan
Rx: Bronchoscopy/Biopsy lung, Bronch: Cytology/Gram/C&S+Fungal/AFB
Rx: Albuterol/Ipratropium, Levofloxacin
Consults: Pulmonary, Oncology
Location: Ward

PLEURAL EFFUSION: *x-ray blunting of costophrenic angles*
Labs: Cr/LDH/Protein, PT/PTT
Pleural Fluid: pH/LDH/Protein/Glucose/Gram/Cells/Cytology/C&S/AFB
Tests: EKG, CXR (PA/Lateral/Decubitus)
Rx: Thoracentesis (Pulmonary consult)
Location: Ward

PNEUMONIA: *may present with RUQ pain*
Labs: ABG, Blood cx, Urine cx, Sputum: Gram/C&S
Tests: EKG, CXR (PA/Lateral)
Rx:

- Community Acquired: Azithromycin (+ Ceftriaxone if dyspnea)
- Nosocomial: Piperacillin + Tazobactim
- Aspiration: Clindamycin
- Group B Strep: Ampicillin + Cefotaxime
- Atypicals (Chlamydia/Mycoplasma/Legionella): Erythromycin

Location: Ward

PULMONARY EMBOLUS:
Labs: ABG, D-dimer, Cardiac enzymes, PT/PTT
Tests: EKG, CXR (PA), V/Q scan

Rx: O_2, ASA, Heparin/Plt/Guaiac, Warfarin/INR/β-HCG
Location: ICU

SINUSITIS:
Rx: Humidified air, Acetaminophen, Pseudoephedrine, Amoxicillin
Location: Office

TENSION PNEUMOTHORAX: *decreased breath sounds, hypotension*
Rx: Needle thoracostomy (stat) → chest tube → CXR
Labs: ABG
Tests: EKG, repeat CXR qam
Location: Ward

GASTROINTESTINAL:

APPENDICITIS:
Labs: ABG, β-HCG, Amylase/Lipase, PT/PTT
Tests: Guaiac, X-ray (abd), CT (abd) for adults, or US (abd) for kids
Rx: Morphine, Cefazolin + Metronidazole, Appendectomy (Surgery consult)
Pre-op: Analgesia,Consent, NPO, Type x cross, CXR/UA
Location: Ward

CHOLECYSTITIS:
Labs: β-HCG, Amylase/Lipase, PT/PTT
Tests: X-ray (abdomen), US (abdomen)
Rx: NG tube, Demerol, Cefuroxime (Surgery consult)
Pre-op: Analgesia, Consent, NPO, IVF, Type x cross, CXR/UA
Location: Ward

COLON CANCER:
Labs: TSH, CEA
Tests: Guaiac, X-ray (abdomen), CT (abdomen)
Rx: Fe sulfate, Docusate, Surgery consult/Oncology consult
Location: Office
Follow-up (2wk): Colonoscopy (GI consult)

CONSTIPATION:
Labs: TSH, Mg/Phos
Tests: EKG, Guaiac, X-ray (abd)
Rx: High fiber diet, Docusate, Metamucil
Location: Office
Follow-up: NPO, Polyethylene Glycol, Colonoscopy w/ consent (if symptomatic)

CROHN'S:
Labs: Stool: culture/O&P
Tests: Guaiac
Rx: Colonoscopy, Biopsy (small intestine), GI consult, Mesalamine
Location: Office

DIVERTICULITIS: *LLQ pain*
Labs: Amylase/Lipase, Blood culture, UA/C&S
Stool: WBC/Culture/O&P
Tests: EKG, Guaiac, X-ray (abdomen), CT (abdomen)
Rx: Acetaminophen (rectal), Ciprofloxacin/Metronidazole, GI consult
Location: Ward
Follow-up (4wk): Colonoscopy

DIVERTICULOSIS: *bloody stool*
Labs: Type x cross, PT/PTT, H/H q6h
Tests: Postural vitals (stat), Guaiac, EKG, Anoscopy
Rx: NG, NPO, Polyethylene Glycol, Colonoscopy (GI consult)
Location: Ward
Follow-up (1wk): high fiber diet, check H/H

DUODENAL ULCER (acute): *vomit blood, black stool*
Labs: PT/PTT, Type x match, H/H q6h
Tests: EKG, EGD, Biopsy (gastric), GI consult
Rx: NG, Gastric Lavage, Pantoprazole
Location: ICU
Follow-up (2wk): check H/H

DUODENAL ULCER (chronic): *weakness, SOB*
Labs: β-HCG, Fe/Ferritin/TIBC, UA, Type x cross, PT/PTT, H. pylori Ab,
Amylase/Lipase
Tests: Guaiac, CXR, EGD, Biopsy (gastric), GI consult
Rx: Fe/Vit C, Omeprazole + Clarithromycin + Amoxicillin (if H. pylori +)
Location: Ward
Follow-up (2wk): Urea breath test

DUODENAL ULCER (perforated): *free air under diaphragm*
Labs: Amylase/Lipase, Type x cross, PT/PTT
Tests: EKG, X-ray (abdomen)
Rx: NG, Ranitidine, Ampicillin + Gentamycin + Metronidazole
Pre-op: Surgery consult (stat), Analgesia, Consent, NPO, CXR/UA
Location: ICU

GASTROENTERITIS:

Labs: UA, Stool: Culture/WBC/Heme
Tests: Guaiac
Rx: Acetaminophen, Fluids, Replace electrolytes
Location: Ward

GIARDIASIS:
Labs: Giardia Ag, Stool: Culture/WBC/Heme, UA
Tests: Rectal exam, Guaiac
Rx: Metronidazole, Hydration, Replace electrolytes
Location: Ward

INTUSSUSCEPTION:
Labs: CBC/CMP
Tests: X-ray (abdomen), US (abdomen)
Rx: NG suction, Barium enema (Pediatric Surgery consult, stat)
Location: Ward

IRRITABLE BOWEL SYNDROME:
Labs: ESR, TSH, 72-hr stool fat, Stool: Culture/WBC/O&P
Tests: Guaiac
Rx: High fiber diet, Lactose-free diet, Loperamide
Location: Office

PANCREATITIS (GALLSTONE):
Labs: Ca/Mg, Amylase/Lipase, PT/PTT, (CBC/CMP)
Tests: EKG, Portable X-ray (abdomen), US (RUQ), ERCP (GI consult)
Rx: NPO/NG tube, Mepiridine/Phenergan
Location: Ward

PANCREATITIS (ALCHOLIC):
Labs: ABG, Ca/Mg, Amylase/Lipase
Tests: CT (abdomen)
Rx: NPO/NG tube, Meperidine/Phenergan, Ca chloride/Mg sulfate prn
Location: Ward

SIGMOID VOLVULUS:
Labs: CBC/CMP
Tests: X-ray (abdomen)
Rx: NG tube, Rectal tube, Sigmoidoscopy (GI consult)
Location: Ward

SPLENIC RUPTURE: *due to trauma*
Labs: ABG, EtOH/Urine tox, UA, Amylase/Lipase, Type x cross, PT/PTT,
Tests: EKG, X-rays (spine/chest/abd/pelvis), CT w/ contrast (abdomen)

Rx: Morphine, Observe if stable, Abd exam q4h, CT next day, Surgery consult
Location: Ward
Discharge Vaccines: Pneumococcal/HIB/Meningococcal

ULCERATIVE COLITIS:
Labs: ESR, Amylase/Lipase, Stool: Culture/WBC/O&P, PT/PTT
Tests: Flexible sigmoidoscopy, Biopsy (small intestine), GI consult
Rx: 5-asa (rectal), Sulfasalazine/Folate, Loperamide, Dicyclomine (tx IBS sx)
Location: Office

RENAL:

ADULT POLYCYSTIC KIDNEY DISEASE:
Labs: Fe/Ferritin/TIBC, Ca/Phos, UA
Tests: EKG, US (kidneys)
Rx: Amlodipine (HTN), Ca acetate (↑Phos), Epo (anemia), Nephrology consult
Location: Ward

CYSTITIS: *bladder infection*
Labs: UA/C&S (stat), β-HCG
Rx: Bactrim (or Amoxicillin if pregnant)
Location: Office

NEPHROTIC SYNDROME: *facial and scrotal swelling (child)*
Labs: UA, Complement, Lipid profile, PT/PTT
Rx: Albumin, Furosemide, Prednisone, Nephrology consult
Location: Ward

PRE-RENAL FAILURE:
Labs: ABG, Ca/Mg/Phos, BUN/Cr >20, UA/C&S/Na/Cr, 24-hr urine protein
Tests: EKG, CXR, US (renal), Daily weights
Rx: Foley, Heparin SQ, Nephrology consult, avoid NSAIDs
Location: Ward

PROSTATITIS:
Labs: Blood cx, UA/Gram/C&S
Tests: Rectal exam
Rx: Acetaminophen, Ampicillin + Gentamycin → Cipro PO
Location: Ward

RENAL CELL CARCINOMA:
Labs: Fe/Ferritin/TIBC, Alk Phos/PTH/Ca, UA
Tests: CXR, US (kidneys), CT (head/chest/abd/pelvis), Bone scan
Rx: Pamidronate (tx high Ca), Surgery/Oncology consults

Location: Ward

UROSEPSIS:
Labs: Blood cx, UA/C&S
Tests: portable CXR
Rx: Acetaminophen, Ceftriaxone IV → Ciprofloxacin PO
Location: ICU

HEMATOLOGY:

DEEP VENOUS THROMBOSIS:
Labs: D-dimer, PT/PTT
Tests: Guaiac, venous Doppler (lower leg)
Rx: Enoxaparin + Warfarin (check β-HCG), avoid Oral Contraceptives
Location: ER → home

FOLATE DEFICIENCY: *fatigue*
Labs: Folate, Vit B_{12}, Retics, MCV
Tests: Guaiac
Rx: Multivitamin, Folate, Fe sulfate, Vit B_{12}, Thiamine
Location: Office

G-6PD DEFICIENCY:
Labs: G-6PD, UA, Blood smear/Retics, Haptoglobin/LDH, PT/PTT, Type x cross
Rx: PRBC, follow Hb/Hct
Location: Ward

HEMOPHILIA A/B: *bleeding at the dentist*
Labs: Factors 8/9, Blood smear, ABO/Rh, PT/PTT/Bleed time
Tests: Head CT (if trauma)
Rx: Factor 8/9, Avoid asa/contact sports
Location: ER
Follow-up (4wk): Genetics consult

IDIOPATHIC THROMBOCYTOPENIC PURPURA:
Labs: Retics, Anti-platelet Ab
Rx: Bone marrow biopsy
1) Prednisone
2) Immunoglobulins (therapy)
3) Splenectomy
Location: ICU

LEAD POISONING:
Labs: Blood lead, Serum tox, Fe/Ferritin/TIBC, Protoporphyrin, Ca, Glucose, UA

Tests: Guaiac
Rx:

- Succimer (if Pb >45)
- EDTA (if Pb>70)

Consults: Hematology
Location: Ward
Discharge: Lead paint assay in home, Lead abatement agency

SICKLE CELL ANEMIA: *with chest pain/fever*
Labs: Type x screen, Blood cx, UA/C&S, Sputum: Gram/Cx
Tests: EKG, CXR, Continuous pulse ox
Rx: O_2, Morphine, Albuterol, Cefuroxime + Azithromycin, PRBC (if hypoxic)
Location: Ward
Discharge: Penicillin prophylaxis until 5y/o, Pneumococcal vaccine

ENDOCRINE:

DIABETIC KETOACIDOSIS:
Labs: ABG, Amylase/Lipase, UA/Urine tox, Osmolarity/Ketones, Glucose/HbA_{1c}, Phos
Tests: EKG, CXR, X-ray (abd)
Rx: ***"I Never Kept Dogs"***

- Insulin drip
- NS
- KCl (20mEq)
- D5 1/2NS (when Glucose=250 and Ketones=0)

Location: ICU

HYPERTHYROID:
Labs: TSH, free T_3/T_4, β-HCG
Tests: EKG, Radioactive iodine uptake
Rx: Propanolol → Methimazole (PTU if pregnant) → ^{131}I → Levothyroxine,
Prednisone (if exophthalmos)
Location: Office
Follow-up (6wk): TSH/Free T_4

HYPOTHYROID:
Labs: TSH, free T_3/T_4, Lipid panel
Tests: EKG, Guaiac
Rx: Levothyroxine, Vit D/Ca carbonate
Location: Office
Follow-up (6wk): TSH

RHEUMATOLOGY:

GOUT: *negative birefringence*
Labs: ESR, Uric acid, UA, BUN/Cr, Blood culture, PT/PT/INR
    Aspirate: Gram/C&S/Crystals/Cell count
Tests: X-ray, Aspirate joint
Rx: Indomethacin → Colchicine → Glucocorticoids (No ASA or diuretics)
Location: Ward

LUPUS:
Labs: ESR, ANA, UA, anti-ds DNA, anti-Smith Ab, Complement, PT/PT
Tests: EKG, CXR
Rx: Prednisone, Rheumatology consult
Location: Office

OSTEOARTHRITIS:
Rx: Acetaminophen, warm compresses
Location: Office

RHABDOMYOLYSIS:
Labs: CPK/Ca/Mg/Pi q4h, urine Myoglobin, Uric acid, PT/PTT, UA
Tests: EKG
Rx: NS → 1/2NS + Mannitol/Sodium Bicarb (until urine pH >6.5)
Location: Ward

RHEUMATOID ARTHRITIS:
Labs: RF, Synovial fluid: Analysis/Gram/C&S
Tests: X-ray (joint), Aspirate joint
Rx: Indomethacin (acute) + Methotrexate (chronic), Exercise
Location: Office

SEPTIC ARTHRITIS:
Labs: ESR, UA, Blood cx, PT/PTT, Throat C&S, Urethral C&S, Rectal C&S, (CBC/BMP), Aspirate:Gram/C&S/Crystals/Cells/GC, HIV, RPR, Hepatitis panel
Tests: X-ray (joint), Aspirate joint (Ortho consult)
Rx: Morphine x 1, Acetaminophen, Ceftriaxone + Azithromycin
Location: Ward

TEMPORAL ARTERITIS: *stiff joints, changes in vision, jaw claudication, H/A*
Labs: ESR/ANA/CRP/RF, UA, TSH, PT/PTT
Tests: EKG, CXR, Head CT, Temporal artery biopsy
Rx: Prednisone (Ranitodine + Vit D/Ca carbonate + Alendronate + DEXA),
  Acetaminophen, Ophthalmology consult
Location: Ward

VERTEBRAL COMPRESSION FRACTURE:
Labs: ESR, Ca, TSH, SPEP (r/o myeloma)
Tests: X-ray (spine)
Rx: Toradol x 1, Naproxen, Vit D/Ca carbonate, Alendronate
Location: ER
Follow-up (1wk): DEXA, neuro exam

OB-GYN:

DYSFUNCTIONAL UTERINE BLEEDING: *diagnosis of exclusion*
Labs: β-HCG, TSH, Prolactin, PT/PTT, CBC
Tests: US (pelvic), Pap
Rx: Oral contraceptives, Fe sulfate
Location: Office

ECTOPIC PREGNANCY: *pelvic pain, vaginal bleeding*
Labs: β-HCG, Blood type/Rh, Type x cross, PT/PTT, (CBC/CMP)
Cervix: C&S/GC, HIV, RPR, Hepatitis panel
Tests: US (transvaginal), Orthostatic vitals
Rx: Methotrexate, RhoGam if Rh(-), Ob-Gyn consult
Location: Ward
Follow-up (4d): check β-HCG

FIBROADENOMA:
Labs: β-HCG, (CBC/BMP)
Tests: US (<35 y/o) or Mammogram (>35y/o), FNA
Rx: Follow q3mo
Location: Office

MENOPAUSE:
Labs: β-HCG, LH/FSH >30, TSH, Prolactin, fasting Lipids
Tests: Pap, Mammogram, DEXA, Guaiac, Colonoscopy
Rx: HRT (Estrogen + Progestins), Vit D/Ca carbonate
Location: Office

OVARIAN CANCER: *ascites, pleural effusion, ovary mass*
Labs: CA-125, UA, Blood type/Rh, PT/PTT, (CBC/CMP)
Tests: EKG, Guaiac, CXR, US (abdomen), CT (C/A/P)
Rx: Paracentesis (for ascites), Ob-Gyn/Oncology consults
Location: Ward

OVARIAN TORSION: *surgical emergency*
Labs: β-HCG, UA, (CBC/BMP)
Tests: US (transvaginal)

Rx: Morphine/Phenergan, Gynecology consult
Location: ICU
Pre-op: Analgesia, Consent, NPO, IVF, Type x cross, PT/PTT, CXR

PELVIC INFLAMMATORY DISEASE: *cervical motion tenderness*
Labs: β-HCG, UA/C&S, Cervix: Gram/C&S/GC, (CBC/BMP)
    HIV, RPR, Hepatitis panel
Tests: Pap
Rx: Acetaminophen, Cefoxitin + Doxycyline
Location: Ward
Follow-up: 1wk

POLYCYSTIC OVARIAN DISEASE: *obese, hirsutism, 2° amenorrhea*
Labs: β-HCG, TSH, Glucose, LH/FSH, Prolactin, Testosterone, DHEAS
Labs: 24-hr urine cortisol, 24-hr urine 17-ketosteroids
Tests: US (pelvis), Pap
Rx: Oral contraceptives (or Clomiphene), Metformin, Spironolactone
Location: Office
Follow-up (1wk): Fasting lipids, Glucose tolerance

PREGNANCY:
"Always Remember… Physicians Check Baby's Growth By Ultrasound. Then, Understanding Girls Have Greater Contractility Risk, Suggest HIV optionally."
Labs: **A**BO/**A**typical Ab screen, **R**h, **P**ap, **C**BC, **B**MP, **G**lucose, β-HCG,
    Ultrasound, **T**B, **U**A, **G**roup B Strep, **H**BsAg, **G**onorrhea, **C**hlamydia,
    Rubella Ab, **S**yphilis, **H**IV (optional)
Rx: Prenatal vitamin, Fe, Folate, Ob-Gyn consult
Location: Office

TOXIC SHOCK SYNDROME:
Labs: Blood cx, UA/C&S, Tampon: C&S, PT/PTT, (CBC/CMP)
Tests: EKG, portable CXR
Rx: Tampon removal, 20L IVF, Acetaminophen, Clindamycin
Location: ICU

TURNER'S SYNDROME:
Labs: Karyotype, LH/FSH, TSH, UA, BUN/Cr, Fasting Glucose
Tests: Echo, US (pelvis), Skeletal survey
Rx: GH (until 12y/o), Estrogen (until puberty), Vit D/Ca carbonate
Location: Office

VAGINITIS:
Labs: Vaginal pH, KOH/Wet mount prep, Vaginal: Gram/C&S/GC, (CBC/CMP)
Tests: Pap

Rx:

- White cheesy (pH<4) = Candida (Rx: Fluconazole, check Glucose)
- Green frothy = Trichomonas (Rx: Metronidazole); sexual partner needs tx
- Fishy smell + clue cells = Bacterial (Rx: Metronidazole or Clindamycin cream)

Location: Office

PEDIATRICS:

ABO INCOMPATIBILITY: *jaundice at birth*
Labs: CRP, Blood type (Mom&baby)/Rh, Direct Coomb's, Bilirubin, (CBC/BMP) CMV titer/Rubella/Toxoplasma, Kleihauer-Betke test
Rx: D5 1/4NS, Phototherapy (>12) + Erythromycin oint(eyes), Transfusion (>20)
Location: ICU

ALCOHOL INTOXICATION (CHILD):
Labs: EtOH, Tox panel (urine/serum), Glucometer, (CBC/BMP)
Tests: Screen for abuse and domestic violence
Rx: IVF, D50 ampule (if low glucose), Naloxone
Location: Ward

ANAPHYLAXIS W/ ANGIOEDEMA:
Labs: ABG, (CBC/BMP)
Tests: EKG, CXR
Rx: Epi SubQ (q15min), Albuterol, Hydrocortisone, Diphenhydramine
Note: If pt on β-blocker, give Glucagon first
Location: ICU (Intubate/Ventilate prn)
Follow-up (6wk): RAST testing, Immunology consult, MedAlert bracelet

CHILD ABUSE BURNS:
Labs: UA, Retics, PT/PTT/Bleeding time, (CBC/BMP)
Tests: Skeletal survey, Bone scan, Head CT
Rx: Admit, complete PE, Silver sulfadiazine cream, Wound dressings
Consults: CPS, Psychiatry, Ophthalmology (check for retinal hemorrhages)
Location: Ward

FOREIGN BODY ASPIRATION:
Labs: CBC, BMP
Tests: CXR, X-ray (neck)
Rx: O_2, Methylprednisolone + Cefazolin → Bronchoscopy (Pulm Consult)
Location: ER

INFECTIOUS DISEASES:

CELLULITIS:

Labs: Blood cx, UA, Wound: Gram/C&S, (CBC/BMP), Glucose/Hb_{A1c}
Tests: EKG, X-ray, Bone scan (if suspect osteomyelitis)
Rx: NPO, Acetaminophen, Clindamycin + Ciprofloxacin, Surgery consult
Location: Ward

FEBRILE NEUTROPENIA: *history of chemotherapy*

Labs: Blood cx, UA/C&S/Gram, Sputum: Gram/Culture, (CBC/CMP)
Tests: CXR (PA/Lateral)
Rx: Ceftazidime
Location: ICU

HEPATITIS A: *dark urine, cigarette aversion, Mexico travel*

Labs: anti-HAV Ab, Retics, PT/PTT, (CBC/CMP)
Rx: avoid Acetaminophen
Location: Office
Follow-up: Check LFT/PT q3days, Public Health Dept

HUMAN IMMUNODEFICIENCY VIRUS:

Labs: ELISA, W. blot, PCR (HIV RNA), CD4 count, Toxoplasmosis Ag
Labs: RPR, PPD, Hepatitis panel
Rx: Didanosine + Indinavir + Zidovudine (if CD4 <350)
Location: Office
Follow-up: Influenza/Pneumococcal vaccine, Public Health Dept

PNEUMOCYSTIS JIROVECIINIA:

Labs: ABG, CD4 count, Sputum: Gram/C&S/AFB, BAL: Silver stain
Tests: EKG, CXR, Broncoscopy/Lavage for PCP (Thoracic Surgery consult)
Rx: Acetaminophen, Bactrim (or Pentamidine if allergic)
Location: Office

TUBERCULOSIS: *travel to India*

Labs: ESR, Cr, Sputum: Gram/C&S/AFB, (CBC/CMP)
Tests: EKG, CXR
Rx: Public Health Dept

- Rifampin (check LFTs)
- Ethambutol (+ Opthalmology consult)
- Pyrazinamide (check uric acid)
- INH + Pyridoxime

Location: Office

NOTES:

PASS Program™ Clues:

"You have been empowered. You now have the eyes of Physio!"

–Francis Ihejirika

Main Concepts:

What electrolytes does the low volume state have?	↑total Na, ↓serum Na (dilutional affect), ↓Cl, ↓K
What pH does the low volume state have?	Alkalotic (except diarrhea, RTA Type II, and DKA) b/c Aldo dumps H^+
What pH do vomiters have?	Alkalotic b/c you vomit out H^+
What pH does diarrhea have?	Acidosis b/c stool has bicarb
What happened if pulse ↑>10 on standing?	Hypovolemic shock
What happened if pulse ↓<5 on standing?	Autonomic dysfunction
What are the symptoms of a low energy state?	**CNS:** mental retardation, dementia **CV:** heart failure, pericardial effusion **Muscle:** weakness, SOB, vasodilation, impotence, urinary retention, constipation **<u>Rapidly Dividing Cells</u>:** • Skin: dry • Cuticles: brittle • Hair: alopecia • Bone marrow: suppressed • Vascular endothelium: breaks down • Lungs: infection, SOB • Kidney: PCT will feel the effect first • GI: N/V/D • Bladder: oliguria • Sperm: decreased • Germ cells: predisposed to cancer • Breasts: atrophic • Endometrium: amenorrhea
What are the most common signs of the low energy state?	Tachypnea and dyspnea
What are the most common symptoms of the low energy state?	Weakness and SOB
What are the most common infections of the low energy state?	UTI and respiratory infections
What is the most common cause of death in the low energy state?	Heart failure
Explain all restrictive lung diseases:	Restrictive: interstitial problem (non-bacterial) • Small stiff lungs (↓VC) • Trouble breathing in => FEV_1/ FVC: > 0.8 • ABG: ↓pO_2 => ↑RR, ↓ pCO_2, ↑pH • CXR: reticulo-nodular pattern, ground-glass

	apperance • Die of cor pulmonale • Ex: NM diseases (breathing out is passive), drugs, autoimmune dz • Tx: Pressure support on ventilator, ↑O_2, ↑RR, ↑inspiratory time
Explain all obstructive lung diseases:	Obstructive: airway problem (bacterial) • Big mucus-filled lungs (↑RV, ↑Reid index = ↑airway thickness/ airway lumen) • Trouble breathing out => FEV_1/ FVC: < 0.8 • ABG:↑pCO_2 => ↑RR, ↓pH • Die of bronchiectasis • Ex: COPD • Tx: Manipulate rate on ventilator, ↑RR, ↑expiratory time, ↑O_2 only if needed
What symptoms does a "more likely to depolarize" state have?	Brain: psychosis, seizures, jitteriness Skeletal muscle: muscle spasms, tetany SM: diarrhea, then constipation Cardiac: tachycardia, arrhythmias
What symptoms does a "less likely to depolarize" state have?	Brain: lethargy, mental status changes, depression Skeletal muscle: weakness, SOB SM: constipation, then diarrhea Cardiac: hypotension, bradycardia
What is the humoral immune response?	B cells and PMNs patrol the blood looking for bacteria
What is the cell-mediated immune response?	T cells and Macrophages patrol the tissues looking for non-bacteria
What are macrophages called in each area of the body?	Blood = Monocytes Brain = Microglia Lung = T1 pneumocytes Liver = Kupffer cells Spleen = RES cells Lymph = Dendritic cells Kidney = Mesangial cells Payers patches = M cells Skin = Langerhans Bone = Osteoclasts CT: • Histiocytes • Giant cells • Epitheloid cells
What is the CBC for every vasculitis?	↓RBC, ↓platelets, ↑WBC, ↑T cells, ↑MP, and

	schistocytes, ↑ESR
What is the time course of the inflammatory response?	1 hr: Swelling Day 1: PMNs show up Day 3: PMNs peak Day 4: MP/T cells show up Day 7: MP/T cells peak, Fibroblasts arrive Day 30: Fibroblasts peak Month 3-6: Fibroblasts leave
What state does estrogen mimic?	The neuromuscular disease state (estrogen is a muscle relaxant)
What does high GABA levels lead to?	Bradycardia, lethargy, constipation, impotence, and memory loss

Neurology:

What is the central nervous system?	Brain and spinal cord; oligodendrocytes
What is the peripheral nervous system?	Everything else; Schwann cells
What is the autonomic nervous system?	Automatic stuff
What is the somatic nervous system?	Moving your muscles
What is the parasympathetic system?	Rest-and-Digest => slows stuff down
How does the parasympathetic system behave?	*"DUMBBELS":* **D**iarrhea **U**rination **M**iosis *"constrict"* **B**radycardia **B**ronchoconstrict **E**rection *"point"* **L**acrimation **S**alivation
What is the sympathetic system?	Fight-or-Flight => speeds stuff up
How does the sympathetic system behave?	*Opposite of Parasympathetics:* Constipation Urinary retention Mydriasis *"eyes wide with fright"* Tachycardia Bronchodilate Ejaculation *"shoot"* Xerophthalmia (dry eyes) Xerostomia (dry mouth)
What is Cushing's triad?	HTN, bradycardia, irregular breathing
What is Budd-Chiari?	Hepatic vein obstruction
What is Arnold-Chiari?	Foramen magnum obstruction
What is Anencephaly?	Notochord did not make contact w/ brain => only have medulla

What is an Encephalocele?	Brain tissue herniation
What is a Dandy Walker malformation?	No cerebellum, distended 4th/lateral ventricles
What is an Arnold-Chiari malformation?	Herniation of cerebellum through foramen magnum **Type I:** cerebellar tonsils (asymptomatic) **Type II:** cerbellar vermis/ medulla => hydrocephalus, syringomyelia (loss of pain/temp)
What is Spina bifida occulta?	Covered by skin w/ tuft of hair
What is Spina bifida aperta?	Has opening *(high AFP)*
What is a Meningocele?	Sacral pocket w/ meninges in it
What is a Meningomyelocele?	Sacral pocket w/ meninges and nerves in it
What is Open-angle glaucoma?	Overproduction of fluid => painless ipsilateral dilated pupil, gradual tunnel vision, optic disc cupping
What is Closed-angle glaucoma?	Obstruction of canal of Schlemm => sudden onset, pain, emergency
What are the Watershed areas?	Hippocampus, splenic flexure
What bug loves the frontal lobe?	Rubella
What bug loves the temporal lobe?	HSV
What bug loves the parietal lobe?	Toxoplasma
What bug loves the hippocampus?	Rabies
What bug loves the posterior fossa?	TB
What bug loves the DCML tract?	Treponema
How do migraines present?	Aura, photophobia, numbness and tingling, throbbing HA, nausea
How do tension headaches present?	"Band-like" pain starts in posterior neck, worse as day progresses, sleep disturbance
How do cluster headaches present?	Rhinorrhea, unilateral orbital pain, suicidal, facial flushing, worse w/ lying down
How does temporal arteritis present?	Pain with chewing, blind in one eye
How does trigeminal neuralgia present?	Sharp, shooting face pain Carbamazepine = Gabapentin
What are the 2 kinds of partial seizures?	Simple (aware), Complex (not aware)
What are the 3 kinds of generalized seizures?	• Tonic-Clonic *"Grand mal"* • Absence *"Petit mal"* Ethosuximide (SLE; Urticaria, Steven Johnson) • Status Epilepticus
How does an epidural hematoma present?	Intermittent consciousness, "lucid interval"
How does a subdural hematoma present?	Headache 4wks after trauma, elderly (loose brain)
How does a subarachnoid hemorrhage present?	"Worst headache of my life", h/o berry aneurysm
What is an Astrocytoma?	Rosenthal fibers, #1 in kids w/ occipital headache
What is an Ependymoma?	Rosettes, in 4th ventricle, hydrocephalus
What is a Craniopharyngioma?	Motor oil biopsy, tooth enamel, Rathke's pouch, ADH problem, bitemporal hemianopsia
What is Glioblastoma multiforme?	Pseudopalisading, necrosis, worst prognosis,

	intralesional hemorrhage
What is a Hemangioblastoma?	Cerebellum, von-Hippel-Lindau
What is a Medulloblastoma?	Pseudorosettes, compresses brain, early **m**orning vomiting
What is a Meningioma?	Parasagittal, psammoma bodies, whorling pattern, best prognosis
What are the most common places to metastasize to the brain?	From lung, breast, skin; see at white-grey junction
What is an Oligodendroglioma?	Fried-egg appearance, nodular calcification
What is a Pinealoma?	Loss of upward gaze, loss of circadian rhythms => precocious puberty
What is a Schwannoma?	CN8 tumor, unilateral deafness
What is Neurofibromatosis?	Café au lait spots (hyperpigmentation) => peripheral nerve tumors, axillary freckle • Type 1 "Von Recklinghausen's": Peripheral (Chr#17), optic glioma, Lisch nodules, scoliosis • Type 2 "Acoustic Neuroma": Central (Chr#22), cataracts, bilateral deafness
What is Sturge-Weber?	**P**ort wine stain (big purple spot) on forehead, angioma of retina
What is Tuberous Sclerosis?	Ashen leaf spots (hypopigmentation), 1° brain tumors, Heart rhabdomyomas, Renal cell CA, Shagreen spots (leathery)

Psychiatry:

How is major depression diagnosed?	*need 5 "SIGE CAPS" >2wks* **S**leep disturbances: wake in am **I**nterest/ Libido loss **G**uilt **E**nergy loss **C**oncentration loss **A**ppetite loss **P**sychomotor agitation **S**uicide: hopelessness
What is Autism?	Repetitive movements, lack of verbal skills and bonding, sx since birth
What is Asperger's?	Good communication, impaired relationships, no mental retardation
What is Rett's?	Only in girls, ↓head growth, lose motor skills, hand-wringing, normal until 5mo
What is Childhood Disintegrative Disorder?	Kid stops walking/ talking
What is Selective Mutism?	Kid talks sometimes
What is Separation Anxiety Disorder?	Kid screams when Mom leaves

What is Conduct Disorder?	Aggressive, disregard for rules, no sense of guilt, harm animals, illegal activity *"bite"*
What is Oppositional Defiant Disorder?	Defiant, noncompliant, directed at authority *"bark"*
What is Attention Deficit Hyperactivity Disorder?	Overactivity, difficulty in school
What is Dysthymia?	Low level sadness >2yr
What is Cyclothymia?	Dysthymia w/ hypomania
What is Double Depression?	Depression followed by dysthymia
What is Bipolar I?	Depression and Mania (psychosis)
What is Bipolar II?	Depression and Hypomania (no psychosis)
What are Loose associations?	Ideas switch subjects, incoherent
What is Tangentiality?	Wanders off the point
What is Circumstantiality?	Digresses, but finally gets back to the point
What is Clanging?	Words that sound alike
What is Word salad?	Unrelated combinations of words
What is Perseveration?	Keeps repeating the same words
What is Neologisms?	New words
What is Delusion?	False belief
What is Illusion?	Misinterprets stimulus
What is Hallucination?	False sensory perception, EtOH withdrawal/Cocaine intox => *formication*
What is Nihilism?	Thinks the world has stopped
What is Loss of ego boundaries?	Not knowing where I end and you begin
What are Ideas of reference?	Believes the media is monitoring you
What is Thought blocking?	Stops mid-sentence
What is Thought broadcasting?	Believes everyone can read his thoughts
What is Thought insertion?	Believes others are putting thoughts into his head
What is Thought withdrawal?	Believes others are taking thoughts out of his head
What is Concrete thinking?	Can't interpret abstract proverbs, just sees the facts
What is Synesthesia?	Smell colors
What is Cataplexy?	Loss of mm. tone due to strong emotions
What is a Paranoid personality disorder?	Suspicious about everything, use projection
What is Schizotypal personality disorder?	"Magical thinking", bizarre behavior
What is Schizoid personality disorder?	"Recluse", don't want to fit in
What is Antisocial personality disorder?	Lie, steal, cheat, destroy property, impulsive w/o remorse, illegal activity
What is Conduct Disorder personality disorder?	<15 y/o antisocial disorder
What is Histrionic personality disorder?	Theatrical, sexually provocative, use repression
What is Borderline personality disorder?	"Perpetual teenager", splitting (love/hate), projection, acting out, self-mutiliation
What is Narcissistic personality disorder?	Pompous, no empathy
What is Dependant personality disorder?	Clingy, submissive, low self-confidence, *regression*

What is Obsessive-Compulsive personality disorder?	Perfectionist, doesn't show feelings, detail-oriented, uses isolation
What is Avoidant personality disorder?	Socially withdrawn, afraid of rejection but wants to fit in
What is Kleptomania?	Steals for the fun of it
What is Pyromania?	Starts fires
What is Intermittent Explosive Disorder?	Loses self-control without adequate reason
What is Pathological Gambling?	Can't stop gambling, affects others
What is Trichotillomania?	Pull out their hair
What is Lewy body dementia?	Stiff, visual hallucinations
What is Normal Pressure Hydrocephalus?	Tr*IAD* = **I**ncontinence, **A**taxia "magnetic gait", **D**ementia
What is Korsakoff psychosis?	Alcoholic Thiamine deficiency
What is Vascular "multi-infarct" dementia?	Sudden onset, uneven progression of deficits, "stair-step" decline
What is Huntington's?	In caudate/putamen, triplet repeat disorder, choreiform movements, Δpersonality
What is Creutzfeldt-Jacob?	Prion induced, die within 1 year, post-cornea transplant
What is Pick's disease?	Frontal lobe atrophy, disinhibition
What is Alzheimer's?	↓ACh in nucleus basalis of Meynert, bad ApoE, amyloid plaques, tangles of tau
What is Parkinson's ?	In substantia nigra, bradykinesia, pill-rolling tremor, shuffling gait, Lewy bodies
What is Somatization?	Think they have a *different* illness all the time (at least 4 organ systems)
What is Hypochondriasis?	Think they have the *same* illness all the time
What is Body Dysmorphic Disorder?	Imagined physical defect
What is Pain disorder?	Prolonged pain not explained by physical causes
What is Conversion?	Neuro manifestation of internal conflict, indifferent to disability
What is Malingering?	Fake illness for ***monetary gain,*** avoids medical treatment
What is Factitious?	Fake illness to get *attention,* seeks medical treatment
What is Munchausen?	Need to be the caregiver, Mom fakes child's illness to get attention
What is Munchausen by proxy?	Mom makes child ill for gain, move a lot
What is Amnesia?	Can't recall important facts
What is Dissociative Fugue?	No past, travel to new place, usually due to trauma
What is Multiple Personality Disorder?	Have 5-10 alters, usually associated w/ incest
What is Depersonalization Disorder?	"Out of body" experiences, deja vu
What is Sublimation?	Substitute acceptable for unacceptable (boxer vs. fighting)

What is Imitation?	Dress like someone else
What is Identification?	Act like someone else
What is Displacement?	Take anger out on someone else
What is Idealization?	Wait for "ideal spouse" while they are beating you up
What is Transference?	Patient views doctor as parent
What is Countertransference?	Doctor views patient as child
What is Acting out?	Expression of impulse, "tantrums"
What is Regression?	Immature behavior
What is Rationalization?	Make excuses for all situations
What is Intellectualization?	Act like a "know-it-all" to avoid feeling emotions
What is Isolation of Affect?	Isolate feelings to keep on functioning
What is Suppression?	Consciously block memory
What is Repression?	Subconsciously block memory
What is Reaction Formation?	Unconsciously act opposite to how you feel (tears of a clown)
What is Undoing?	Doing *exact opposite* of what you used to do to fix a wrong
What is Compensation?	Doing something *different* of what you used to do
What is Sadism?	Gives pain
What is Masochism?	Receives pain
What is Exhibitionism?	Exposure to others
What is Voyeurism?	Watching other people without their permission
What is Telephone Scatalogia?	Phone sex
What is Frotteurism?	Rub penis against fully clothed women
What is a Transvestite?	Dress up as opposite sex, no identity crisis
What is a Transsexual?	Gender identity crisis *"man trapped in a woman's body"*
What is a Fetish?	Objects (vibrators, dildos, shoes)
What is a Pedophile?	Children (watching child pornography)
What is a Necrophile?	Corpses
What is a Beastophile?	Animals
Can you die during EtOH withdrawal?	Yes
Can you die during opiod withdrawal?	No; just very painful

Cardiology:

What organs have resistance in series?	Liver, kidney
What organs have resistance in parallel?	All the rest
What organ has the highest A-VO_2 difference at rest?	Heart
What organ has the highest A-VO_2 difference after exercise?	Muscle
What organ has the highest A-VO_2 difference after meal?	Gut
What organ has the highest A-VO_2	Brain

difference during a test?	
What organ has the lowest $A\text{-}VO_2$ difference?	Kidney
Where does Type A thoracic aortic dissection occur?	Ascending aorta (occurs in cystic medial necrosis, syphilis)
Where does Type B thoracic aortic dissection occur?	Descending aorta (occurs in trauma, atherosclerosis)
What layers does a true aortic aneurysm occur?	Intima, media, and adventitia
What layers does a pseudo aortic aneurysm occur?	Intima and media
What is pulse pressure?	Systolic – Diastolic pressure
What vessel has the thickest layer of smooth muscle?	Aorta
What vessels have the most smooth muscle?	Arterioles
What vessels have the largest cross-sectional area?	Capillaries
What vessel has the highest compliance?	Aorta
What vessels have the highest capacitance?	Veins and venules
What is your max heart rate?	220 – age
What is Stable angina?	Pain with exertion (atherosclerosis)
What is Unstable angina?	Pain at rest (transient clots)
What is Prinzmetal's angina?	Intermittent pain (coronary artery spasm)
What is Amyloidosis?	Stains Congo red, Echo Apple-green birefringence
What is Hemochromatosis?	Fe deposit in organs => hyperpigmentation, arthritis, DM
What is Cardiac tamponade?	Pressure equalizes in all 4 chambers, quiet precordium, no pulse or BP, Kussmaul's sign, pulsus paradoxus (↓ >10mm Hg BP w/ insp)
What is a Transudate?	An effusion with mostly water *Too much water:* • Heart failure • Renal failure *Not enough protein:* • Cirrhosis (can't make protein) • Nephrotic syndrome (pee protein out)
What is an Exudate?	An effusion with mostly protein *Too much protein:* • Purulent (bacteria) • Hemorrhagic (trauma, cancer, PE) • Fibrinous (collagen vascular dz, uremia, TB) • Granulomatous (non-bacterial)

What is Systole?	Squish heart, ↓blood flow to coronary aa., more extraction of O_2 (Phase 1 Korotkoff)
What is Diastole?	Fill heart, ↑blood flow to coronary aa., less extraction of O_2 (Phase 5 Korotkoff)
What are the only arteries w/ deoxygenated blood?	Pulmonary arteries and umbilical arteries
What murmur has a Waterhammer pulse?	AR
What murmur has Pulsus tardus?	AS
What cardiomyopathy has Pulsus alternans?	Dilated cardiomyopathy
What disease has Pulsus bigeminus?	Idiopathic Hypertrophic Subaortic Stenosis
What murmur has an irregularly irregular pulse?	A Fib
What murmur has a regulary irregular pulse?	PVC
What sound radiates to the neck?	AS
What sound radiates to the axilla?	MR
What sound radiates to the back?	PS
What disease has a boot-shaped x-ray?	Right ventricle hypertrophy
What disease has a banana-shaped x-ray?	IHSS
What disease has an egg-shaped x-ray?	Transposition of the great arteries
What disease has a snowman-shaped x-ray?	Total Anomalous Pulmonary Venous Return
What disease has a "3" shaped x-ray?	Coarctation of the aorta
What is Osler-Weber-Rendu?	AVM in lung, gut, CNS => sequester platelets => telangiectasias
What is Von Hippel-Lindau?	AVM in **h**ead, retina => renal cell CA risk
When do valves make noise?	When valves close
What valves make noise during systole?	Mitral and tricuspid
What murmurs occur during systole?	Holosystolic, ejection murmur or click
What are the holosystolic murmurs?	TR, MR, or VSD
What are the systolic ejection murmurs?	AS, PS, or HCM
What valves make noise during diastole?	Aortic and pulmonic
What are the diastolic murmurs?	Blowing and Rumbling
What are the diastolic blowing murmurs?	AR or PR
What are the diastolic rumbling murmurs?	TS or MS
What are the continuous murmurs?	PDA or AVMs
What has a friction rub while breathing?	Pleuritis
What has a friction rub when holding breath?	Pericarditis
What does a mid-systolic click tell you?	Mitral valve prolapse
What does an ejection click tell you?	A/P stenosis
What does an opening snap tell you?	M/T stenosis
What does S_2 splitting tell you?	Normal on inspiration (b/c pulmonic valve closes later)

What does wide S_2 splitting tell you?	↑O_2, ↑RV volume, or delayed pulmonic valve opening
What does fixed wide S_2 splitting tell you?	ASD
What does paradoxical S_2 splitting tell you?	AS (or left bundle branch block)
What is cor pulmonale?	Pulmonary HTN => RV failure
What is Eisenmenger's?	Pulmonary HTN => reverse L-R to R-L shunt
What is Transposition of the great arteries?	Aorticopulmonary septum did not spiral
What is Tetrology of Fallot?	• Overriding Aorta: aorta sits on IV septum over the VSD; pushes on PA • Pulmonary Stenosis "Tet spells" • RV hypertrophy => boot-shaped heart • VSD (L to R shunt)
What is Total Anomalous Pulmonary Venous Return?	All pulmonary veins to RA, snowman x-ray
What is Truncus Arteriosus?	Spiral membrane not develop => one A/P trunk, mixed blood
What is Ebstein's Anomaly?	Tricuspid prolapse, Mom's Li increases risk
What can Lithium do to Mom?	Nephrogenic Diabetes Insipidus
What is Cinchonism?	Hearing loss, tinnitus, thrombocytopenia

Pulmonary:

What is the difference between a carotid body and a carotid sinus?	Carotid body: chemoreceptor Carotid sinus: baroreceptor
What color is air on an x-ray?	Black "radiolucent"
What color is fluid/solid on x-ray?	White "radiopaque"
What disease has a steeple sign on neck film?	Croup
What disease has a thumb sign on neck film?	Epiglottitis
What is a "blue bloater"?	Bronchitis
What is a "pink puffer"?	Emphysema
What diseases have pulmonary eosinophilia?	Aspergillosis, Strongyloides
What drugs cause pulmonary eosinophilia?	Nitrofurantoin, Sulfonamides
What are the risk factors for lung cancer?	Smoking, Radon, 2nd hand smoke, pneumoconiosis (except anthracosis)
What diseases have hemoptysis?	Bronchiectasis, bronchitis, pneumonia, TB, lung cancer
Where is a Bokdalek hernia?	Back of diaphragm
Where is a Morgagni hernia?	Middle of diaphragm
What diseases have respiratory alkalosis?	Restrictive Lung Dz (anxiety, pregnancy, Gram – sepsis, PE)
What diseases have respiratory acidosis?	Obstructive Lung Dz

What diseases have metabolic alkalosis?	Low Volume State (vomiting, diuretics, GI blood loss)
What diseases have metabolic acidosis?	Acid production (MUDPILES, RTA II, diarrhea)
What does stridor tell you?	Extrathoracic narrowing => narrows when breathe in => neck x-ray
What does wheezing tell you?	Intrathoracic narrowing => narrows when breathe out => chest x-ray
What does rhonchi tell you?	Mucus in airway => obstructive lung disease
What does grunting tell you?	Blows collapsed alveoli open => restrictive lung disease
What does dull percussion tell you?	Something b/w airspace and chest wall absorbing sound (fluid or solid)
What does hyperresonance tell you?	Air in lungs
What does tracheal deviation tell you?	Away from pneumothorax OR toward atelectasis *"air-phobic"*
What does fremitus, egophony, and bronchophony tell you?	Consolidation=> *pathognomonic for pneumonia*
What is Restrictive lung disease?	Small stiff lungs, trouble breathing in
What is Obstructive lung disease?	Big mucus-filled lungs, trouble breathing **o**ut
What is Epiglottitis?	X-ray thumb sign, drooling
What is Croup?	X-ray steeple sign, barking cough
What is Tracheitis?	Look toxic, grey pseudomembrane, leukocytosis
What is Asthma?	Wheeze on expiration, IgE, eosinophils
What is Bronchiolitis?	Asthma in kids <2 y/o
What is Sinusitis?	Teeth pain worse with bending forward
What is Bronchiectasis? Obstructive Lung	Bad breath, purulent sputum, hemoptysis CF
What is Bronchitis?	Lots of sputum, "**b**lue **b**loater"
What is Emphysema?	* Restrictive to obstructive pattern, "pink puffer"
What is Laryngomalacia?	Epiglottis roll in from side-to-side
What is Pneumonia?	Consolidation of airway
What is Pneumothorax?	Decreased breath sounds on one side
How do you detect a Pulomary Embolus?	Tachypnea, increased V/Q scan, EKG: $S_1Q_3T_3$
What is Tamponade?	Decreased breath sounds/BP; increased JVD
What is Tracheomalacia?	Soft cartilage, stridor since birth
What is Cystic Fibrosis?	Meconium ileus, steatorrhea,bronchiectasis
What is Aspergillosis?	Allergy to mold, dead plants, compost piles
What is Asbestosis?	Shipyard workers, pipe fitters, brake mechanics, insulation install
What is Silicosis?	Sandblasters, glassblowers
What is Byssinosis?	Cotton workers, chest tightness
What is Beryliosis?	Radio, TV welders
What is Anthracosis?	Coal workers, massive fibrosis
What is Sarcoidosis?	Non-caseating granulomas, eggshell calcification of

	lymph nodes
What is Carcinoid syndrome?	Flushing, wheezing, diarrhea
What is Small cell CA? Smoking perihilar	At carina, malignant, Cushing's, SIADH, SVC syndrome
What is Large cell CA?	Large stuff
What is Squamous cell CA?	Smoker, high PTH, high Ca^{2+}
What is Brochealveolar CA?	Looks like pneumonia; due to pneumoconiosis periphery non smokers

Gastrointestinal:

What disease has a corkscrew x-ray?	Esophageal spasm
What disease has an apple core x-ray?	Cancer
What disease has a stacked coin x-ray?	Intussusception
What disease has a thumbprint x-ray?	Toxic megacolon
What disease has an abrupt cutoff x-ray?	Volvulus
What disease has a barium clumping x-ray?	Celiac sprue
What disease has a bird's beak x-ray?	Achalasia
What disease has a string sign x-ray?	Pyloric stenosis
What diseases have solid dysphagia?	Schatzki's rings, stricture, cancer
What diseases have solid and liquid dysphagia?	Esophageal spasm, scleroderma, achalasia
What is Barrett's Esophagus?	Metaplasia, ↑AdenoCA risk
What are Esophageal Varices?	Vomit blood everywhere, portal HTN
What is Mallory-Weiss?	Tear LES mucosa, chronic vomiters
What is Boerhaave's?	Tear all layers of esophagus, left-sided pneumo/pain/effusion
What is Achalasia?	Lost LES Auerbach's, bird's beak, Chaga's, choke on solids
What is Hirschsprung's?	Lost rectum Auerbach's, no meconium passage
What is a Zenker's diverticulum?	Cough undigested food from above UES, halitosis
What is a Traction diverticulum?	Eat big bolus => gets stuck above LES
What is Plummer-Vinson syndrome?	Esophageal webs, spoon nails, Fe-deficiency anemia
What are Schatzki rings?	Esophageal webs in lower esophagus
What is a TE fistula?	Choke w/ each feeding
What is an Esophageal atresia w/ TE fistula?	Vomit w/ 1st feeding, huge gastric bubble
What is Duodenal atresia?	Bilious vomiting w/1st feed, double bubble, **D**own's
What is Pyloric stenosis?	Projectile vomiting (3-4 wk old), RUQ olive mass
How does Choanal atresia present?	Turns blue with feeding
How is the Tetrology of Fallot presentation different?	Turns blue with crying
What makes Scleroderma unique?	↓LES pressure
What makes Esophageal spasms unique?	↑Peristalsis
What makes Achalasia unique?	↓Peristalsis and ↑LES pressure

What disease has a RUQ olive mass?	Pyloric stenosis
What disease has a RLQ sausage mass?	Intussusception
What is a Bezoar?	Mass of hair or vegetables => antrum obstruction
What is Gastritis type A?	Upper GI bleed, anti-parietal cell Ab
What is Gastritis type B?	Upper GI bleed, spicy foods, H. pylori
What is a Duodenal ulcer?	Too much acid: pain after meal/ at night, type O blood, H. pylori, pain relieved by eating
What is a Gastric ulcer?	Broken mucus layer: pain during meal, NSAIDs, type A blood
What is a Sliding hiatal hernia?	Fundus slides from esophageal hiatus to thorax => sucks acid into thorax
What is a Rolling hiatal hernia?	Fundus sticks through hole in diaphragm, strangulates bowel *"rolls through a hole"*
What is Ménétrier's disease?	Protein-losing, thick stomach rugal folds
What defines Constipation?	< 3 BM per week
What defines Diarrhea?	>200g per day
What is Osmotic diarrhea?	Watery
What is Secretory diarrhea?	Laxative use
What is Inflammatory diarrhea?	Blood, pus
What is Celiac sprue?	Jejunum, wheat allergy, villous atrophy, anti-gliadal Ab
What is Tropical sprue?	Ileum celiac sprue
What is Mesenteric ischemia?	Pain out of proportion to exam
What bugs cause bloody diarrhea?	***"CASES"*** **Campylobacter** **Amoeba (E. histolytica)** **Shigella** E. **coli** **Salmonella**
What is the difference b/w 1° Biliary Cirrhosis and 1° Sclerosing Cholangitis?	**1° Biliary Cirrhosis:** anti-mitochondrial Ab, bile ductules destroyed, xanthelasma **1° Sclerosing Cholangitis:** p-ANCA Ab, bile duct inflammation, beading, onion skinning, associated w/ UC
What is Ascending Cholangitis?	Common duct stone gets infected
What are the signs of alcoholic cirrhosis?	Spider angioma, palmar erythema, Dupuytren's contractions, gynecomastia
What is Hepatorenal Syndrome?	Pts w/ liver disease build up liver toxins that cause renal failure
What is Cholangitis?	Inflammation of bile duct => Charcot's triad, Reynold's pentad
What is Cholecystitis?	Inflammation of gall bladder => Murphy's sign
What is Cholelithiasis?	Formation of gallstones => RUQ colic
What is Choledocholithiasis?	Gallstone obstructs bile duct

What is Cholestasis?	Obstruction of bile duct => pruritis, ↑alkaline phosphatase, jaundice
What is Conjugated bilirubin?	Water soluble "direct"
What is Unconjugated bilirubin?	Fat soluble "indirect"
What is the most common type of gallstone?	Cholesterol (can't see on x-ray)
What type of gallstones can be seen on x-ray?	Ca-bilirubinate
What is a Xanthoma?	Cholesterol buildup (elbow or Achilles)
What is a Xanthelasma?	Triglyceride buildup (under eye)
What does high cholesterol cause?	Atherosclerosis
What do high triglycerides cause?	Pancreatitis
What is Type 1 Hyperlipidemia?	Bad Liver LL (CM)
What is Type 2a Hyperlipidemia?	Bad LDL or B-100 receptors: trapped in ER (LDL only)
What is Type 2b Hyperlipidemia?	Less LDL/VLDL receptors (LDL/VLDL)
What is Type 3 Hyperlipidemia?	Bad Apo E (IDL/VLDL)
What is Type 4 Hyperlipidemia?	Bad Adipose LL (VLDL only)
What is Type 5 Hyperlipidemia?	Bad C2 (VLDL/CM) b/c C2 stimulates LL
What is Crigler-Najjar?	Unconjugated bilirubin, usually in infants
What is Gilbert's syndrome?	Glucuronyl transferase is saturated => stress unconjugugated bilirubin
What is Rotor's?	Bad bilirubin storage => conjugated bilirubin
What is Dubin-Johnson?	Bad bilirubin excretion => black liver
What is Cullen's sign?	Bleed around umbilicus => hemorrhagic pancreatitis
What is Turner's sign?	Bleed into flank => hemorrhagic pancreatitis
What tests are used for following pancreatitis?	• Amylase – sensitive, breaks down carbs • Lipase – specific, breaks down TGs
What does Ranson's criteria tell you?	Poor prognosis for pancreatitis pts
What is Ranson's criteria at presentation?	*"WAGLA"* **W**BC: >16K/μL (infection) **A**ge: >55 (usually multiple illnesses) **G**lucose: >200 mg/dL (islet cells are fried) **L**DH: >350 IU/L (cell death) **A**ST: >250 IU/L (cell death)
What is Ranson's criteria at 48hr?	*"BuCH was a SOB"* **B**UN: ↑ >5mg/dL (↓renal blood flow) **C**a: <8 mg/dL (saponification) **H**ct: drops >10% (bleed into pancreas) **S**equester >6 L fluid => 3rd spacing p**O**$_2$: <60mm Hg (fluid/protein leak → ARDS) **B**ase deficit >4mEq/L (diarrhea =>

	pancreatic enzymes are dead)
What is Carcinoid syndrome?	Diarrhea, flushing, wheezing
What produces Currant Jelly sputum?	Klebsiella
What produces Currant Jelly stool?	Intussusception
What is Gardener's syndrome?	Familial polyposis w/ bone tumors
What is Turcot's syndrome?	Familial polyposis w/ brain tumors
What is Familial polyposis?	100% risk of colon cancer, APC defect => annual colonoscopy at 5y/o
What is Peutz-Jegher syndrome?	Hyperpigmented mucosa => dark gums/vagina
What is Crohn's disease?	IBD w/ cobblestones, melena, creeping fat, fistulas
What is Ulcerative colitis?	IBD w/ pseudopolyps, hematochezia, lead pipe colon, toxic megacolon
What is Intussusception?	Currant jelly stool, stacked coin enema, sx come and go
How does Diverticulosis present?	Bleeds
How does Diverticulitis present?	Hurts
How does Spastic Colon present?	Intermittent severe cramps
How does IBS present?	Alternating diarrhea/constipation
How do External Hemorrhoids present?	Pain
How do Internal Hemorrhoids present?	No pain
What is Pseudomembranous Colitis?	Overgrowth of C. difficile due to normal flora being killed off, usually by Clindamycin use
What is Whipple's disease?	T. whippleii destroy GI tract, then spread causing malabsorption, arthralgia
What color is an upper GI bleed?	Black
What color is a lower GI bleed?	Red
What adds color to stool?	Bilirubin
What is the default color of stool?	Clay-colored
What is the default color of urine?	Tea-colored

Renal:

Which part of the nephron concentrates urine?	Medulla
What is Goldblatt's kidney?	Flea-bitten kidney (blown capillaries)
What is Uremia?	Azotemia + symptoms
What is Azotemia?	$\uparrow$BUN/Cr
What is Nephritic kidney disease?	$\uparrow$Size of fenestrations => vasculitis hematuria <3.5g protein
What is Nephrotic kidney disease?	Lost BM charge due to deposition on heparin sulfate => massive proteinuria and lipiduria
What is seen in RPGN (Rapidly Progressive Glomerulonephritis)?	Crescents
What is Post-Strep GN?	Subepithelial, IgG/C_3/C_4 deposition, ASO Ab

What is Interstitial Nephritis?	Urine eosinophils Drugs
What is Lupus Nephritis?	Subepithelial
What is MGN (Membranoglomerulonephritis)?	Deposition
What is MPGN (Membranoproliferative Glomerulonephritis)?	Tram-tracks (type II has low C_3)
What is MCD (Minimal Change Disease)?	Kids, fused foot processes, no renal failure, loss of charge barrier
What is FSGS (Focal Segmental Glomerulosclerosis)?	AA, HIV pts
What are the vasculitis w/ low C_3?	*"PMS in Salt Lake City"** **P**ost-strep GN **M**PGN Type II **S**BE **S**erum sickness **L**upus **C**ryoglobulinemia
What is the most common cause of kidney stones?	Dehydration
What are the most common type of kidney stones?	Calcium pyrophosphate
What type of kidney stones have coffin-lid crystals?	Triple phosphate
What type of kidney stones have rosette crystals?	Uric acid
What type of kidney stones have hexagonal crystals?	Cystine
What type of kidney stones have envelope or dumbbell-shaped crystals?	Oxalate
What disease has Aniridia?	Wilm's tumor
What disease has Iridocyclitis?	Juvenile rheumatoid arthritis
What is Phimosis?	Foreskin scarred at penis head (foreskin stuck smooshed up)
What is Paraphimosis?	Foreskin scarred at penis base (retraction of foreskin => strangulates penis)
What is Urge incontinence?	Urgency leads to complete voiding (detrusor spasticity → small bladder vol)
What is Stress incontinence?	Weak pelvic floor muscles (estrogen effect)
What is Overflow incontinence?	Runs down leg but can't complete empty bladder
What structures have one-way valves?	Urethra, ejaculatory duct
What structures have fake sphincters?	Ureters, LES, Ileocecal valve
What has WBC casts?	Nephritis
What has WBC casts only?	Pyelonephritis (sepsis)

bullous pemphigus = sloughs

What has WBC casts + eosinophils?	Interstitial nephritis (allergies)
What has WBC casts + RBC casts?	Glomerulonephritis
What has Fat casts?	Nephrotic syndrome
What has Waxy casts?	Chronic renal failure
What has Tubular casts?	ATN
What has Muddy brown casts?	ATN
What has Hyaline casts?	Normal sloughing
What has Epithelial casts?	Normal sloughing
What has Crescents?	RPGN
How do you measure afferent renal function?	Creatinine (or inulin)
How do you measure efferent renal function?	BUN (or PAH)
What is the afferent arteriole's job?	Filter
What is the efferent arteriole's job?	Secrete
How do you test afferent arteriole function?	GFR
How do you test efferent arteriole function?	RPF
What is pre-renal failure?	Low flow to kidney (BUN:Cr >20)
What is renal failure?	Damage glomerulus (BUN:Cr <20)
What is post-renal failure?	Obstruction (haven't peed in last 4 days)
What is the job of the proximal tubule?	Reabsorb glucose, amino acids, salt, bicarb,
What is the job of the thin ascending limb?	Reabsorbs water
What is the job of the thick ascending limb?	Make the concentration gradient by reabsorbing Na, K, Cl, Mg, Ca without water
What is the job of the early distal tubule?	Concentrate urine by reabsorbing NaCl (hypotonic) Thiazide
What is the job of the late distal tubule and collecting duct?	Final concentration of urine by reabsorbing water, excretion of acid (isotonic) Aldosterone cortex ADH medulla
What does the macula densa do?	Measures osmolarity
What does the J-G apparatus do?	Measures volume
What is Fanconi's syndrome?	Old tetracycline use => urine phosphates, glucose, amino acids
What is Bartter's syndrome?	Baby w/ defective triple transporter (low Na, Cl, K w/ normal BP)
What is Psychogenic polydipsia?	No concentrating ability → cerebral edema
What is Hepatorenal syndrome?	High urea from liver → increase glutaminase → NH_4^+ → GABA → kidney stops working
What is Type 1 RTA?	Distal renal tubular acidosis: H/K in CD is broken → *high urine pH* (UTI, stones, Li)
What is Type 2 RTA?	Proximal RTA: bad CA → lost all bicarb → *low urine pH* (multiple myeloma)
What is Type 3 RTA?	RTA I + II → *normal urine pH (5-6)*
What is Type 4 RTA?	Infarct J-G → no renin → no Aldo → *high K* (DM, NSAIDs, ACE-I, Heparin)
What is Central Pontine Myelinolysis?	Due to correcting Na faster than 0.5mEq/hr

Hematology:

What is a Neutrophil?	The Phagocyte (has anti-microbials, most abundant)
What is an Eosinophil?	The Parasite Destroyer, Allergy Inducer
What is a Basophil?	The Allergy Helper (IgE receptor => histamine release)
What is a Monocyte?	The Destroyer => MP (hydrolytic enzymes, coffee-bean nucleus)
What is a Lymphocyte?	The Warrior => T, B, NK cells
What is a Platelet?	The Clotter (no nuclei, smallest cells)
What is a Blast?	Baby Hematopoietic cell
What is a Band?	Baby Neutrophil
What does high WBC and high PMNs tell you?	Stress demargination
What does high WBC and <5% blasts tell you?	Leukemoid reaction, seen in burn pts (extreme demargination looks like leukemia)
What does high WBC and >5% blasts tell you?	Leukemia
What does high WBC and bands tell you?	Left shift => have infection
What does high WBC and B cells tell you?	Bacterial infection
What diseases have high eosinophils?	*"NAACP"* **N**eoplasm (lymphoma) **A**llergy/ **A**sthma **A**ddison's disease (no cortisol → relative eosinophilia) **C**ollagen vascular disease **P**arasites
What diseases have high monocytes (>15%)?	*"STELS"* **S**yphilis: chancre, rash, warts **T**B: hemoptysis, night sweats **E**BV: teenager sick for a month **L**isteria: baby who is sick **S**almonella: food poisoning
What do high retics (>1%) tell you?	RBC being destroyed peripherally
What do low retics tell you?	Bone marrow not working right (↓production)
What is Poikilocytosis?	Different shapes
What is Anisocytosis?	Different sizes
What is the RBC lifespan?	120 days
What is the platelet lifespan?	7 days
What does –penia tell you?	Low levels (usually due to virus or drugs)
What does –cytosis tell you?	High levels
What does –cythemia tell you?	High levels
What is the difference between plasma and serum?	**Plasma:** no RBC **Serum:** no RBC or fibrinogen

What is Chronic Granulomatous Disease?	NADPH oxidase deficiency → recurrent Staph/Aspergillus infections (Nitroblue Tetrazolium stain negative)
What does MPO deficiency cause?	Catalase + infections
What is Chediak Higashi? *Can you say that real fast 3 times in a row?...*	Lazy lysosome syndrome: lysosomes are slow to fuse around bacteria
What organ can make RBCs if the long bones are damaged?	Spleen => splenomegaly
What causes a shift to the right in the Hb curve?	*"All CADETs face right"** • ↑**C**O_2 • **A**cid/**A**ltitude • 2,3-**D**PG • **E**xercise • **T**emp
How does CO poison Hb?	Co**mpetitive inhibitor of O_2 on Hb => cherry-red lips, pinkish skin hue**
How does Cyanide poison Hb?	**Non-competitive inhibitor of O_2 on Hb => almond breath**
What is MetHb?	Hb w/ Fe^{3+}
What is Acute Intermittent Porphyria?	↑Porphyrin, urine δ-ALA, porphobilinogen => abdominal pain, neuropathy, red urine
What is Porphyria Cutanea Tarda?	**Sunlight => skin blisters w/ porphyrin deposits, Wood's lamp=orange-pink**
What is Erythrocytic Protoporphyria?	**Porphyria cutanea tarda in a baby**
What is Sickle cell disease?	**Homozygous HbS: ($\beta_{Glu6 \rightarrow Val}$) => vaso-occlusion, necrosis, dactylitis (painful fingers/toes) at 6mo, protects against malaria**
What is Sickle cell trait?	**Heterozygous HbS => painless hematuria, sickle with extreme hypoxia (can't be a pilot, fireman, diver)**
What is Hb C disease?	($\beta_{Glu6 \rightarrow Lys}$), still charged => no sickling
What is α-thalassemia?	• **1 deletion:** Normal • **2 deletions "trait":** Microcytic anemia 3 deletions: **Hemolytic anemia, Hb H=β_4** • **4 deletions:** Hydrops fetalis, Hb Bart=γ_4
What is β-thalassemia?	• **1 deletion "β minor":** ↑HbA_2 and HbF • **2 deletions "trait/intermedia/major":** • only HbA_2 and HbF => hypoxia at 6 mo
What is Cooley's anemia?	**See w/ β thalassemia major (no HbA => excess RBC production); baby making blood from everywhere => frontal bossing,**

	hepatosplenomegaly, long extremities
What is Virchow's triad?	Thrombosis risk factors: 1) Turbulent blood flow *"slow"* 2) Hypercoagulable *"sticky"* 3) Vessel wall damage *"escapes"*
What does acute hypoxia cause?	Shortness of breath
What does chronic hypoxia cause?	Clubbing of fingers/toes
What is intravascular hemolysis?	RBC destroyed in blood vv. → low haptoglobin (binds free floating Hb)
What is extravascular hemolysis?	RBC destroyed in spleen (problem w/ RBC membrane) => splenomegaly
What enzymes need lead (Pb)?	• δ-ALA dehydrase • Ferrochelatase
What does EDTA bind?	X^{2+}
What disease has a smooth philthrum?	Fetal alcohol syndrome
What disease has a long philthrum?	William's
What disease has sausage digits?	Pseudo-hypoparathyroidism, psoriatic arthritis
What disease has 6 fingers?	Trisomy 13
What disease has 2-jointed thumbs?	Diamond-Blackfan
What disease has painful fingers?	Sickle cell disease
What are the Microcytic Hypochromic anemias? *"FAST Lead"*	
• **F**e deficiency	↑TIBC, menses, GI bleed, koilonychia
• **A**nemia of chronic disease	↓TIBC
• **S**ideroblastic anemia	↓δ-ALA synthase, blood transfusions
• α-**T**halassemia	AA, Asians (Chr.16 deletion)
• β-**T**halassemia	Mediterraneans (Chr.11 point mutation)
• **Pb** poisoning	↓δ-ALA dehydrogenase, ↓ferrochelatase, x-ray blue line, eating old paint chips
What are the Megaloblastic anemias?	
• Vit B_{12} deficiency	Tapeworms, vegans, type A gastritis, pernicious anemia
• Folate deficiency	Old food, glossitis
• Alcohol	Fetal alcohol syndrome: smooth philthrum, stuff doesn't grow
What are the Intravascular Hemolytic anemias?	*IgM*
• G-6PD deficiency	Sulfa drugs, moth balls, fava beans, sudden drop in Hb
• Cold autoimmune	Mononucleosis, mycoplasma infections, RBC agglutination
What are the Extravascular Hemolytic	*IgG*

anemias?	
• Spherocytosis	Defective spherin or ankyrin, + osmotic fragility test
• Warm autoimmune	Anti-Rh Ab, dapsone, PTU, anti-malarials, sulfa drugs
• Paroxysmal cold autoimmune	Bleeds after cold exposure, Donath-Landsteiner Ab
• Sickle cell anemia	Crew haircut x-ray, avascular necrosis of femur, short fingers
What are the Production Anemias?	
• Diamond-Blackfan	No RBCs, 2-jointed thumbs
• Aplastic anemia	Pancytopenia, autoimmune, benzene, AZT, CAM, radiation
What is Basophilic Stippling?	Lots of immature cells, ↑mRNA (Pb poisoning)
What is a Bite cell = Basket cell?	Unstable Hb inclusions (G6-PD deficiency)
What is a Burr cell = Echinocyte?	Pyruvate kinase deficiency, Liver dz, Post-splenectomy
What is Cabot's ring body?	Vit B_{12} deficiency, Pb poisoning
What is a Doehle body?	PMN leukocytosis (infection, steroids, tumor)
What is a Drepanocyte?	Sickle cell anemia
What is a Helmet cell?	Fragmented RBC (Hemolysis: DIC, HUS, TTP)
What is a Heinz body?	Hb precipitates and sticks to cell membranes (G-6PD deficiency)
What is a Howell-Jolly body?	Spleen or bone marrow should have removed nuclei fragments (hemolytic anemia, spleen trauma, cancer)
What is a Pappenheimer body?	Fe ppt inside cell (sideroblastic anemia)
What is a Pencil cell = Cigar cell?	Fe deficiency anemia
What is Rouleaux formation?	Multiple myeloma
What is a Schistocyte?	Broken RBC (DIC, artificial heart valves)
What is a Sideroblast?	Macrophages pregnant w/Fe (genetic or multiple transfusions)
What is a Spherocyte?	Old RBC
What is a Spur cell = Acanthocyte?	Lipid bilayer dz
What is a Stomatocyte?	Liver dz
What is a Target cell = Codocyte?	Less Hb (Thallasemias or Fe deficiency)
What is a Tear drop cell = Dacrocyte?	RBCs squeezed out of marrow (hemolytic anemia, bone marrow cancer)
What is the Clotting Cascade?	How you stop bleeding
What do platelet problems cause?	Bleeding from skin and mucosa
What do clotting problems cause?	Bleeding into cavities
What causes increased PTT and bleeding time?	von Willebrand disease and Lupus
What is Bernard-Soulier?	Baby w/ bleeding from skin and mucosa, **b**ig platelets (low GP1b)
What is Glanzmann's?	baby w/ bleeding from skin and mucosa (low

	GP2b3a)
How does Factor 13 deficiency present?	Umbilical stump bleeding (1st time baby has to stabilize a clot)
What is Factor V Leiden?	Protein C can't break down Factor 5 => more clots
How does von Willebrand Disease present?	Heavy menstrual bleeding
What are the types of VWD?	**Type 1 (AD):** ↓VWF production **Type 2 (AD):** ↓VWF activity (+ Ristocetin aggregation test) **Type 3 (AR):** No VWF
What is Hemophilia A?	Defective Factor 8 (< 40% activity) => bleed into cavities (head, abdomen, etc.)
What is Hemophilia B?	Factor 9 deficiency => bleed into joints (knee, etc.)
What diseases have low LAP?	CML, PNH
What has high LAP?	Leukemoid reaction
What is the difference between acute and chronic leukemias?	**Acute:** started in bone marrow, squeezes RBC out of marrow **Chronic:** started in periphery, not constrained => will expand
What is the difference between myeloid and lymphoid leukemias?	**Myeloid:** ↑RBC, WBC, platelets, MP (↓lymphoid cells) => bone marrow biopsy **Lymphoid:** ↑NK, T, B cells (↓myeloid cells) => do lymph node biopsy
What defines ALL?	<15y/o males, bone pain, PAS stain ⊕, TdT ⊕
What defines AML?	15-30y/o males, Sudan Stain, Auer rods
What defines CML?	30 50y/o females, t(9,22) "Philadelphia chromosome", bcr-abl, ↓LAP
What defines CLL?	>50 y/o males w/ lymphadenopathy, "soccer ball" nuclei, smudge cells
What defines Hodgkin's lymphoma?	EBV, may have Reed-Sternberg cells
What are the B cell Non-Hodgkin's lymphomas?	**Follicular:** t(14,18), bcl-2 **Burkitt:** t(8,14), c-myc, starry sky MP • American kids: abdominal mass • African kids: jaw mass
What are the T cell Non-Hodgkin's lymphomas?	Mycosis Fungoides: total body rash Sezary syndrome: cerebreform cells
What is Polycythemia Rubra Vera?	Hct >60%, ↓Epo, Budd-Chiari, plethora "pruritis after bathing"
What is Essential Thrombocythemia?	Very high platelets, stainable Fe, ↓c-mpl
What is Myelofibrosis?	Megakaryocytes, fibrotic marrow => teardrop cells, extramedular hematopoiesis
What are plasma neoplasms?	Produce lots of Ab
What is Waldenstrom Macroglobulinemia?	IgM, hyperviscous

What is Monoclonal Gammopathy of Undetermined Significance?	Old person w/ gamma spike
What is Multiple Myeloma?	Serum M prot (IgG), urine Bence-Jones protein, rouleaux, punched-out lesions
What is Heavy Chain Disease?	↑IgA
What is Histocytosis X?	Kid w/ eczema, skull lesions, diabetes insipidus, exophthalmos
What does the Coombs test tell you?	Ab involved
What does the direct Coombs test tell you?	On surface => hemolytic anemias
What does the indirect Coombs test tell you?	In serum
What is type and cross?	You know you can use that blood, save it for specific pt
What is type and match?	Type it and wait
What is forward typing?	Uses **Ab** to detect Ag ***"Fabulous"***
What is backward typing?	Uses Ag to detect Ab
What does blood type A tell you?	Have the A antigen
What does blood type O tell you?	Have no antigens, universal donor
What does blood type AB tell you?	Have both antigens, universal recipient
What does Rh + tell you?	Has D antigen
What does Rh – tell you?	Does not have D antigen
What is Hemolytic Disease of the Newborn?	Rh – Mom's placenta tears, 100cc baby's blood sees Mom/produces Ab, attacks fetus
What is RHOGAM?	Anti-D IgG
When do you give RHOGAM?	1st Dose: 28 wk gestation (of 2nd child) 2nd Dose: 72 hrs post-delivery (Rh+ baby)
What is the most common transplant?	Blood
What is a Syngenic transplant?	Twin to twin
What is an Autograft?	Self to self transplant
What is an Allograft?	Human to human transplant
What is a Xenograft?	1 species to another species
What is Hyperacute rejection?	Within 12 hrs *(preformed Ab)*
What is Acute rejection?	4 days to years later *(T-cells, MP)*
What is Chronic rejection?	> 7 days (Fibroblasts)
What is Graft vs. Host disease?	Bone marrow transplants reject *(T_K, MP)*
What are Immunoprivileged sites?	No lymphatic flow => no Ag => easy to transplant (brain, cornea, thymus, testes)
What is INR?	Measured PT/ Control PT

Endocrinology:

What is Necrosis?	Non-programmed cell death = noisy, inflammation, nucleus destroyed first
What is Apoptosis?	Programmed cell death = quiet, no inflammation, nucleus guides it => destroyed last
What is Pyknosis?	Nucleus turns into blobs *"pick blobs"*
What is Karyohexis?	Nucleus fragments
What is Karyolysis?	Nucleus dissolves
What is a Somatotrope?	GH
What is a Gonadotrope?	LH, FSH
What is a Thyrotrope?	TSH
What is a Corticotrope?	ACTH
What is a Lactotrope?	PRL
What receptors do protein hormones use?	Cell membrane receptors
What receptors do steroid hormones use?	Nuclear membrane receptors
What are the steroid hormones?	*"PET CAD"* *Note: thyroid hormone acts like a steroid* **P**rogesterone $\mathbf{E_2}$ **T**estosterone **C**ortisol **A**ldo Vit **D**
What does Endocrine mean?	Secretion into blood
What does Exocrine mean?	Secretion into non-blood
What is Autocrine?	Works on itself
What is Paracrine?	Works on it's neighbor
What is Merocrine?	Cell is **m**aintained => exocytosis
What is Apocrine?	**Ap**ex of the cell is secreted
What is Holocrine?	The **whole** cell is secreted
What organs do not require insulin?	*"BRICKLE"* **B**rain **R**BC **I**ntestine **C**ardiac, Cornea **K**idney **L**iver **E**xercising muscle
What does GnRH do?	Stimulates LH, FSH
What does GRH do?	Stimulates GH
What does CRH do?	Stimulates ACTH
What does TRH do?	Stimulates TSH
What does PRH do?	Stimulates PRL

What does DA do?	Inhibits PRL
What does SS do?	Inhibits GH
What does ADH do?	Conserves water, vasoconstricts
What does oxytocin do?	Milk letdown, baby letdown
What does GH do?	IGF-1 release from liver
What does TSH do?	T_3,T_4 release from thyroid
What does LH do?	Testosterone release from testis, E_2 and Progesterone release from ovary
What does FSH do?	Sperm or egg growth
What does PRL do?	Milk production
What does ACTH do?	Cortisol release from adrenal gland
What does MSH do?	Skin pigmentation
What are the stress hormones?	Epi: immediate Glucagon: 20min Insulin: 30min ADH: 30min Cortisol: 2-4hr GH: 24hr
What does ADH do?	Concentrates urine
What is Diabetes Insipidus?	Too little ADH => urinate a lot
What is Central DI?	Brain not making ADH
What is Nephrogenic DI?	Blocks ADH receptor, can be caused by Li and Domecocycline
What does the Water Deprivation test tell you?	Water deprivation => DI *fails to concentrate urine*
What does giving DDAVP tell you?	DDAVP => Central DI *concentrates >25%*
What is SIADH?	Too much ADH => expand plasma vol => pee Na
What is the difference b/w DI and SIADH?	DI has dilute urine, SIADH has concentrated urine
What is Psychogenic Polydipsia?	Pathologic water drinking => low plasma osmolarity
What does Aldosterone do?	Reabsorbs Na, secretes H^+/ K^+
What is a Neuroblastoma?	Adrenal medulla tumor in kids, dancing eyes/feet, secretes catecholamines
What is a Pheochromocytoma?	Adrenal medulla tumor in adults, 5 P's
What does the Zona Glomerulosa make?	Aldosterone "salt"
What does the Zona Fasiculata make?	Cortisol "sugar"
What does the Zona Reticularis make?	Androgens "sex"
What is Conn's syndrome?	High Aldo (tumor), Captopril test makes it worse
What does ANP do?	Inhibits Aldo, dilates renal artery (afferent arteriole)
What does Calcitonin do?	Inhibits osteoclasts => low serum Ca^{2+}
What is MEN I?	"Wermer's": **P**ancreas, **P**ituitary, **P**arathyroid adenoma (high gastrin) *"PPP"*

What is MEN II?	"Sipple's": Pheo, Medullary thyroid cancer, PTH
What is MEN III?	"MEN IIb": Pheo, Medullary thyroid cancer, Oral/GI neuromas
What does CCK do?	Gallbladder contraction, bile release
What does Cortisol do?	Gluconeogenesis by proteolysis => thin skin
What is Addison's disease?	Autoimmune destruction of adrenal cortex => hyperpigmentation, ↑ACTH
What is Waterhouse Friderichsen?	Adrenal hemorrhage
What is Cushing's syndrome?	High cortisol (pituitary tumor or adrenal tumor or small cell lung CA)
What is Cushing's disease?	High ACTH (pituitary tumor)
What is Nelson's syndrome?	Hyperpigmentation after adrenalectomy
If the low-dose dexamethasone test suppresses, what does that tell you?	Normal, obese, or depressed
If the low-dose dexamethasone test does not suppress, what does that tell you?	Cushing's => do high dose test
If the high-dose dexamethasone test suppresses, what does that tell you?	Pituitary tumor => ACTH (call brain surgeon)
If the high-dose dexamethasone test does not suppress, what does that tell you?	• Adrenal adenoma => Cortisol (call general surgeon) • Small cell lung cancer => ACTH (call thoracic surgeon)
What are the survival hormones?	Cortisol: permissive under stress TSH: permissive under normal
What does Epinephrine do?	Gluconeogenesis, glycogenolysis
What does Erythropoietin do?	Makes RBCs
What does Gastrin do?	Stimulates parietal cells => IF, H^+
What does Growth hormone do?	Growth, sends somatomedin to growth plates, gluconeogenesis by proteolysis
What is a Pygmie?	No somatomedin receptors
What is Achondroplasia = Laron Dwarf?	Abnormal FGF receptors in extremities
What is a Midget?	↓Somatomedin receptor sensitivity
What is Acromegaly?	Adult bones stretch "my hat doesn't fit", coarse facial features, large furrowed tongue, deep husky voice, jaw protrusion, ↑IGF-1 b/c of GH tumor
What is Gigantism?	childhood acromegaly
What does GIP do?	Enhances insulin action => post-prandial hypoglycemia
What does Glucagon do?	Gluconeogenesis, glycogenolysis, lipolysis, ketogenesis
What does Insulin do?	Pushes glucose into cells
What is Type I DM?	Anti-islet cell Ab, GAD Ab, Coxsackie B, low insulin, DKA, polyuria, polydipsia, polyphagia
What is Type II DM?	Insulin receptor insensitivity, high insulin, HONK coma, acanthosis nigricans

How does DKA present?	Kussmaul respirations, fruity breath (acetone), altered mental status
What is the Dawn phenomenon?	Morning hyperglycemia 2° to GH
What is the Somogyi Effect?	Morning hyperglycemia 2° to evening hypoglycemia
What is Factitious Hypoglycemia?	Insulin injection (↑insulin, ↓C-peptide)
What is an Insulinoma?	Tumor (↑insulin, ↑C-peptide)
What is Erythrasma?	Rash in skin folds, coral-red Wood's lamp
What is Syndrome X = Metabolic Syndrome?	"Pre-DM"=> HTN, dyslipidemia, hyperinsulinemia, acanthosis nigricans
What are foot ulcer risk factors?	• DM/ Glycemic control • Male smoker • Bony abnormalities • Previous ulcers
What conditions cause weight gain?	• Obesity • Hypothyroidism • Depression • Cushing's • Anasarca
What does Motilin do?	stimulates segmentation (1° peristalsis, MMC)
What does Oxytocin do?	Milk ejection, baby ejection
What does PRL do?	Milk production
What does PTH do?	Chews up bone
What does Vit. D do?	Builds bone
What do parathyroid chief cells secrete?	PTH
What do stomach chief cells secrete?	Pepsin
What is the difference between Norepinephrine and Epinephrine?	NE: Neurotransmitter Epi: Hormone
What is 1° hyperparathyroidism?	Parathyroid adenoma
What is 2° hyperparathyroidism?	Renal failure
What is Familial Hypocalciuria Hypercalcemia?	↓Ca excretion
What if both serum Ca and PO_4 decrease?	Vit D deficiency
What if serum Ca and PO_4 change in opposite directions?	PTH problem • High Ca => hyperPTH • Low Ca => hypoPTH
What is the most common cause of 1° hypoparathyroidism?	Thyroidectomy
What is Pseudohypoparathyroidism?	Bad kidney PTH receptor, ↓urinary cAMP
What is Pseudopseudohypoparathyroidism?	G-protein defect, no Ca^{2+} problem
What is Hungry Bone syndrome?	Remove PTH → bone sucks in Ca^{2+}
What does Secretin do?	Secretion of bicarb, inhibit gastrin, tighten pyloric sphincter

What does Somatostatin do?	Inhibits secretin, motilin, CCK
What do T_3 and T_4 do?	Growth, differentiation
What disease has Exophthalmos?	Grave's
What disease has Enophthalmos?	Horner's
What are the Hyperthyroid diseases?	
• **Grave's**	Exophthalmos, pretibial myxedema, TSHr Ab
• **DeQuervain's**	Viral, painful jaw
• **Silent thyroiditis**	Post-partum
• **Plummer's**	Benign adenoma, old person
• **Jod-Basedow**	Transient hyperthyroidism due to ↑I
What are the Hypothyroid diseases?	
• **Hashimoto's**	Antimicrosomal Ab = TPO Ab
• **Reidel's struma**	Woody neck
• **Cretin**	Freaky features, hypothyroid Mom and Baby
• **Euthyroid sick syndrome**	Low T_3 syndrome
• **Wolff-Chaikoff**	Transient hypothyroidism
What is Plummer's syndrome?	Hyperthyroid adenoma
What is Plummer-Vinson syndrome?	Esophageal webs
What does Testosterone do?	Makes external male genitalia
What does Müllerian Inhibiting Factor do?	Makes internal male genitalia
What do TPO and Thymosin do?	Help T cells mature
What does VIP do?	Inhibits secretin, motilin, CCK
How does a VIPoma present?	Watery diarrhea
How does a SSoma present?	Constipation
What are the hormones with disulfide bonds?	*"PIGI"* • **P**RL • **I**nsulin • **G**H • **I**nhibin
Which hormones have the same α subunits?	• LH, FSH • TSH • β-HCG
What hormones produce acidophils?	*"GAP"* • **G**H • **P**RL
What hormones produce basophils?	*"B FLAT"* • **F**SH • **L**H • **A**CTH • **T**SH

Dermatology:

What are 1st degree burns?	Red (epidermis)
What are 2nd degree burns?	Blisters (hypodermis)
What are 3rd degree burns?	Painless neuropathy (dermis)
What diseases have palm and sole rashes?	*"TRiCKSSS"* **T**oxic Shock Syndrome **R**ocky mountain spotted fever **C**oxsackie A: Hand-Foot-Mouth disease **K**awasaki **S**carlet fever **S**taph Scaled Skin **S**yphilis
What is Erythema Multiforme?	Target lesions (viral, drugs)
What is Stevens Johnson syndrome?	Erythema Multiforme Major (mouth, eye, vagina)
What is Toxic Epidermal Necrolysis?	Stevens Johnson w/ skin sloughing
What is Pemphi*gus* vulgaris?	Ab against desmosomes => circular immunofluorescence, in epider***mis***, oral lesions, + Nikolsky sign
What is Bullous Pemphigoid?	Ab against hemidesmosomes => linear immunofluoresence, subepidermal, "floating" keratinocytes, eosinophils
What is Eczema?	Dry flaky dermatitis in flexor creases *"itch that rashes"*
What is Nummular dermatitis?	Circular eczema
What is Spongiotic eczema?	Weeping eczema: scratching causes oozing *"like a sponge"*
What is Lichenification?	Scratching => thick leathery skin
What is Pityriasis Rosea?	Herald patch that follows skin lines (Tx: sunlight) *"C-mas tree pattern"*
What is Lichen Planus?	Polygonal pruritic purple papules
What is Scabies?	Linear excoriation "burrows" in webs of fingers, toes, belt line (Sarcoptes feces)
What does UV-A cause?	**A**ging
What does UV-B cause?	**B**urns and cancer
What are the ABCD's that indicate worse prognosis of skin cancer?	• **A**ssmetric • Irregular **B**orders • **C**olor differences • >4mm **D**iameter
What does the Clark level tell you?	Invasion of melanoma
What does Breslow's classification tell you?	Depth of melanoma
Where are Malignant Melanomas usually found?	Males: back Females: leg
What is the precursor of a Malignant Melanoma?	Hutchison's freckle

What are the types of Malignant Melanomas?	**Superficial spreading**: most common, flat brown **Nodular**: worst prognosis, black, dome-shape **Lentigo maligna melanoma:** elderly pts, fair-skin **Acral lentigous: A**IDS pts, dark skin
Where are Squamous Cell carcinomas usually found?	Flat flaky stuff on lower face, keratin pearls
What is the precursor of Squamous Cell carcinoma?	Actinic keratosis (red scaly plaque)
What are the types of Squamous Cell carcinoma?	**Bowen's disease:** SCC in situ on uncircumcised penis dorsum **Verrucous carcinoma:** wart on foot
Where are Basal Cell carcinomas usually found?	Pearly papules on upper face, palisading nuclei, good progosis
What are the types of Basal Cell carcinomas?	**Nodular:** waxy nodule w/ central necrosis **Superficial:** red scaly plaques, like eczema **Pigmented:** looks like melanoma **Sclerosing:** yellow waxy plaques
What is Acne Rosacea?	Blush all the time, worse w/ stress/alcohol
What is a Brown Recluse Spider Bite?	Painful black necrotic lesion
What is Cellulitis?	Warm red leg
What is Cutaneous Anthrax?	Painless black necrotic lesion
What is a Decubitus Ulcer?	Bedsore
What is a DVT?	Blood clot in veins, associated w/ hypercoagulable state
What is Erysipelas?	Shiny red, raised, does not blanch, usually on facc, assoc w/ Strep pyogenes
What is Ichthyosis?	Gradual lizard skin
What is Miliaria?	"Heat rash": burning, itching papules on trunk
What is Molluscum Contagiosum?	Fleshy papules w/ central dimple, pox virus (STD)
What is Psoriasis?	Silver scales on extensors, nail pitting, differentiated too fast, worse w/ stress • Auspitz sign: remove scale => pinpoint bleeding • Koebner's phenomenon: lesions at sites of skin trauma
What is a Pyogenic Granuloma?	Vascular nodule at site of previous injury
What is Seborrheic Dermatitis?	Dandruff in eyebrows, nose, behind ears
What is Seborrheic Keratosis?	Rubbery warts with aging, greasy
What is Thrombophlebitis?	Vein inflammation w/ thrombus
What is Vitiligo?	White patches, anti-melanocyte Ab
What is Xeroderma Pigmentosa?	Bad DNA repair
What is Erythema Chronicum Migrans?	Lyme disease (solitary lesion that spreads)
What is Erythema Infectiosum?	Fifth disease "slapped cheeks" due to Parvovirus B19
What is Erythema Marginatum?	Rheumatic fever (red margins)

What is Erythema Multiforme?	Target lesions due to HSV, Phenytoin, Barbs, Sulfas
What is Erythema Multiforme Major?	Stevens Johnson syndrome (> 1 mucosal surface)
What is Erythema Nodosum?	Fat inflammation (painful red nodules on legs), sarcoidosis
What is Erythema Toxicum?	Newborn benign rash (looks like flea bites w/ eosinophils)

Muscle:

Where is CK-MB found?	Heart
Where is CK-MM found?	Muscle
Where is CK-BB found?	Brain
Why should you wait 30min after a meal before swimming?	All blood in gut and skeletal mm. have ran out of ATP
How does a neurogenic muscle disease present?	Distal weakness + Fasciculations
How does a myopathic muscle disease present?	Proximal weakness + pain
What is a light chain composed of?	Actin
What is a heavy chain composed of?	Myosin
What band of the sarcomere does not change length?	A band
Where are the T-tubules located?	Cardiac muscle: Z line Skeletal muscle: A-I junction
What is Duchenne's MD?	Dystrophin frameshift, Gower sign, calf pseudohypertrophy
What is Becker's MD?	Dystrophin missense, sx >5 y/o
What is Myotonic Dystrophy?	Bird's beak face, can't let go when shake hands
What is Myasthenic syndrome = Lambert-Eaton?	Gets stronger as day goes by, stronger w/ EMG, not small cell CA
What is Myasthenia gravis?	Gets weaker as day goes, stronger w/ Edrophonium, weaker w/ EMG, rule out thymoma
What is Multiple sclerosis?	Anti-myelin Ab, young woman w/ vision problems, sx come&go
What is Metachromatic Leukodystrophy?	Arylsulfatase deficiency, kid with MS presentation
What is Ataxia Telangectasia?	Spider veins, IgA deficiency
What is Friedreich's ataxia?	Retinitis pigmentosa, scoliosis
What is Adrenal Leukodystrophy?	Carnitine shuttle problem, adrenal failure
What is Guillain-Barre?	Ascending paralysis, 2 wk after URI
What is ALS?	Middle age male w/ Fasciculations, descending paralysis, no sensory problems
What is Werdnig-Hoffman?	Fasciculations in a newborn, no anterior horns
What is Polio?	Asymmetric Fasciculations in child, 2 wk after gastroenteritis

What is Choreoathetosis?	Dance-like movements, wringing of hands, quivering voice
What is Atonic cerebral palsy?	No muscle tone, floppy

Bone:

What is Slipped Capital Femoral Epiphysis?	Obese boys w/ dull achy pain
What is Legg-Calves-Perthes?	Limp (femur head avascular necrosis)
What is Osgood-Schlatter?	Knee pain (tibial tubercle avascular necrosis)
What is Septic arthritis?	Joint pain (Staph aureus)
What is Ankylosing Spondylitis?	Ligament ossification => vertebral body fusion, ↓lumbar curve, stiffer in morning, kyphosis, uveitis, HLAB-27
What is Cauda Equina Syndrome?	"Saddle anesthesia": can't feel butt, thighs, perineum
Where does bone cancer metastasis occur from?	Breast, prostate, lung, kidney
What is Costochondritis?	Painful swelling of chest joint-bone attachments, worse w/ deep breath
What is Disk Herniation?	Straight leg raise => shooting pain
What is Lumbar Stenosis?	MRI *"hourglass"*, low back pain
What is Ochronosis = Alkaptonuria?	Kids w/ OA, black urine, homogentisic acid oxidase deficiency
What is Osteitis Fibrosis cystica?	Inflammation of bone w/ holes
What is Osteogenesis Imperfecta?	Blue sclera, multiple broken bones
What is Osteomalacia = Rickets?	Soft bones (waddling gait) Craniotabes (soft skull) Rachitic rosary (costochondral thickening) Harrison's groove Pigeon breast (sternum protrusion)
What is Osteomyelitis?	Infected bones
What is Osteonecrosis = Legg-Calve-Perthes?	Wedge-shaped necrosis of femur head
What is Osteopenia?	Lost bone mass
What is Osteopetrosis?	↓Osteoclast activity => marble bones (obliterate own bone marrow)
What is Osteoporosis?	Loss of bone matrix (not calcification) => compression fractures
What is Osteosclerosis?	Thick bones
What is Paget's Disease?	"My hat doesn't fit", **p**aramyxovirus, ↑osteoclasts/blasts, fluffy bone, osteosarcoma, ↑CO heart failure, deafness, ↑alkaline phosphatase alone

Rheumatology:

What is Rheumatoid Factor?	An IgM against IgG F_c
What are Tophi?	Gout crystals + giant cells
What is Podagra?	Big toe inflammation
What is CREST syndrome?	**C**alcinosis **R**aynaud's **E**sophageal dysmotility **S**clerodactyly **T**elangiectasia
What Ab is associated with CREST?	Anti-**ce**ntromere Ab
What diseases have Raynaud's syndrome?	• Scleroderma • Takayasu's • RA • SLE
What platelet count do most types of vasculitis have?	Low platelets
Which vasculitis has a high platelet count?	Kawasaki disease
Which vasculitis has a normal platelet count?	Henoch-Schönlein purpura
What is Osteoarthritis?	Pain worse w/ activity, PIP/DIP joints
What is Rheumatoid Arthritis?	Pain worse in morning, MCP/PIP joints
What is Stills' Disease?	Juvenile RA
What is Pseudogout?	Pyrophosphate crystals in knees/wrists
What is Gout?	Urate crystals in big toe, (-) birefringence
What is Myositis?	1 muscle hurts
What is Polymyositis?	>1 muscle weak
What is Dermatomyositis?	Myositis + rash
What is Fibromyalgia	Hurt all the time, 11 trigger points
What is Polymyalgia Rheumatica?	Weak shoulders, temporal arteritis
What is SLE = Lupus?	Meet 4 criteria: *"DOPAMIN RASH"* **D**iscoid rash **O**ral ulcers **P**hotosensitivity **A**rthritis **M**alar rash **I**mmunologic disorder: Anti-ds DNA, Sm, Cardiolipin Ab **N**eurologic disorder: seizure or psychosis **R**enal failure: *die of this* **A**NA **S**erositis: pleuritis/pericarditis (Libman-Sacks endocarditis) **H**emolytic anemia

What is Scleroderma?	Tight skin, fibrosis
What is Takayasu's Arteritis?	Pulseless Asian women, aorta inflammation
What is Polyarteritis Nodosa?	p-ANCA Ab, attacks gut/kidney, Hep B
What is Wegener's Granulomatosis?	c-ANCA Ab, attacks ENT, lungs, kidney
What is Goodpasture's?	Anti-GBM Ab, attacks lung/kidney, RPGN
What is Reiter's syndrome?	Males that can't see, pee, or climb a tree
What is Sjögren's syndrome?	Females that have dry eyes/mouth, RA
What is Behçet's syndrome?	Oral and genital ulcers, uveitis
What is Churg-Strauss?	Asthma, eosinophils, multi-organ involved
What is Felty's syndrome?	RA, leukopenia, splenomegaly
What is Kawasaki's disease?	*"CRASH"* **C**onjunctivitis **R**ash (palm/sole) **A**neurysm (coronary artery) → MI in kids **S**trawberry tongue **H**ot (fever > 102°F for at least 3 days + cervical lymphadenopathy)

Gynecology:

What does the Seminal Vesicle give to sperm?	Food (fructose) and clothes (semen)
What do the Bulbourethral = Cowper's glands secrete?	Bicarbonate (neutralize lactobacilli)
What does the Prostate secrete?	*"The prostate HAZ it"* **H**yaluronidase **A**cid phosphatase **Z**inc
What is the Capacitation reaction?	Zn used to peel semen off
What is the Acrosomal reaction?	Sperm release enzymes to eat corona radiata
What is the Crystalization reaction?	Wall formed after 1 sperm enters (to prevent polyspermy)
Where does Testosterone come from?	Adrenal gland and testicles
Where does DHT come from?	Testicles (at puberty)
What is a Pseudohermaphrodite?	External genitalia problem
What is a True Hermaphrodite?	Internal genitalia problem => has both sexes
What is a Female Hermaphrodite?	Impossible b/c the default is female
What is a Female Pseudohermaphrodite?	XX with low 21-OHase => high testosterone
What is a Male Hermaphrodite?	XY with no MIF
What is a Male Pseudohermaphrodite?	XY that has low 17-OHase => low testosterone
What is Hirsutism?	**H**airy
What is Virilization?	Man-like

What is Testicular Feminization = Androgen Insensitivity Syndrome?	Bad DHT receptor → XY w/ blind pouch vagina
What is McCune-Albright?	Precocious sexual development, polyostotic fibrous dysplasia "whorls of CT", "Coast of Maine" pigmented skin macules
What is Cryptochordism?	Testes never descended => sterility after 15mo, seminomas
Which stage of the menstrual cycle has the highest estrogen levels?	Follicular stage (has proliferative endothelium)
What stage of the menstrual cycle has the highest temperature?	Ovulatory stage
What stage of the menstrual cycle has the highest progesterone levels?	Luteal stage (has secretory endothelium)
What form of estrogen is highest at menopause?	**E_1:** Estr<u>one</u> (made by fat)
What form of estrogen in highest in middle-age females?	**E_2:** Estra<u>di</u>ol (made by ovaries)
What form of estrogen is highest at pregnancy?	**E_3:** Es<u>tri</u>ol (made by placenta)
What states have increased estrogen?	Pregnancy, liver failure, p450 inhibition, obesity
What is Adenomyosis?	Growth of endometrium → myometrium, enlarged "boggy" uterus w/ cystic areas
What does DES taken by Mom cause in her daughter?	• Adenomyosis → menorrhagia • Clear Cell CA of vagina • Recurrent abortions
What is Kallman's syndrome?	No GnRH, anosmia (can't smell)
What is Polycystic Ovarian Syndrome?	↑Cysts: no ovulation → no progesterone (↑endometrial CA) → can't inhibit LH, obese, hairy, acne
What is Savage's syndrome?	Ovarian resistance to FSH/LH
What is Turner's syndrome (XO)?	High FSH, low E_2, ovarian dysgenesis
What does the Progesterone challenge test tell you?	Bleeds => she has estrogen Not bleed => she has no E_2 or ovaries • ↑FSH → ovary problem • ↓FSH → pituitary problem
What is Sheehan syndrome?	Post-partum hemorrhage → pituitary, hyperplasia infarcts → no lactation
What is Asherman's syndrome?	Previous D&C → uterine scars
What is Oligomenorrhea?	Too few periods
What is Polymenorrhea?	Too many periods
What is the most common cause of post-coital vaginal bleeding?	Cervical cancer

What is the most common cause of post-coital vaginal bleeding in pregnant women?	Placenta previa
What is the most common cause of vaginal bleeding in post-menopause women?	Endometrial cancer
What is Chronic Pelvic Pain?	Endometriosis until proven otherwise
What is Dysfunctional Uterine Bleeding?	Diagnosis of exclusion, usually due to anovulation
What is Dysmenorrhea?	PG-F causes painful menstrual cramps (teenagers miss school/work)
What is Endometriosis?	Painful cyclical heavy menstrual bleeding => "powder burns, chocolate cysts" due to ectopic endometrial tissue
What is Kleine regnung?	Scant bleeding at ovulation
What is Menorrhagia?	Heavy menstrual bleeding
What is Fibroids = Leiomyoma?	Benign uterus SM tumor • Submucosal type => bleeding • Subserosal type => pain
What is Metrorrhagia?	Bleeding or spotting in between periods
What is Mittelschmerz?	Pain at ovulation
What causes Syphilis?	Treponema pallidum (spirochete)
What is Herpes?	ds DNA virus
What is HPV?	ds DNA virus
What is Chlamydia?	Obligate intracellular bacteria
What causes Gonorrhea?	Gram – diplococcus
What causes Chancroid?	H. ducreyi
What causes Lymphogranuloma Venereum?	Chlamydia trachomatis
What causes Granuloma Inguinale?	C. granulomatosis
What causes Epididymitis?	Chlamydia
What is Condyloma Lata?	Flat fleshy warts, ulcerate, 2° Syphilis
What is Condyloma Accuminata?	Verrucous "cauliflower" warts, koilocytes, HPV 6,11
How does Herpes present?	**1°:** Painful grouped vesicles on red base **2°:** Painful solitary lesion
How does Syphilis present?	**1°:** Painless chancre (1-6 wks) **2°:** Rash, condyloma lata (6 wks) **3°:** Neuro, cardio, bone (6yrs)
How does Chancroid present?	Painful w/ necrotic center, Gram – rod, "school of fish" pattern
How does Lymphogranuloma Venereum present?	Painless ulcers → abscessed nodes → genital elephantiasis
How does Granuloma Inguinale present?	Spreading ulcer, Donovan bodies, granulation test
How does Chlamydia present?	Cervicitis (yellow pus), conjunctivitis, PID
How does Gonorrhea present?	Palmar pustules, arthritis, urethral discomfort

What is Epididymitis?	Unilateral scrotal pain decreased by support
What causes Congenital blindness?	CMV
What causes Neonatal blindness?	Chlamydia
What is Lichen simplex chronicus?	Raised white lesions, chronic scratching
What is Lichen sclerosis?	Paperlike vulva, itching, cancer risk
What is Hidradenoma?	Sweat gland cysts
What causes non-bacterial fetal infections?	*"TORCHS"* **T**oxoplasma: multiple ring-enhancing lesions, cat urine, parietal lobe **O**thers **R**ubella: cataracts, hearing loss, PDA, meningoencephalitis, pneumonia, "blueberry muffin" rash **C**MV: spastic diplegia of legs, hepatosplenomegaly, blindness, central calcifications **H**SV-2: temporal lobe hemorrhagic encephalitis, need C/S prophylaxis **S**yphilis: Rhagade's (lip fissure), saber shin legs, Hutchison's razor teeth, mulberry molars
What is Paget's disease of the breast?	Rash and ulcer around nipple, breast cancer
What is Lobular carcinoma?	Cells line up single file, contralateral primary
What is a Comedocarcinoma?	Multiple focal areas of necrosis, "blackheads"
What is Inflammatory carcinoma?	Infiltrates lymphatics, pulls on Cooper's ligaments, *"peau d'orange"*
What is Cystosarcoma phylloides?	"Exploding mushroom", firm, rubbery, moveable, good prognosis
What is Intraductal papilloma?	Nipple bleeding, most common breast CA
What is Ductal carcinoma?	Worst prognosis breast cancer
What is Sarcoma Botyroides?	Vagina cancer, ball of grapes
What is a Sister Mary Joseph Nodule?	Ovarian CA spread to umbilicus
What is Meig's syndrome?	Pleural effusion, ovarian fibroma, ascites
What are the side effects of estrogen?	Weight gain, breast tenderness, nausea, HA
What are the side effects of progesterone?	Acne, depression, HTN

Obstetrics:

Why do pregnant women get anemia?	Dilutional effect; RBC rises 30% but volume rises 50%
What are the degrees of vaginal lacerations?	**1st Degree:** Skin **2nd Degree:** Muscle **3rd Degree:** Anus **4th Degree:** Rectum
What is Vernix?	Cheesy baby skin
What is Meconium?	Green baby poop
What is Lochia?	Endometrial slough
What is normal blood loss during a vaginal delivery?	500mL
What is normal blood loss during a C-section?	1L
How do you treat A1 Gestational DM?	Diet
How do you treat A2 Gestational DM?	Insulin
What are identical twins?	Egg split into perfect halves "monochorionic"
What are fraternal twins?	Multiple eggs fertilized by different sperm
What is Ovarian Hyperstimulation Syndrome?	Weight gain and enlarged ovaries after clomiphene use
Who makes the Trophoblast?	Baby
Who makes the Cytotrophoblast?	Mom => GnRH, CRH, TRH, Inhibin
Who makes the Syncitotrophoblast?	Mom and baby => HCG, HPL
When does implantation occur?	1 week after fertilization
When is β-HCG found in urine?	2 weeks after fertilization
What is the function of Estrogen?	Muscle relaxant, constipation, ↑protein production, irritability, varicose veins
What is the function of Progesterone?	↑appetite, ↑acne, dilutional anemia, quiescent uterus, pica, hypoTN, melasma
What makes progesterone <10wk old?	Corpus luteum
What makes progesterone >10wk old?	Placenta
What is the function of β-HCG?	Maintains corpus luteum, sensitizes TSHr => act hyperthyroid (to ↑BMR)
What makes β-HCG?	Placenta
How fast should β-HCG rise?	Doubles every 2 days until 10 wks (when placenta is fully formed)
What is the function of AFP?	Regulates fetal intravascular volume
What is the function of HPL?	Blocks insulin receptors => sugar stays high (baby's stocking up on stuff needed for the journey)
What is the function of Inhibin?	Inhibits FSH => no menstruation
What is the function of Oxytocin?	Milk ejection, baby ejection
What is the function of Cortisol in pregnancy?	Decreases immune rejection of baby, lung maturation

What are the thyroid hormone levels during pregnancy?	↑TBG => ↑bound T_4, normal free T_4 levels
When can you first detect fetal heartones?	Week 20
When can you tell the sex of a fetus by US?	Week 16
What does an AFI <5 indicate?	Oligohydramnios (cord compression)
What does an AFI >20 indicate?	Polyhydramnios (DM)
How fast should fundal height change?	Uterus grows 1cm/wk
What is the Pool Test?	Fluid in vagina
What is Ferning?	Estrogen crystallizes on slide
What is Nitrazine?	Shows presence of amniotic fluid
What is the risk of chorionic villus sampling?	Fetal limb defects
What is the risk of amniocentesis?	Abortion (2% risk)
What is a normal biophysical profile?	>8
What is a biophysical profile?	*"Test the Baby, MAN!"* **T**ones of the heart **B**reathing **M**ovement: BPD, HC, AC, FL **A**FI **N**on-stress test (normal = "reactive")
What pelvis types are better for vaginal delivery?	Gynecoid, anthropoid
What pelvis types will need C/S?	Platypelloid, android
How do you predict a due date with Nägele's Rule?	9 months from last menses → add 1 wk
Why is Nägele's Rule inaccurate?	B/c it does not start from ovulation date
How do you correct Nägele's Rule for cycles >28 days?	Add x days if cycle is x longer
How much weight should a pregnant woman gain?	1lb/wk
When should intercourse be avoided during pregnancy?	3rd trimester b/c PG-F in semen may cause uterine contractions
What are the Leopold maneuvers?	1) Feel fundus 2) Feel baby's back 3) Feel pelvic inlet 4) Feel baby's head
What is Stage I of labor?	Up to full dilation 1) Latent Phase (<20h): Contractions → 4 cm cervical dilation 2) Active Phase (<12h): 4-10 cm cervical dilation (1cm/ hr)
What is Stage II of labor?	Full dilation → delivery Station 0: Baby above pelvic rim (most uteri are

	anteverted) 1. Engage 2. Descend 3. Flex head 4. Internal rotation 5. Extend head 6. Externally rotate 7. Expulsion: LDA most common presentation
What is Stage III of labor?	Delivery of placenta (due to PG-F) Blood gush → cord lengthens → fundus firms
How do you monitor baby's HR?	Doppler, scalp electrode
How do you monitor uterus?	Tocodynamics, uterine pressure catheter
What Bishop's score predicts delivery will be soon?	>8
What are Braxton-Hicks contractions?	Irregular contractions w/ closed cervix
What is a Vertex presentation?	Posterior fontanel (triangle shape) presents first, normal
What is a Sinciput presentation?	Anterior fontanel (diamond shape) presents first
What is a Face presentation?	Mentum anterior → forceps delivery
What is a Compound presentation?	Arm or hand on head → vaginal delivery
What is a Complete breech?	Butt down, thighs and legs flexed
What is a Frank breech?	Butt down, thigh flexed, legs extended (~pancake)
What is a Footling breech?	Butt down, thigh flexed, one toe is sticking out of cervical os
What is a Double Footling breech?	Two feet sticking out of cervical os
What is a Transverse Lie?	Head is on one side, butt on the other
What is Shoulder Dystocia?	Head out, shoulder stuck
Can you try vaginal delivery on a woman who has had a Classic Horizontal C/S previously?	No, must have C/S for all future pregnancies
Can you try vaginal delivery on a woman who has had a Low Transverse C/S previously?	Yes
What is early deceleration?	Normal, due to head compression
What is late deceleration?	Uteroplacental insufficiency b/c placenta can't provide O_2/nutrients
What is variable deceleration?	Cord compression
What is increased beat-to-beat variability?	Fetal hypoxemia
What is decreased beat-to-beat variability?	Acidemia
What is Pre-eclampsia?	Ischemia to placenta => HTN (>140/90)
What is the treatment for Pre-eclampsia?	Delivery
What is HELLP syndrome?	Hepatic injury causing: • **H**emolysis • **E**levated **L**iver enzymes

	• Low Platelets
What is Eclampsia?	HTN + seizures
What are the symptoms of eclampsia?	H/A, changes in vision, epigastric pain
What is the treatment for eclampsia?	4mg Mg sulfate as seizure prophylaxis
What is Chorioamnionitis?	Fever, uterine tenderness, ↓fetal HR
What are the symptoms of Amniotic Fluid Emboli?	Mom just delivered baby and has SOB → PE, death (amniotic fluid → lungs)
What is Endometritis?	Post-partum uterine tenderness
What is an incomplete molar pregnancy?	2 sperm + 1 egg (69, XXY), has embryo parts
What is a complete molar pregnancy?	2 sperm + no egg (46, XX – both paternal), no embryo
What is Pseudocyesis?	Fake pregnancy w/ all the signs and symptoms
What is the most common cause of 1st trimester maternal death?	Ectopic pregnancy
What is the most common cause of 1st trimester spontaneous abortions?	Chromosomal abnormalities
What are the most common causes of 3rd trimester spontaneous abortions?	Anti-cardiolipin Ab, placenta problems, infection, incompetent cervix
What is a threatened abortion?	Cervix closed, baby intact (Tx: bed rest)
What is an incvitable abortion?	Cervix open, baby intact (Tx: cerclage = sew cervix shut until term)
What is an incomplete abortion?	Cervix open, fetal remnants (Tx: D&C to prevent placenta infection)
What is a complete abortion?	Cervix open, no fetal remnants (Test: β-HCG)
What is a missed abortion?	Cervix closed, no fetal remnants (Tx: D&C)
What is a septic abortion?	Fever >100°F, malodorous discharge
What is Placenta Previa?	Post-coital bleeding, placenta covers cervical os; ruptures placental arteries
What is Vasa Previa?	Placenta aa. hang out of cervix
What is Placenta Accreta?	Placenta attached to superficial lining
What is Placenta Increta?	Placenta invades into myometrium
What is Placenta Percreta?	Placenta perforates through myometrium
What is Placenta Abruptio?	Severe pain, premature separation of placenta
What is Velamentous Cord Insertion?	Fetal vessels insert between chorion and amnion
What is a Uterus Rupture?	Tearing sensation, halt of delivery
What is an Apt test?	Detects HbF in vagina
What is Wright's stain?	Detects nucleated fetal RBC in Mom's vagina
What is a Kleihauer-Betke test?	Detects percentage of fetal blood in maternal circulation (dilution test)
What is maternity blues?	Post-partum crying, irritability
What is post-partum depression?	Depression >2wks
What is post-partum psychosis?	Hallucinations, suicidal, infanticidal

Pediatrics:

What are the newborn screening tests?	1. "Please Check BBefore Going HHome"* 2. **P**KU 3. **C**ongenital adrenal hyperplasia 4. **B**iotinidase, β-thalassemia 5. **G**alactosemia 6. **H**ypothyroidism, **H**omocystinuria
What is VATER syndrome?	7. **V**ertebral abnormality 8. **A**nal 9. **TE** fistula 10. **R**enal
What is an average IQ?	85-100
What are the most common causes of mental retardation in the US?	EtOH, Fragile X, Down's
What is Rubeola also known as?	Measles
What is Rubella also known as?	German measles
What is the APGAR test?	*Test at 1 and 5min (normal>7)* **A**ppearance (color) **P**ulse **G**rimace **A**ctivity **R**espiration
What is the most common eye infection the first day of life?	Clear discharge due to silver nitrate
What is the most common eye infection the first week of life?	Gonorrhea => purulent discharge
What is the most common eye infection the second week of life?	Chlamydia
What is the most common eye infection the third week of life?	Herpes
What are the causes of Hyperbilirubinemia?	Sepsis, ABO incompatibility, hypothyroidism, breastfeeding
What is the cause of symmetrically small babies?	Chromosomal abnormality or TORCHS
What is the cause of asymmetrically small babies?	Poor blood supply spares brain => small body, normal head
What is the cause of large babies?	DM or twin-twin transfusions
What is Milia?	Neonatal whiteheads on malar area
What is Nevus Flammeus?	"Stork bites" on back of neck, look like flames
What is Seborrheic Dermatitis?	Red rash w/oily skin and dry flaky hairline
What are Hemangiomas?	Flat blood vessels
What are Mongolian spots?	Melanocytes on lower back

What is Erythema Toxicum?	White wheal on red area, has eosinophils
What is Subgaleal Hemorrhage?	Prolonged jaundice in newborns due to birthing trauma
What is Caput Succedaneum?	Under scalp (edema crosses suture lines)
What is Cephalohematoma?	Under bone (blood not cross suture lines)
What is an Epstein's pearl?	White pearls on hard palate
What is persistent eye drainage since birth usually due to?	Blocked duct
What are wide sutures due to?	Hypothyroidism, Down's
What causes a Cleft Lip?	Medial nasal prominence did not fuse
What causes a Cleft Palate?	Maxillary shelves did not fuse
What is the most common cause of no red reflex?	Cataracts
What is the most common cause of a white reflex?	Retinoblastoma
What is the sign of a Clavicle Fracture?	Asymmetric Moro reflex
What is an Omphalocele?	Intestines protrude out of umbilicus covered by peritoneum
What is Gastroschisis?	Abdominal wall defect, off-center
What is a Nephroblastoma?	Kidney "Wilm's" tumor, hemihypertrophy, aniridia
What is a Neuroblastoma?	Adrenal medulla tumor, hypsarrhythmia, opsoclonus, ↑VMA
What is Polyhydramnios?	Too much amniotic fluid, baby can't swallow
What are the most common causes of Polyhydramnios?	NM problem: **Werdnig-Hoffman** GI problem: **Duodenal atresia**
What is Oligohydramnios?	Too little amniotic fluid, baby can't pee
What are the most common causes of Oligohydramnios?	Abd muscle problem: **Prune Belly** Renal agenesis: **Potter's syndrome**
What is Fifth disease?	Erythema infectiosum "slapped cheeks", arthritis in mom, aplastic anemia
What is Sixth disease?	Roseola, exanthema subitum (*fever disappears, then rash appears*)
What is Hand-Foot-Mouth disease?	Mouth ulcers => won't eat or drink, palm/sole rash, Coxsackie A virus
What is Measles = Rubeola?	1) Cough, Coryza, Conjunctivitis 2) Koplik spot 3) Morbilliform blotchy rash
What is Molloscum Contagiosum?	Flesh-colored papules w/ central dimple
What is Mumps?	Parotiditis => red Stenson's duct
What is Otitis Media?	Fluid in middle ear
What is Pityriasis Rosea?	Herald patch → migrates along skin lines "C-mass tree" appearance
What is Rubella = German 3-day measles?	Trunk rash, lymphadenopathy behind ears, don't look sick

How is Smallpox different from Chickenpox?	Smallpox is on face, same stage of development, fever
What is Varicella = Chickenpox?	1) Red macule 2) Clear vesicle on red dot 3) Pus 4) Scab
What is Zellweger's?	Neonatal seizures
What is the most common cause of delayed speech development?	Hearing loss
What are the signs of child abuse?	• Multiple ecchymoses • Retinal hemorrhage • Epidural/Subdural hemorrhage • Spiral fractures (twisted) • Multiple fractures in different healing stages
What should you rule out when child abuse is suspected?	• Osteogenesis imperfecta • Bleeding disorders • Fifth disease • Mongolian spots

Surgery:

What causes fever post-op Day 1?	Wind: Pneumonia, Atelectasis
What causes fever post-op Day 3?	Water: UTIs
What causes fever post-op Day 5?	Wound: Staph aureus
What causes fever post-op Day 7?	Walk: DVT
What causes fever post-op Day 10?	Abdominal abscesses
What causes fever anytime?	H_2 blockers, Antibiotics
What do you check for in an unresponsive patient?	*"ABC"* **A**irway: call, listen for noise → can pt talk? **B**reathing: listen to chest → intubate **C**irculation: color and capillary refill
What is an Indirect inguinal hernia?	Goes through the inguinal canal into scrotum
What is a Direct inguinal hernia?	Goes directly through abdominal wall in Hesselbach's triangle
What is potency?	How long a drug will keep the patient knocked out
What is Minimum Alveolar Concentration?	Concentration needed to push the drug into the brain

Biochemistry:

What is the most common intracellular buffer?	Protein
What is the most common extracellular buffer?	Bicarbonate
What is a Zwitterion?	A molecule with one negative and one positive end
What is the Isoelectric Point?	The pH at which there is no net charge
What is the rate limiting enzyme in Glycolysis?	PFK-1
What is the rate limiting enzyme in Gluconeogenesis?	Pyruvate carboxylase
What is the rate limiting enzyme in the HMP shunt?	G-6PD
What is the rate limiting enzyme in Glycogenesis?	Glycogen synthase
What is the rate limiting enzyme in Glycogenolysis?	Glycogen phosphorylase
What is the rate limiting enzyme in FA synthesis?	AcCoA carboxylase
What is the rate limiting enzyme in β-oxidation?	CAT-1
What is the rate limiting enzyme in Cholesterol synthesis?	HMG CoA reductase
What is the rate limiting enzyme in Ketogenosis?	HMG CoA synthase
What is the rate limiting enzyme in Purine synthesis?	PRPP synthase
What is the rate limiting enzyme in Pyrimidine synthesis?	Asp transcarbamoylase
What is the rate limiting enzyme in TCA cycle?	Isocitrate dehydrogenase
What is the rate limiting enzyme in Urea cycle?	CPS-I
What is the rate limiting enzyme in Heme synthesis?	δ-ALA synthase
What are the catabolic pathways that create energy?	***"ABC"*** **AcetylCoA production** **β-oxidation** **Citric acid cycle**
What are the anabolic pathways that store energy?	*"EFGH"* **ER** **Fatty acid synthesis** **Glycolysis** **HMP shunt**
What are the anabolic + catabolic	***"HUG"***

pathways?	**Heme synthesis** **Urea cycle** **Gluconeogenesis**
What does an Isomerase do?	Creates an isomer
What does an Epimerase do?	Creates an epimer, which differs around 1 chiral carbon
What does a Mutase do?	**Moves sidechain from one carbon to another (intrachain)**
What does a Transferase do?	**Moves sidechain from one substrate to another (interchain)**
What does a Kinase do?	Phosphorylates using ATP
What does a Phosphorylase do?	Phosphorylates using Pi
What does a Carboxylase do?	Forms C-C bonds (w/ ATP and biotin)
What does a Synthase do?	Consumes 2 substrates
What does a Synthetase do?	Consumes 2 substrates, uses ATP
What does a Phosphatase do?	Breaks phosphate bond
What does a Hydrolase do?	Breaks a bond with water
What does a Lyase do?	Cuts C-C bonds w/ ATP
What does a Dehydrogenase do?	Removes H with a cofactor
What does a Thio do?	Breaks S bonds
What is Diffusion?	From high to low concentration
What is Active Transport?	Goes against concentration gradient
What is Zero-order kinetics?	Metabolism independent of concentration
What is 1st-order kinetics?	Constant drug percentage metabolism over time, depends on drug concentration
What is Efficacy?	Max effect regardless of dose (lower w/ non-competitive antagonist)
What effects Efficacy?	V_{max}
What is Potency?	Amount of drug needed to produce effect (lower w/ comp antagonist)
What affects Potency?	K_m
What is K_d?	Concentration of drug that binds 50% of receptors
What is EC_{50}?	Concentration of drug that produces 50% of maximal response
What is Competitive Inhibition?	Fights for active site, no ΔV_{max}, potency decreases
What is Non-competitive Inhibition?	Binds a regulatory site, no ΔK_m, efficacy decreases, $\downarrow V_{max}$
What is an Endothermic Reaction?	Consumes heat
What is an Exothermic Reaction?	Gives off heat
What is the Peak level?	11. 4 hrs after dose (too high => decrease dose)
What is the Trough level?	2 hrs before dose (too high => give less often)
What is $t_{1/2}$?	Half-life, the time it takes for the body to use half of the drug ingested

What is von Gierke?	G-6Pase deficiency => hypoglycemia, hepatosplenomegaly
What is Pompe's?	Cardiac α-1,4-glucosidase deficiency => DIE early
What is Cori's?	Debranching enzyme deficiency => *short* branches of glycogen
What is Anderson's?	Branching enzyme deficiency => *long* chains of glycogen
What is McArdle's?	Muscle phosphorylase deficiency => muscle cramps w/ exercise
What is Essential Fructosuria?	Fructokinase deficiency => excrete fructose (still have hexokinase)
What is Fructosemia?	"Fructose intolerance" (Aldolase B deficiency) => liver damage
What does a Galactokinase deficiency cause?	Cataracts
What does Galactosemia cause?	Cataracts, mental retardation, liver damage
What does the Citrate shuttle do?	FA transport out of the mitochondria
What does the Carnitine shuttle do?	FA transport into the mitochondria
What lysosomal diseases have a cherry-red macula?	Tay-Sachs, Niemann-Pick
What lysosomal diseases have a Gargoyle-face?	Gaucher's, Hurler's
What is Tay-Sachs?	Hexosaminidase A deficiency => blindness, incoordination, dementia
What is Sandhoff's?	Hexosaminidase A/B deficiency
What is Gaucher's?	Glucocerebrosidase deficiency => wrinkled tissue MP, bone pain
What is Niemann-Pick?	Sphingomyelinase deficiency => zebra bodies
What is Fabry's?	α-galactosidase deficiency => corneal clouding, attacks *baby's* kidneys, X-linked
What is Krabbe's?	β-galactosidase deficiency => globoid bodies
What is Metachromatic Leukodystrophy?	Arylsulfatase deficiency => childhood MS
What is Hunter's?	Iduronidase deficiency, milder form
What is Hurler's?	Iduronidase deficiency, worse form
What is Lesch-Nyhan?	(HGPRT deficiency) => gout, neuropathy, self-mutilation
What do white diaper crystals suggest?	Excess orotic acid
What does biotin donate methyl groups for?	Carboxylation
What does THF donate methyl groups for?	Nucleotides
What does SAM donate methyl groups for?	All other reactions
What is the difference b/w Heterochromatin and Euchromatin?	Heterochromatin = tightly coiled Euchromatin = loose (10nm fibers)
What are the Purines?	A, G

What are the Pyrimidines?	C, U, T
What is a silent mutation?	Changes leave the same amino acid
What is a point mutation?	Changes 1 base
What is a transition?	Changes 1 purine to another purine
What is a transversion?	Changes 1 purine to a pyrimidine
What is a frameshift mutation?	Insert or delete 1-2 bases
What is a missense mutation?	Mistaken amino acid substitution
What is a nonsense mutation?	Early stop codon
What does a Southern blot detect?	DNA
What does a Northern blot detect?	RNA
What does a Western blot detect?	Protein
What are the essential amino acids?	*"PVT TIM HALL"* 12. **P**he 13. **V**al 14. **T**rp 15. **T**hr 16. **I**le 17. **M**et 18. **H**is 19. **A**rg 20. **L**eu 21. **L**ys
What are the essential fatty acids?	Linolenic Linoleic
What are the acidic amino acids?	Asp, Glu
What are the basic amino acids?	Lys, Arg
What are the sulfur-containing amino acids?	Cys, Met
What are the O-bond amino acids?	Ser, Thr, Tyr
What are the N-bond amino acids?	Asn, Gln
What are the branched amino acids?	Leu, Ile, Val
What are the aromatic amino acids?	Phe, Tyr, Trp
What is the smallest amino acid?	Gly
What are the ketogenic amino acids?	Lys, Leu
What are the glucogenic + ketogenic amino acids?	*"PITT"* **P**he, **I**so, **T**hr, **T**rp
What are the glucogenic amino acids?	All the rest
What amino acids does Trypsin cut?	Lys, Arg
What amino acids does β-ME cut?	Cys, Met
What amino acids does Acid Hydrolysis denature?	Asn, Gln
What amino acids does Chymotrypsin cut?	Phe, Tyr, Trp
What amino acid turns yellow on Nurhydrin reaction?	Pro

What does Carboxypeptidase cut?	Left of any amino acid on the carboxy terminal
What does Aminopeptidase cut?	Right of N terminus
What does CNBr cut?	Right of Met
What does Mercaptoethanol cut?	Right of Cys, Met
What does Elastase cut?	Right of Gly, Ala, Ser
What does Trypsin cut?	Arg, Lys
What does Chymotrypsin cut?	Phe, Tyr, Trp
What does α_1-AT do?	Inhibits trypsin from getting loose
What is PKU?	No Phe → Tyr (via Phe Hydroxylase): Nutrasweet sensitivity, mental retardation, pale, blond hair, blue eyes, musty odor
What is Albinism?	No Tyr → Melanin (via Tyrosinase)
What is Maple Syrup Urine disease?	**Defective metabolism of branced aa (Leu, Iso, Val) => aa leak out**
What is Homocystinuria?	No Homocystine → Cys: *"COLA" stones* **C**ystine, **O**rnithine, **L**ysine, **A**rginine
What is Pellagra?	Niacin deficiency: • Dermatitis, Diarrhea, Dementia, Death
What is Hartnup's?	No Trp => Niacin + Serotonin • Presents like Pellagra • Can mimic corn-rich diet
What causes anterior leg bowing?	Neonatal syphilis
What causes lateral leg bowing?	Rickets
What are the names of the B vitamins?	*"The Rich Never Lie about Panning Pyrite Filled Creeks"* Vit B_1 = **Thiamine** Vit B_2 = **R**iboflavin Vit B_3 = **N**iacin Vit B_4 = **L**ipoic acid Vit B_5 = **P**antothenic acid Vit B_6 = **P**yridoxine Vit B_9 = **F**olate Vit B_{12} = **C**obalamin
What does Vit A do?	Night vision, CSF production, PTH
What does Vit B_1 do?	Dehydrogenases, transketolase (PPP) cofactors
What does Vit B_2 do?	FAD cofactor
What does Vit B_3 do?	NAD cofactor
What does Vit B_4 do?	Glycolysis, no known diseases
What does Vit B_5 do?	Part of AcetylCoA, no known diseases
What does Vit B_6 do?	Transaminase cofactor, myelin integrity
What does Vit B_9 do?	Nuclear division
What does Vit B_{12} do?	Cofactor for HMT and MMM MYELIN
What does Vit C do?	Collagen synthesis

X-LINKED DOM: Vit D Res. Rick
Pseudo hypoparathyroidism
Pyruvate Dehydrogenase

What does Vit D do?	Mineralization of bones, teeth
What does Vit K do?	Clotting
What does Biotin do?	Carboxylation
What does Ca^{2+} do?	Neuronal function, atrial depolarization, SM contractility
What does Cu^{2+} do?	Collagen synthesis
What does Fe^{2+} do?	Hb function, electron transport
What is Bronze pigmentation?	Fe deposit in skin
What is Bronze cirrhosis?	Fe deposit in liver
What is Bronze diabetes?	Fe deposit in pancreas
What is Hemosiderosis?	Fe overload in bone marrow
What is Hemochromatosis?	Fe deposit in organs
What does Mg^{2+} do?	PTH and kinase cofactor
What does Zn^{2+} do?	Taste buds, hair, sperm function
What does Cr do?	Insulin function
What does Mb do?	Purine breakdown (xanthine oxidase)
What does Mn do?	Glycolysis
What does Se do?	Heart function => dilated cardiomyopathy
What does Sn do?	Hair
What is Kwashiorkor?	Malabsorption, big belly (ascites), protein deficiency
What is Marasmus?	Starvation, skinny, calorie deficiency
Where does the Pre label send stuff to?	ER
Where does the Pro label send stuff to?	Golgi
Where does the Mannose-6-P send stuff to?	Lysosome
Where does the N-terminal sequence send stuff to?	Mitochondria
What are the 4 types of collagen?	***"SCAB"*** **Type I: Skin, bone** **Type II: Connective tissue, aqueous humor** **Type III: Arteries** **Type IV: Basement membrane**
How does Scleroderma present?	Tight skin
How does Ehlers Danlos present?	Hyperstretchable skin
How does Marfan's present?	Hyperextensible joints, arachnodactyly, wing span longer than height, Aortic root dilatation, aortic aneurysm, mitral valve prolapse, Dislocated lens from *bottom* of eye → look up
How does Homocystinuria present?	• Dislocated lens from *top* → look down
How does Kinky hair disease present?	Hair looks like copper wire (Cu deficiency)
How does Scurvy present?	Bleeding gums, bleeding hair follicles
How does Takayasu arteritis present?	Asian female with very weak pulse
How does Osteogenesis Imperfecta present?	Shattered bones, blue sclera

Folate
Pantothenic Acid
Biotin
Vit K
B12

Genetics:

What is the typical incidence of rare things?	1-3%
What is the typical incidence with 1 risk factor?	10%
What is the typical incidence with 2 risk factors?	50%
What is the typical incidence with 3 risk factors?	90%
What does Autosomal Dominant usually indicate?	Structural problem, 50% chance of passing it on
What does Autosomal Recessive usually indicate?	Enzyme deficiency, 1/4 get it, 2/3 carry it
What are the X-linked Recessive deficiencies?	*"Lesch-Nyhan went Hunting For Pirates and Gold Cookies"** **L**esch**N**yhan (HGPRT def.): self mutilation, gout, neuropathy **Hu**nter's (iduronidase def.) **F**abry's (α-galactosidase def.): corneal clouding, attacks baby's kidneys ~~**P**DH deficiency~~ **G-**6PD deficiency: infxns, hemolytic anemia **C**hronic Granulomatous Dz: NADPH oxidase deficiency
Where did X-linked Recessive diseases come from?	From maternal uncle or grandpa
What are the X-linked Dominant diseases?	• Na-resistant rickets (kidney leaks phosphorus): waddling gait • Pseudohypoparathyroidism: sausage digits, osteodystrophy
Where did X-linked Dominant diseases come from?	Dad → daughter
What are the Mitochondrial diseases?	• Leber's = atrophy of optic nerve • Leigh's = subacute necrotizing encephalomyelopathy
Where did Mitochondrial diseases come from?	Mom → all kids
Why do we stop CPR after 20-30min?	The brain has irreversible cell injury
Why do we only have 6hrs to use t-PA?	The body has irreversible cell injury
What is Turner's?	(X,O): web neck, cystic hygroma, shield chest, coarctation of aorta, rib notching
What is Klinefelter's?	(47, XXY): tall, gynecomastia, infertility, ↓testosterone
What is XXX syndrome?	(47, XXX): normal female w/ two Barr bodies
What is XYY syndrome?	(47, XYY): tall aggressive male

What is Trisomy 13?	**P**atau's, **p**olydactyly, high arch **p**alate, **p**ee problem, holoprosencephaly
What is Trisomy 18?	Edward's, rocker bottom feet
What is Trisomy 21?	Down's, simian crease, wide $1^{st}/2^{nd}$ toes, macroglossia, Mongolian slant of eyes, Brushfield spots, retardation
What disease has a Dinucleotide repeat?	HNPCC
What diseases have Trinucleotide repeats?	Huntington's, Fragile X, Myotonic Dystrophy, Friedreich's Ataxia
What is Angelman's?	"Happy puppet syndrome", ataxia
What is Prader-Willi?	Hyperphagia, hypogonadism, almond-shaped eyes
What is Kallman's?	Anosmia, small testes
What is Anaplasia?	Regress to infantile state
What is Atrophy?	Decreased organ or tissue size
What is Desmoplasia?	Cell wraps itself w/ dense fibrous tissue
What is Dysplasia "carcinoma in situ"?	Lose contact inhibition (cells crawl on each other)
What is Hyperplasia?	Increased cell number
What is Hypertrophy?	Increased cell size
What is Metaplasia?	Change from one adult cell type to another
What is Neoplasm?	New growth
What is Benign?	Well circumscribed, freely movable, maintains capsule, obeys physiology, hurts by compression, slow growing
What is Malignant?	Not well circumscribed, fixed, no capsule, doesn't obey physiology, hurts by metastasis, rapidly growing (outgrows blood supply → hunts for blood → secretes angiogenin and endostatin to inhibit blood supply of other tumors)
What are the fastest killing cancers?	Pancreatic cancer, Esophageal cancer
What does Adeno- tell you?	Glandular
What does Leiomyo- tell you?	Smooth muscle
What does Rhabdomyo- tell you?	Skeletal muscle
What does Hemangio- tell you?	Blood vessel
What does Lipo- tell you?	Fat
What does Osteo- tell you?	Bone
What does Fibro- tell you?	Fibrous tissue
What does –oma tell you?	Tumor
What does –carcinoma tell you?	Cancer
What does –sarcoma tell you?	Connective tissue cancer
What is a Hamartoma?	Abnormal growth of normal tissue
What is a Choristoma?	Normal tissue in the wrong place
What is the most common anterior mediastinum tumor?	Thymoma
What is the most common middle mediastinum tumor?	Pericardial

What is the most common posterior mediastinum tumor?	Neuro tumors
What organs have the most common occurrence of metastasis?	*"BBLLAP"* **B**rain (grey-white jxn) **B**one (bone marrow) **L**ung **L**iver (portal vein, hepatic artery) **A**drenal gland (renal arteries) **P**ericardium (coronary arteries)
What cancers have psammoma bodies?	**P**apillary (thyroid) **S**erous (ovary) **A**denocarcinoma (ovary) **Me**ningioma **Me**sothelioma
What cancer has CA-125?	Ovarian
What cancer has CA-19?	Pancreatic
What cancer has S-100?	Melanoma
What cancer has BRCA?	Breast
What cancer has PSA?	Prostate
What cancer has CEA?	Colon, Pancreatic
What cancer has AFP?	Liver, Yolk sac
What cancer has Rb?	Ewing's sarcoma, Retinoblastoma
What cancer has Ret?	Medullary thyroid cancer
What cancer has Ras?	Colon
What cancer has bcl-2?	Follicular lymphoma
What cancer has c-myc?	Burkitt's lymphoma
What cancer has L-myc?	Small cell lung carcinoma
What cancer has N-myc?	Neuroblastoma => pseudorosettes
What cancer has Bombesin?	Neuroblastoma
What cancer has β-HCG?	Choriocarcinoma
What cancer has 5-HT?	Carcinoid syndrome
What has t(9,22)?	CML (bcr-abl gene)
What has t(14,18)?	Follicular lymphoma (bcl-2 gene)
What has t(8,14)?	Burkitt's lymphoma (c-myc gene)
What has t(15,17)?	AML M3
What has t(11,22)?	Ewing's sarcoma
What has HLA A3, A6?	Hemochromatosis
What has HLA B5?	Behcet's
What has HLA B13?	Psoriasis without arthritis
What has HLA B27?	Psoriasis Ankylosing spondylitis Reiter's
What has HLA DR2?	Goodpasture's, MS

What has HLA DR3?	Celiac sprue
What has HLA DR4?	Pemphigus vulgaris
What has HLA DR5?	Pernicious anemia

Microbiology:

How do you know it's an anaerobe?	Gas, bad odor
What makes an adequate sputum sample?	<10 epithelial cells and >25 leukocytes per low power field
What bug causes Acute Bacterial Endocarditis?	Staph aureus (attacks healthy valves)
What bug causes Subacute Bacterial Endocarditis?	Strep viridans (attacks damaged valves)
How do you diagnose Rheumatic Fever?	*Need at least 2 "SPECC":* **S**ubcutaneous nodules **P**olyarthritis **E**rythema Marginatum (red margins) **C**horea (Sydenham's) **C**arditis **(MS > AS > TS)** *"MAT"*
What are the only Gram – diplococci?	Neisseria
What is the only Gram + diplococci?	Strep pneumo
What bugs cause heart block infections?	*"Don't TeLL Chaga"* • **D**iphtheria • **T**yphoid fever (Salmonella typhii) • **L**egionella • **L**yme disease • **C**hagas disease (Whipple's)
What are the IgA protease bugs?	Strep pneumo. H. influenza Neisseria catt.
What do the IgA protease bugs cause?	Sinusitis, otitis media, pneumonia, bronchitis
What are the simple Gram – bugs?	*"simple HE"* • **H**. influenza • **E**. coli
What are the atypical (no cell wall) bacteria?	*"CLUMsy"* • **C**hlamydia • **L**egionella • **U**reaplasma • **M**ycoplasma
What is the only Gram + endotoxin?	Listeria
What are the Gram – exotoxins?	H. influenza, Bordetella pertussis
What are the spore forming Gram + bacteria?	Bacillus anthracis, Clostridium botulinum

What bugs cause walking pneumonia?	*"Come My Love for a walk"* 0-2 mo: **C**hlamydia pn. 10-30 y: **M**ycoplasma pn. >40 y: **L**egionella pn. AIDS/Premies: Pneumo Jirovecii
What causes Cryoglobulinemia?	*"I AM HE"* **I**nfluenza **A**denovirus **M**ycoplasma **H**ep B, C **E**BV
What are the Silver Stainers?	*"HaLV-C = Comma"** **H.** pylori **L**isteria **V**ibrio **C**ampylobacter
What bacteria cause diarrhea <8hrs after introduction?	*"CBS"* • Clostridium • Bacillus cereus • Staph aureus
What are the Gram – capsules?	*"Some Killers Have Pretty Nice Capsules"* **S**almonella **K**lebsiella **H**. influenza B **P**seudomonas **N**eisseria **C**itrobacter
What is the Gram + capsule?	Strep pneumo
What are the Urease + bugs?	*"Urease PPUNCH"* **P**roteus **P**seudomonas **U**reoplasma **N**ocardia **C**ryptococcus **H.** Pylori
What bugs cause Monocytosis?	*"STELS"* **S**yphilis **T**B **E**BV **L**isteria **S**almonella
What are the Big Mama Anaerobes?	Bacteroides fragilis Strep. bovis

	C. septicum
What bugs disrupt G proteins?	Pertussis: inhibits Gi Cholera: stimulates Gs E. coli: stimulates Gs
What bacteria cause UTIs and prostatitis?	E. coli Proteus Klebsiella Pseudomonas
What bug is associated w/ cat scratch?	Bartonella henselae
What bug is associated w/ cat saliva?	Pasturella multocida
What bug is associated w/ cat pee?	Toxoplasmosis
What bug is associated w/ cat caca?	Toxocara cati
What diseases are caused by the Ixodes Tick?	• Lyme disease – bull's eye lesion • Babesiosis – hemolytic anemia • Ehrlichiosis – puncture near eye
What bugs cause hospital abscesses?	**Day 1-3:** Staph aureus – lots of O_2 **Day 4-7:** Strep viridans – no enzymes **Day >7:** Anaerobes – PMNs
What kind of patients does Staph and Pseudomonas like to attack?	DM Cystic fibrosis pts Burn pts Neutropenic pts
What diseases have granulomas?	TB Sarcoidosis Syphilis Histiocytosis X The -ellas
What diseases provide malaria protection?	**Sickle cell:** AAs **G-6PD:** Mediterraneans
What bug causes multiple liver abscesses?	Entamoeba histolytica
What bug causes multiple cerebral abscesses?	Citrobacter
What has multiple lung aneurysms?	Osler-Weber-Rendu
What bugs cause Loeffler syndrome?	*"NASSA"* **N**ecator americanus **Ancylostoma** duodenale **S**chistosomiasis **S**trongyloides **A**scaris lumbricoides
What is the symptom of Loeffler syndrome?	Pulmonary eosinophilia
What is PIE syndrome?	**P**ulmonary **I**nfiltrate w/ **E**osinophilia
What diseases have PIE syndrome?	Aspergillus, Loeffler's, Churg-Strauss
What disease does Coxsackie A cause?	Hand-Foot-Mooth disease

What diseases does Coxsackie B cause?	Endocarditis, DM type 1
What bugs cause Meningitis?	**0-2mo:** *"baby BEL"** • Group **B** Strep • **E**. coli • **L**isteria **>2mo:** Strep pneumo **10-21y/o:** N. meningitidis
What are the CSF lab values for Meningitis?	• PMNs → bacteria • T cells/MP → non-bacterial • Normal glucose → viral
What does the Hep B Core Ag tell you?	Gone before pt has symptoms (2 mo)
What does the Core Ab tell you?	Past infection
What does the Surface Ag tell you?	Current infection
What does the Surface Ab tell you?	Vaccination has occured
What does the E Ag tell you?	Transmissibility/ infectivity
What does the E Ab tell you?	Low transmissibility
What antibody shows current Hep A infection?	IgM
What antibody shows past Hep A infection?	IgG
What type of people does Hep E like to infect?	Pregnant women, Asians
What is associated w/ Staph aureus?	Cellulitis, dairy products, gold pigment, endocarditis, styes, hordeolum (on eyelid)
What is associated w/ Strep pyogenes?	*"LINES"*, ASO, hyaluronic acid capsule
What is associated w/ Staph epidermidis?	Central lines, VP shunts, white pigment
What is associated w/ Propionibacterium acnes?	Acne, progesterone => propionic acid
What is associated w/ Pasteurella Multicoda?	Cat/dog saliva => cellulitis w/ lymphadenitis
What is associated w/ Staph saprophyticus?	Female UTIs, honeymooner's cystitis
What is associated w/ Hemophilus aegyptus?	Swollen eyeball w/ pus
What is associated w/ Francisella tularensis?	Ulcers at rabbit or deer tick bite site (Tx: Streptomycin)
What is associated w/ Pseudomonas aeruginosa?	Otitis externa, ecthyma granulosum, loves rubber, whirlpool folliculitis, ADP ribosylates EF-2
What is associated w/ Streptococcus pneumoniae?	Otitis media, red bulging tympanic membrane
What is associated w/ Fusobacterium?	Painful mouth ulcers, gum pus, Vincent's angina
What is associated w/ Strep mutans?	Dental cavities
What is associated w/ Strep viridans?	Subacute bacterial endocarditis, green pigment
What is associated w/ Strep salivarius?	Cold agglutinin test

What is associated w/ Strep pyogenes?	Rheumatic fever, strain 12 => PSGN
What is associated w/ Group B strep?	Pharyngitis
What is associated w/ Corynebacterium diphtheriae?	Chinese letters, gray pseudomembrane, suffocation, ADP ribosylates EF_2, heart block, Elek test
What is associated w/ Strep pneumonia?	Gram + diplococci, rusty sputum, IgA protease, vaccine covers 23 strains
What is associated w/ H. influenza?	Gram – pleomorphic rod, Factors V/X, 5C-ribose capsule, B-type: meningitis, epiglottitis, sepsis
What is associated w/ N. catarrhalis?	Loves mucus → attacks respiratory tract
What is associated w/ Chlamydia psittaci?	Parakeets, parrots
What is associated w/ Chlamydia pneumonia?	Staccato coughing, elementary and reticulate bodies
What is associated w/ Mycoplasma pneumonia?	Bullous myringitis, ground-glass CXR, P_1 protein, erythema multiforme
What is associated w/ Legionella pneumonia?	AC ducts, silver stains lung, CYEA, Pontiac fever, Legionnaire's pneumonia
What is associated w/ Pneumocystis Jiroveciinia?	AIDS/premies, rusty sputum, silver stains lung, ↑LDH (Tx: steroids)
What is associated w/ Clostridium botulinum?	Canned food, honey, inhibits ACh release, resp failure, stool toxin (Tx: Penicillin)
What is associated w/ Clostridium tetani?	Rusty nail wounds, inhibits Gly release, resp failure, risus sardonicus, **t**ennis racquet shape
What is associated w/ Bordetella pertussis?	Whooping cough, ADP ribosylates Gi
What is associated w/ Bacillus anthracis?	**B**lack eschar, woolsorter's lung disease, D-Glu, **b**oxcar-like spore (Tx: Ciprofloxacin)
What is associated w/ H. Pylori?	Duodenal ulcers
What is associated w/ Salmonella?	Raw chicken&eggs, turtles, rose typhoid spots, H_2S, sickle cell osteomyelitis
What is associated w/ Campylobacter jejuni?	Raw chicken&eggs, very bloody diarrhea, inactivated by gastric juice, Guillain-Barre, seagull-shape
What is associated w/ Clostridium perfringens?	Holiday ham&turkey, DM gas gangrene, soil/feces
What is associated w/ Bacillus cereus?	Fried rice, pre-formed toxin
What is associated w/ Listeria monocytogenes?	Raw cabbage, hot dogs, spoiled milk, migrant workers => heart block, meningitis, abortions
What is associated w/ Vibrio parahaemolyticus?	Raw fish
What is associated w/ Actinomyces?	Face fistulas, sulfur granules
What is associated w/ Vibrio vulnificus?	Raw oysters, cellulitis in swimmer's cuts
What is associated w/ Shigella?	day care outbreaks => seizures, destroys 60S ribosome, infects M cells
What is associated w/ Vibrio cholera?	Rice water diarrhea, ADP ribosylates Gs
What is associated w/ Clostridium difficile?	Explosive diarrhea = pseudomembranous colitis

What is associated w/ Yersinia enterolitica?	Presents like appendicitis + Reiter's, bloody diarrhea
What is associated w/ Strep bovis?	Colon cancer, black pigment
What is associated w/ Clostridium septicum?	Colon cancer
What is associated w/ ETEC?	Traveler's => rice-water diarrhea
What is associated w/ EHEC?	Hemorrhagic => renal failure
What is associated w/ EIEC?	Inflammatory => loose stool
What is associated w/ EPEC?	Pathogenic => newborn diarrhea
What is associated w/ E. coli?	Raw hamburger, ADP ribosylates Gs, verotoxin, HUS
What is associated w/ Proteus mirabilis?	Staghorn calculus, UTI (Tx: Norfloxacin)
What is associated w/ Klebsiella pneumonia?	Alcoholics, currant jelly sputum, UTI, lung fissures
What is associated w/ Staph saprophyticus?	Young & college girls UTIs
What is associated w/ Enterococcus?	UTI, infective endocarditis post-GI/GU procedure
What is associated w/ N. meningitides?	DIC, pili, releases toxin in log phase
What is associated w/ Brucella?	farmers/vets, spiking fever 5x/day, animal placentas
What is associated w/ R. rikettsii?	Ticks => Rocky mountain spotted fever
What is associated w/ R. akari?	Mites => fleshy papules
What is associated w/ R. typhi?	Fleas, starts in armpit
What is associated w/ R. prowazekii?	Lice, starts on body
What is associated w/ R. tsutsugamushi?	Mites => scrub typhus
What is associated w/ Coxiella burnetii?	Dusty barn => Q fever
What is associated w/ Yersinia pestis?	Rats, fleas => Bubonic plague
What is associated w/ Bartonella henselae?	Cat scratch => single painful lymph node, silver stains
What is associated w/ Bacteroides fragilis?	Post-op **b**owel abscess, grows in blood clots
What is associated w/ Chlamydia trachomatis?	Neonatal inturned eyelashes/cornea ulcers
What is associated w/ Neisseria gonorrhea?	Pili, urethritis, tenovitis, septic arthritis

Immunology:

Which cytokine do macrophages stimulate?	IL-1
Which cytokines do T_H cells stimulate?	Everything other than IL-1
Which enlarged lymph nodes are most likely malignant?	• Supraclavicular • Epitrochlear (above elbow) • Inguinal
What is CD8?	T_K (killer) or T_S (suppressor) cell; responds to MHC-1 (self)
Which of your cells express MHC-1?	All except RBCs and platelets
What is CD4?	T_H (helper) cell; responds to MHC-2 (non-self)
What type of immunity does TH_1 provide?	Cell-mediated
What type of immunity does TH_2 provide?	Humoral
What do B cell deficiency patients die of?	Bacterial infection

What is Common Variable Hypogammaglobulinemia?	Kids w/ B cells don't differentiate into plasma cells => low Ab
What is Bruton's Agammaglobulinemia?	Young adults w/ defective Tyr kinase => no Ab, X-linked
What is Job-Buckley Syndrome?	Redheaded females, stuck in IgE stage
What is Multiple Myeloma?	Multiple osteolytic lesions, IgG, κ light chains, rouleaux
What is Heavy Chain Disease?	IgA and multiple myeloma of GI tract
What is Selective IgG_2 Deficiency?	Recurrent encapsulated infections
What is Selective IgA Deficiency?	Transfusion anaphylaxis, mucous membrane infections
What is Hairy Cell Leukemia?	Fried egg/sunburst appearance, TRAP +
What is Ataxia Telangiectasia?	Low IgA, neuro problems
What is Hyper IgM syndrome?	High IgM, low all other Ab
What do T cell deficiency patients die of?	Viral infection
What is DiGeorge Syndrome?	No thymus/inferior parathyroids, low Ca^{2+}
What is Chronic Mucocutaneous Candidiasis?	T cell defect against Candida albicans, chronic fatigue
What is SCID?	No thymus, frayed long bones, baby dies by 18mo
What is Wiscott-Aldrich?	Low IgM, low platelets, high IgA, eczema, petechiae
What does the CD4 count tell you?	Status of HIV (normal =1,000)
What does the viral load tell you?	Progression of HIV
What organs have the most CD4 receptors?	Blood vessels, brain, testicles, cervix, rectum
What are the 3 tests used to screen for HIV?	• ELISA: detects IgG Ab to p24 Ag • Western blot: see >2 proteins • PCR: detects virus (use in babies)
What is the definition of AIDS?	CD4 <200/μL or clinical symptoms
What are the live vaccines?	*"Bring Your Own Very Small Virus + MMR"* • **B**CG • **Y**ellow Fever • **O**PV (Sabin) = oral polio • **V**aricella • **S**mallpox • Rota**V**irus • **M**easles = Rubeola • **M**umps • **R**ubella = German 3-day measles
What is 1° Biliary Cirrhosis?	Anti-mitochondrial Ab, pruritis, females
What is 1° Sclerosing Cholangitis?	p-ANCA Ab, bile duct inflammation, onion skinning, IBD
What is Type I Autoimmune Hepatitis?	Anti-SM Ab, young women
What is Type II Autoimmune Hepatitis?	Anti-LKM Ab, kids

What is Bullous pemphigoid?	Anti-hemidesmosome Ab, skin bullae
What is Celiac sprue?	Anti-gliaden Ab, eating wheat => steatorrhea
What is CREST syndrome?	Anti-centromere Ab
What is Dermatitis herpetiformis?	Anti-BMZ Ab, Anti-endomysial Ab, vesicles on anterior thigh
What is Dermatomyositis?	Anti-Jo-1 Ab, myositis + rash
What is DM type 1?	Anti-islet cell, anti-GAD Ab, polyuria, polydipsia, weight loss
What is Drug-induced Lupus?	Anti-**h**istone Ab, *"**H**IPPPE"* causes it
What is Gastritis Type A?	Anti-parietal cell Ab, atrophic gastritis, adenoCA
What is Goopasture's?	Anti-GBM Ab, attacks lung and kidney, RPGN
What is Grave's disease?	Anti-TSHr Ab, hyperthyroid, bug eyes, pretibial myxedema
What is Guillain-Barre?	Anti-**g**anglioside Ab, ascending paralysis, 2 wk after URI
What is Hashimoto's?	Anti-microsomal Ab = TPO, hypothyroid
What is ITP?	Anti-platelet Ab, Anti-GP2b3a Ab, thrombocytopenia
What is Mixed CT disease?	Anti-RNP Ab
What is Mononucleosis?	Heterophile Ab, teenager w/ sore throat
What is MPGN TypeII?	Anti-C_3 convertase Ab = C_3 nephritic factor
What is Multiple sclerosis?	Anti-myelin Ab, middle aged female w/ vision problems
What is Myasthenia gravis?	Anti-AChr Ab, female w/ ptosis, weaker as day goes by
What is Paroxysmal hemolysis?	Donath Landsteiner Ab, bleeds when cold
What is Pemphigus vulgaris?	Anti-desmosome Ab, skin sloughs off when touched
What is Pernicious anemia?	Anti-IF Ab, Vit B_{12} def => megaloblastic anemia
What is Polyarteritis Nodosa?	**p**-ANCA Ab, attacks gut and kidney, HepB
What is Post-Strep GN?	ASO Ab, nephritic w/ complement deposition
What is Rheumatoid Arthritis?	Rheumatoid Factor, pain worse in the morning
What is Scleroderma?	Anti-**Scl**70 = anti-TopoI, fibrosis, tight skin
What is Sjögrens?	Anti-SSA Ab, dry "sand in" eyes, dry mouth, arthritis
What is SLE?	Anti-ds DNA/ Sm/ Cardiolipin, rash, photosensitivity, oral ulcers, RF
What is SLE cerebritis?	Anti-neuronal Ab, Anti-ribosomal Ab
What is Vitiligo?	Anti-melanocyte Ab, white patches
What is Warm hemolysis?	Anti-Rh Ab, bleeds at body temp
What is Wegener's?	c-ANCA Ab, attacks ENT, lungs, kidney
What is HSP?	IgA disease 2 weeks after common cold => Berger's
What is Berger's?	IgA disease 2 weeks after vaccination => Serum sickness
What is Alport's?	IgA disease 2 weeks after diarrhea => HSP, Polio

Biostatistics:

What is the Mean?	Average
What is the Median?	Middle value
What is the Mode?	Most frequent value
What does Sensitivity tell you?	People that have the disease w/ + test
What does Specificity tell you?	People that don't have the disease w/ - test
What does Positive Predictive Value tell you?	Probability of having a disease w/ + test
What does Negative Predictive Value tell you?	Probability of not having a disease if have a negative test
What does Incidence tell you?	New cases (rate per unit time)
What does Prevalence tell you?	Total cases (at one time)
What does the Odds Ratio tell you?	Diseased are x times more likely to see risk factor
What does a Confidence Interval = 95% tell you?	95% sure it lies within the interval
What does Relative Risk tell you?	Risk of getting disease w/ known exposure
What is NNT?	**N**umber **N**eeded to **T**reat to change 1 life
What does a p value <0.5 tell you?	Random chance that you will be wrong 1 time out of 20
What does a Null hypothesis tell you?	Nothing's happenin'
What does Power tell you?	Probability of detecting a true intervention
What is a type I error?	P value error, false negative *"too optimistic"*
What is a type II error?	Power error, false positive *"too pessimistic"*
What is Accuracy?	Validity *"truth"*
What is Precision?	Reliability *"keep making the same mistake"*
What are the phases of clinical trials?	Phase I: Toxicity *"hurt pt?"* Phase II: Efficacy *"help pt?"* Phase III: Comparison *"any better?"* Phase IV: Post-marketing surveillance *"can they screw it up?"*
What is a Cohort study?	Prospective study that provides incidence (new cases), uses RR
What is a Case Control study?	Restrospective study that provides prevalence (total cases), uses Odds Ratio
What is a Cross Sectional study?	Provides prevalence "snapshot"
What is a Case Report?	Describes an unusual pt
What is a Case Series report?	Describes several unusual pts
What are Consensus Panels?	Panel of experts provides a recommendation
What is Clinical wisdom?	"I think...." paper
What is a Meta-analysis?	Tries to combine data from many trials

Clinical Skills:

What does pain radiating to the jaw indicate?	MI
What does pain radiating to the left shoulder indicate?	MI or spleen rupture
What does pain radiating to the scapula indicate?	Gall bladder problems
What does pain radiating to the trapezius indicate?	Pericarditis
What does pain radiating to the groin indicate?	Ureter stones or osteoarthritis of hip
What is Blumberg's sign?	Rebound tenderness => peritonitis
What is Brudzinski's sign?	**B**end neck, see knee flexion => meningitis
What is Chadwick's sign?	Purple vagina => pregnancy
What is Chandelier sign?	Cervical motion tenderness => PID
What is Chovstek's sign?	Tap facial nerve => muscle spasm
What is Courvoisier's sign?	Palpable gall bladder => cancer (pancreatic)
What is Cozen's sign?	Dorsiflex wrist causes pain => lateral epicondylitis *"tennis elbow"*
What is Cullen's sign?	Bleeding around umbilicus => hemorrhagic pancreatitis
What is Gottran's sign?	Scaly patches on MCP, PIP joints => dermatomyositis
What is Hamman's sign?	Crunching sound on auscultation => subcutaneous emphysema
What is Homan's sign?	Pain in calf w/ dorsiflexion => DVT
What is Kehr's sign?	Left shoulder pain => spleen rupture
What is Kernig's sign?	Flexing **k**nee causes resistance or pain => meningitis
What is Kussmaul's sign?	Big neck veins w/ inspiration => pericardial tamponade
What is Levine sign?	Hold clenched fist over sternum => angina
What is Murphy's sign?	Pressed gallbladder causes pt to stop breathing => cholecystitis
What is Obturator sign?	Inward rotation of hip causes pain => appendicitis
What is Phalen's test?	Flexing wrists for 60s causes paresthesia => carpal tunnel
What is Psoas sign?	Extending hip causes pain => appendicitis
What is Rovsing's sign?	Palpating LLQ causes RLQ pain => appendicitis
What is Tinel's test?	Percussing median nerve causes 1st-3rd finger pain => carpal tunnel
What is Trousseau's sign?	BP check causes carpal spasm => migratory thrombophlebitis
What is Turner's sign?	Bleeding into flank => hemorrhagic pancreatitis

Pharmacology:

What drugs act as haptens on platelets?	asa Heparin Quinidine
What drugs act as haptens on RBCs?	*"PAD PACS"* • **P**enicillamine • α-Me**D**opa • **D**apsone • **P**TU • **A**ntimalarials • **C**ephalosporins • **S**ulfa drugs
What drugs wipe out the bone marrow?	*"ABCV"* • **A**ZT • **B**enzene • **C**hloramphenicol • **V**inblastine
What drugs wipe out granulocytes?	Carbamazapine Clozapine Ticlopidine
What drugs cause pulmonary fibrosis?	*"BBAT"* • **B**usulfan • **B**leomycin • **A**miodarone • **T**ocainide
What drugs dilate veins and arteries?	ACE-I and Nitrates
What are the anti-inflammatory actions of steroids?	• Stabilizes: mast cells/endothelium • Inhibits: MP migration/ PLA • Kills: T cells/eosinophils
What do opioid mu receptors cause?	**M**ental side effects
What do opioid kappa receptors cause?	Pain relief
What drugs have sulfur in them?	ACE-I Dapsone Celocoxib Loop diuretics Thiazide diuretics Sulfonylureas Sulfonamides
What are the short-acting DM Type 1 drugs?	Lispro Regular Asparte

What are the intermediate-acting DM Type 1 drugs?	NPH Lente
What are the long-acting DM Type 1 drugs?	Glargine Ultralente
What drugs cause SLE?	*"HIPPPE"* **Hydralazine** **I**NH **P**rocainamide **P**henytoin **P**enicillamine **E**thosuximide
What antibody is associated with drug-induced lupus?	Anti-**h**istone Ab
What drugs cause labor induction?	• PGE_2 – ripens cervix • Oxytocin – increase contractions • Pitocin – increase contractions
What drugs cause labor halt = Tocolytics?	• Hydration (stop ADH=oxytocin) • $MgSO_4$ – decrease contractions • Terbutaline – decrease contractions • Ritodrine – ↑edema
What drugs can be Nephrotoxic?	Water soluble drugs
What drugs can be Hepatotoxic?	Fat soluble drugs
How do you treat Acetaminophen overdose?	N-acetylcysteine
How do you treat Benzodiazepine overdose?	Flumazenil
How do you treat β-blocker overdose?	Glucagon
How do you treat CCB overdose?	Calcium Chloride + Glucagon
How do you treat CO poisoning?	100% O_2
How do you treat Cu, Gold, Cd, Mercury, Arsenic poisoning?	Penicillamine
How do you treat Cyanide poisoning?	Amyl Nitrite
How do you treat Ethylene Glycol, MeOH overdose?	EtOH, Fomepizole
How do you treat Fe overdose?	Deferoxamine
How do you treat Fibrinolytic overdose?	Aminocaproic acid
How do you treat Heparin overdose?	Protamine sulfate
How do you treat Lithium overdose?	Sodium polystyrene sulfonate
How do you treat Nitrite overdose?	Methylene blue
How do you treat Organophosphate poisoning?	Atropine + Pralidoxime
How do you treat Opioid overdose?	Naloxone

How do you treat Pb poisoning?	Succimer
How do you treat Salicylate overdose?	Charcoal
How do you treat TCA overdose?	Bicarbonate (prevent arrhythmias)
How do you treat Theophylline overdose?	Esmolol
How do you treat Warfarin overdose?	Vit K
What drugs causes myositis?	*"RIPS"* • **R**ifampin • **I**NH • **P**rednisone • **S**tatins
What drugs cause disulfiram reactions?	Metronidazole Cephalosporins Procarbazine
What drugs cause dysgeusia?	Metronidazole Clarithromycin Li
What antibiotics are in a triple antibiotic?	*"Brand New Potient"* **B**acitracin **N**eomycin **P**olymyxin D
What is the 1-dose treatment for Chlamydia?	Azithromycin
What are the 1-dose treatments for Gonorrhea?	*"Try to fix the fox with floxs"* • Cef**tri**axone • Ce**fix**eme • Ce**fox**itin • Ciprofloxacin • Ofloxacin • Gatifloxacin
What bugs are treated with vancomycin?	MRSA Staph. epidermidis Enterococcus
What is the treatment for malaria?	Quinine – tinnitus Mefloquine – good liver penetration Primaquine – prevents relapse, high bug resistance Chloraquine – kills RBCs
What should you always remember when taking a test?	You have been empowered… You now have the eyes of physio!

REFERENCE LABS:

	REFERENCE RANGE	SI REFERENCE INTERVALS
BLOOD, PLASMA, SERUM		
* Alanine aminotransferase (ALT), serum	8-20 U/L	8-20 U/L
Amylase, serum	25-125 U/L	25-125 U/L
* Aspartate aminotransferase (AST), serum	8-20 U/L	8-20 U/L
Bilirubin, serum (adult) Total // Direct	0.1-1.0 mg/dL // 0.0-0.3 mg/dL	2-17 µmol/L // 0-5 µmol/L
* Calcium, serum (Ca^{2+})	8.4-10.2 mg/dL	2.1-2.8 mmol/L
* Cholesterol, serum	Rec:<200 mg/dL	<5.2 mmol/L
Cortisol, serum	0800 h: 5-23 µg/dL // 1600 h: 3-15 µg/dL	138-635 nmol/L // 82-413 nmol/L
	2000 h: ≤ 50% of 0800 h	Fraction of 0800 h: ≤ 0.50
Creatine kinase, serum	Male: 25-90 U/L	25-90 U/L
	Female: 10-70 U/L	10-70 U/L
* Creatinine, serum	0.6-1.2 mg/dL	53-106 µmol/L
Electrolytes, serum		
Sodium (Na^+)	136-145 mEq/L	136-145 mmol/L
* Potassium (K^+)	3.5-5.0 mEq/L	3.5-5.0 mmol/L
Chloride (Cl^-)	95-105 mEq/L	95-105 mmol/L
Bicarbonate (HCO_3^-)	22-28 mEq/L	22-28 mmol/L
Magnesium (Mg^{2+})	1.5-2.0 mEq/L	0.75-1.0 mmol/L
Estriol, total, serum (in pregnancy)		
24-28 wks // 32-36 wks	30-170 ng/mL // 60-280 ng/mL	104-590 nmol/L // 208-970 nmol/L
28-32 wks // 36-40 wks	40-220 ng/mL // 80-350 ng/mL	140-760 nmol/L // 280-1210 nmol/L
Ferritin, serum	Male: 15-200 ng/mL	15-200 µg/L
	Female: 12-150 ng/mL	12-150 µg/L
Follicle-stimulating hormone, serum/plasma	Male: 4-25 mIU/mL	4-25 U/L
	Female: premenopause 4-30 mIU/mL	4-30 U/L
	midcycle peak 10-90 mIU/mL	10-90 U/L
	postmenopause 40-250 mIU/mL	40-250 U/L
Gases, arterial blood (room air)		
pH	7.35-7.45	[H^+] 36-44 nmol/L
P_{CO_2}	33-45 mm Hg	4.4-5.9 kPa
P_{O_2}	75-105 mm Hg	10.0-14.0 kPa
* Glucose, serum	Fasting: 70-110 mg/dL	3.8-6.1 mmol/L
	2-h postprandial: < 120 mg/dL	< 6.6 mmol/L
Growth hormone - arginine stimulation	Fasting: < 5 ng/mL	< 5 µg/L
	provocative stimuli: > 7 ng/mL	> 7 µg/L
Immunoglobulins, serum		
IgA	76-390 mg/dL	0.76-3.90 g/L
IgE	0-380 IU/mL	0-380 kIU/L
IgG	650-1500 mg/dL	6.5-15 g/L
IgM	40-345 mg/dL	0.4-3.45 g/L
Iron	50-170 µg/dL	9-30 µmol/L
Lactate dehydrogenase, serum	45-90 U/L	45-90 U/L
Luteinizing hormone, serum/plasma	Male: 6-23 mIU/mL	6-23 U/L
	Female: follicular phase 5-30 mIU/mL	5-30 U/L
	midcycle 75-150 mIU/mL	75-150 U/L
	postmenopause 30-200 mIU/mL	30-200 U/L
Osmolality, serum	275-295 mOsmol/kg H_2O	275-295 mOsmol/kg H_2O
Parathyroid hormone, serum, N-terminal	230-630 pg/mL	230-630 ng/L
* Phosphatase (alkaline), serum (p-NPP at 30EC)	20-70 U/L	20-70 U/L
* Phosphorus (inorganic), serum	3.0-4.5 mg/dL	1.0-1.5 mmol/L
Prolactin, serum (hPRL)	< 20 ng/mL	< 20 µg/L
* Proteins, serum		
Total (recumbent)	6.0-7.8 g/dL	60-78 g/L
Albumin	3.5-5.5 g/dL	35-55 g/L
Globulin	2.3-3.5 g/dL	23-35 g/L
Thyroid-stimulating hormone, serum or plasma	0.5-5.0 µU/mL	0.5-5.0 mU/L
Thyroidal iodine (^{123}I) uptake	8%-30% of administered dose/24 h	0.08-0.30/24 h
Thyroxine (T_4), serum	5-12 µg/dL	64-155 nmol/L
Triglycerides, serum	35-160 mg/dL	0.4-1.81 mmol/L
Triiodothyronine (T_3), serum (RIA)	115-190 ng/dL	1.8-2.9 nmol/L
Triiodothyronine (T_3) resin uptake	25%-35%	0.25-0.35
* Urea nitrogen, serum	7-18 mg/dL	1.2-3.0 mmol/L
* Uric acid, serum	3.0-8.2 mg/dL	0.18-0.48 mmol/L

REFERENCE LABS:

	REFERENCE RANGE	SI REFERENCE INTERVALS
BODY MASS INDEX (BMI)		
Body mass index	Adult: 19-25 kg/m^2	
CEREBROSPINAL FLUID		
Cell count	0-5/mm^3	0-5 x 10^6/L
Chloride	118-132 mEq/L	118-132 mmol/L
Gamma globulin	3%-12% total proteins	0.03-0.12
Glucose	40-70 mg/dL	2.2-3.9 mmol/L
Pressure	70-180 mm H_2O	70-180 mm H_2O
Proteins, total	<40 mg/dL	<0.40 g/L
HEMATOLOGIC		
Bleeding time (template)	2-7 minutes	2-7 minutes
Erythrocyte count	Male: 4.3-5.9 million/mm^3	4.3-5.9 x 10^{12}/L
	Female: 3.5-5.5 million/mm^3	3.5-5.5 x 10^{12}/L
Erythrocyte sedimentation rate (Westergren)	Male: 0-15 mm/h	0-15 mm/h
	Female: 0-20 mm/h	0-20 mm/h
Hematocrit	Male: 41%-53%	0.41-0.53
	Female: 36%-46%	0.36-0.46
Hemoglobin A_{1c}	≤ 6%	≤ 0.06
Hemoglobin, blood	Male: 13.5-17.5 g/dL	2.09-2.71 mmol/L
	Female: 12.0-16.0 g/dL	1.86-2.48 mmol/L
Hemoglobin, plasma	1-4 mg/dL	0.16-0.62 mmol/L
Leukocyte count and differential		
Leukocyte count	4500-11,000/mm^3	4.5-11.0 x 10^9/L
Segmented neutrophils	54%-62%	0.54-0.62
Bands	3%-5%	0.03-0.05
Eosinophils	1%-3%	0.01-0.03
Basophils	0%-0.75%	0-0.0075
Lymphocytes	25%-33%	0.25-0.33
Monocytes	3%-7%	0.03-0.07
Mean corpuscular hemoglobin	25.4-34.6 pg/cell	0.39-0.54 fmol/cell
Mean corpuscular hemoglobin concentration	31%-36% Hb/cell	4.81-5.58 mmol Hb/L
Mean corpuscular volume	80-100 μm^3	80-100 fL
Partial thromboplastin time (activated)	25-40 seconds	25-40 seconds
Platelet count	150,000-400,000/mm^3	150-400 x 10^9/L
Prothrombin time	11-15 seconds	11-15 seconds
Reticulocyte count	0.5%-1.5%	0.005-0.015
Thrombin time	<2 seconds deviation from control	<2 seconds deviation from control
Volume		
Plasma	Male: 25-43 mL/kg	0.025-0.043 L/kg
	Female: 28-45 mL/kg	0.028-0.045 L/kg
Red cell	Male: 20-36 mL/kg	0.020-0.036 L/kg
	Female: 19-31 mL/kg	0.019-0.031 L/kg
SWEAT		
Chloride	0-35 mmol/L	0-35 mmol/L
URINE		
Calcium	100-300 mg/24 h	2.5-7.5 mmol/24 h
Chloride	Varies with intake	Varies with intake
Creatinine clearance	Male: 97-137 mL/min	
	Female: 88-128 mL/min	
Estriol, total (in pregnancy)		
30 wks	6-18 mg/24 h	21-62 μmol/24 h
35 wks	9-28 mg/24 h	31-97 μmol/24 h
40 wks	13-42 mg/24 h	45-146 μmol/24 h
17-Hydroxycorticosteroids	Male: 3.0-10.0 mg/24 h	8.2-27.6 μmol/24 h
	Female: 2.0-8.0 mg/24 h	5.5-22.0 μmol/24 h
17-Ketosteroids, total	Male: 8-20 mg/24 h	28-70 μmol/24 h
	Female: 6-15 mg/24 h	21-52 μmol/24 h
Osmolality	50-1400 mOsmol/kg H_2O	
Oxalate	8-40 μg/mL	90-445 μmol/L
Potassium	Varies with diet	Varies with diet
Proteins, total	<150 mg/24 h	<0.15 g/24 h
Sodium	Varies with diet	Varies with diet
Uric acid	Varies with diet	Varies with diet

INDEX[*]

[*] This index was created with **TExtract™**

B

C

D

E

F

G

H

I

K

L

M

N

O

P

Q

R

S

T

U

V

W

X

Y